DROUGHT MANAGEMENT PLANNING IN WATER SUPPLY SYSTEMS

# Water Science and Technology Library

VOLUME 32



*The titles published in this series are listed at the end of this volume.*

# DROUGHT MANAGEMENT PLANNING IN WATER SUPPLY SYSTEMS

Proceedings from the UIMP International Course
held in Valencia, December 1997


edited by

ENRIQUE CABRERA
*Fluid Mechanics Group, Polytechnic University,*
*Valencia, Spain*

and

JORGE GARCÍA-SERRA
*Fluid Mechanics Group, Polytechnic University,*
*Valencia, Spain*

SPRINGER-SCIENCE+BUSINESS MEDIA, B.V.

A C.I.P. Catalogue record for this book is available from the Library of Congress.

DOI 10.1007/978-94-017-1297-2

---

*Printed on acid-free paper*

# TABLE OF CONTENTS

**DROUGHT MANAGEMENT IN AN URBAN CONTEXT**



**PRACTICAL CASES**

# ACKNOWLEDGEMENTS

This book contains the valuable contributions to the course "Management of Droughts in Urban Water Distribution Systems" of relevant international experts. Our first acknowledgment goes to them, the genuine protagonists of the book.

This publication has been given the support from the Conseller de Cultura, Eduación y Ciencia of the Valencian Government, Mr. Francisco E. Camps, and from its General Director, also responsible of the recently created Valencian Public Organism of Research (OPVI), Ms. Carmen Martorell. We are grateful to them for such a support and for the preface of the book.

The activity was held within the framework of the Universidad Internacional Menéndez y Pelayo of Valencia. The work of its Director, Mr. José Sanmartín, and of its co-directors, Mr. Luis Moreno and Ms. Mabel López, helped to make easy what is not simple: organizing with success an international course of high level. Their help is also greatly appreciated.

The Course also counted with the support of Aguas de Valencia and Iberdrola. Their support to this type of initiatives of the Fluid Mechanics Group has become customary. It is, without doubt, one of our better stimuli to try to reach the most brilliant results. Their prestige and the confidence they put in our Group are worth it. Also, they jointly introduce this book. Because of this and many other things, we want to express our sincere gratitude to these Companies.

The course was part of the Master "Management and Efficient Use of Water", within the project AGUA, Agua y su Gestión Urbana y Ambiental (Water and its urban and environmental management), financed by the DG V of the EU, through its operative program ADAPT. The promoter of the Project AGUA is the Fluid Mechanics Group of the Polytechnic University of Valencia that, at the same time, is the platform of the Valencian Research Institute of Efficient Use of Water, which has recently been created within the framework of the OPVI. In fact, this publication, closely related with efficient use of water, is the first activity developed in this framework.

We must also thank Kluwer Academic Publishers and especially their publisher Petra van Steenbergenfor her assistance, careful presentation and production of the book.

It would not be fair to conclude this section without explicitly mention that this work is the fruit of the endeavor of all the members of the Fluid Mechanics Group. In fact, the Editors have acted exclusively as the visible head of a wide group integrated by Angeles Alvarez, Miguel Andreu, Francisco Arregui, Carlos Balmaseda, Carmelo Cabezuelo, Enrique Cabrera, Quique Cabrera, Ramón Cañadas, Ricardo Cobacho, José Luis Diago, Ramón Dolz, Vicent Espert, Vicente Fuertes, Francisco García, Marta García, Jorge García-Serra, Pedro Iglesias, Francisco J. Izquierdo, Joaquín Izquierdo, Gonzalo López, P. Amparo López, José M. Llorens, Javier Martínez, Ana Mut, Pablo Navarro, Francisco Pastor, Rosario Perelló, Rafael Pérez, José V. Ribelles, Verónica Romero, Manuel Sánchez and Manuel Zaera. The production of this book would not have been possible without the contribution of the whole Group.

Fluid Mechanics Group
Valencia, July, 1998

# PREFACE

The Valencian Public Organism of Research (OPVI), recently created, has as its first target the promotion and dissemination of Research within the fields of Science, Technology and the Humanities. Under this directive, the OPVI will contribute to propel the social and economic development of the Valencian society in a definite way. Without doubt, Water Management is one of the technological areas having greater relevance and interest. For one thing, our millennial tradition emblematically represented by the *Tribunal de las Aguas* of Valencia accounts for it. For the other, it represents a subject of singular transcendence for the harmonic and sustainable future development of our Community. It is obvious that Water Management directly and substantially conditions all the pillars of our economy and, thus, our future. Namely, urban, agricultural, industrial, tourism and leisure sectors. This justifies the fact that one of the first Institutes launched by the OPVI is the Institute of Efficient Use of Water.

The book we are prefacing is one of the first products of this Institute within the frame of the OPVI. A panel of specialists from all over the world has contributed to it by addressing with rigor a problem of special current importance: the Management of Droughts within the Urban Water Supply Systems. One must not forget that during five long years, 1991 through 1995, up to twelve million Spanish citizens suffered temporal service interruptions of water in their homes. The discomfort and even sanitary risk that this implies needs no explanation. Taking for granted that new and more severe droughts are to come, it is clear that these measures, so far away from modern quality standards, should not be enforced again.

The book in your hands offers technology, experiences and guidelines to prepare and adapt Urban Water Supply Systems to cope with events, which are both natural and undesirable, in a modern and efficient way. This perfectly meets the OPVI objectives. Consequently, we take great pleasure in greeting and prefacing its birth with the strong desire that it be useful both for technicians developing their activity within Water Supply Systems and for politicians that are responsible for taking decisions to spur modernization.

We fervently extend this desire to other communities or countries that, having climatic conditions similar to ours, must face periodically these episodes called droughts.

Francisco E. Camps Ortiz
Conseller de Cultura, Educación y Ciencia
Vicepresidente del Consejo Rector del OPVI.

## FOREWORD

Introducing a book like this in your hands lends a good opportunity to think over a subject that is systematically ignored during hydraulic bonanza but deeply worries under adverse conditions. We refer to the lack or shortage of water, a good not evenly distributed.

The current climatic conditions, with peculiar phenomena like El Niño, produce natural disasters of far-reaching consequences for the life in different areas of our Planet. Even though we are located on a geographical area of relative stability, the cycle drought - abundant rainfall affects us periodically.

The book we are introducing constitutes a detailed study by experts devoting long hours to delve into these phenomena. They work to sensitize technicians, politicians and citizens to the necessity of taking decisions able to reduce the problems that are derived from the bad use or the uneven distribution of this valuable good called water.

The twenty-first century will constitute the framework where the conflicts derived from the ownership and the use of water will be resolved. Professor Cabrera emphasizes on the structural drawbacks of the water distribution systems. And referring to urban water distribution systems he puts the stress on price policies conditioning the network efficiency to the price: *the cheaper the supply, the worse efficiency*, quoting his own words. And strongly rejects the emergency solutions, which he calls *botch-ups*.

The lack of knowledge of the water supply system, the lack of professional management and the lack of knowledge of the evolution of the available hydraulic resources compared with the demand, are the main objectives of Professor Cabrera's reflections.

In short, we face a *water use culture* demanding more flexibility, imagination and coordination. A culture well distant from the current Spanish water administration structure, in which up to nine different official organisms, ranging from the Ministry of Environment to Municipal Governments, interfere. There is no doubt this new culture needs more flexibility, not favored by this complex organization. This is especially evident during water shortage or drought periods.

On the other hand, the work developed by the Professors and Researchers of the Fluid Mechanics Group of the Polytechnic University of Valencia deserves our better attention and support.

Treating the varied information implicit in a modern concept of management of a water distribution system needs organization and systematization. Thus, the different departments within the system can have access to it in a controlled and organized way. The network elements, the economic information and the spatial information are dealt with within the Geographical Information System (GIS) on an integrated basis. This system stores and links data of spatial nature with thematic data. With this modern tool, water supply management becomes more flexible and accurate. It is worth to mention leakage detection as one of the applications of high profitability. Among other functions, it also allows to elaborate rehabilitation plans with higher precision and, consequently, more cost-effective in terms of both time and money.

The use of models to simulate the performance of a system is necessary practice to forecast the network behavior. In this book it is shown that modeling is more than a recommended practice. It must go beyond the academic redoubt and be assumed with rigor by those who are responsible of water management.

The contribution within the book of a panel of experts from all over the world reveals the high level of interest and prestige risen by the Fluid Mechanics Group. After analyzing their activities, their concern to propose imaginative solutions to old problems and to share work and experience within this field are easily verifiable.

Elaborating scientific methods to fight the droughts to come is praiseworthy work that we always will back. These books far from becoming old library pieces must be used as permanent reference for those in charge of water management.

This introduction must not go beyond this point. It only aims to spur those interested in these subjects to deepen into the book. It has two good points: it is a compilation of the work developed by experts and it radiates the stress all they put into water. This resource usually ignored by the citizen -its final user-, but that causes him great concern when he lacks of it, since it is both indispensable and irreplaceable.

**AGUAS DE VALENCIA, S. A.** **IBERDROLA, S. A.**

# WATER SUPPLY SYSTEMS IN DROUGHT PERIODS. THE CASE OF SPAIN

CABRERA, E.; ESPERT, V.; LÓPEZ, P.A.
*Fluid Mechanics Group.*
*Universidad Politécnica. Valencia (Spain).*

## 1. Introduction

From a hydrological perspective, to speak of drought management in Spain while we are still in this excessively generous autumn of 1997 might appear a topic of scarce relevance. Especially since it has been preceded by two years of rainfall whose mean is very superior to the average annual values. Consequently, the current circumstances allow us to consider concluded the extraordinary dry period that was existent in Spain between 1991 and 1995. However it is now a timely moment, without any sense of hurry or need for agitation, to reflect on what happened in order to avoid the repetition of those events.

It would be unreasonable to forget a drought that provoked, over several years, temporary restrictions and interruptions in drinking water supply to more than ten million Spanish citizens and which was threatening to exhaust, in the summer of 1996, the scarce water reserves that remained. Happily it rained during the following winter. But, on this point we should insist, it would be irresponsible to forget not only the third world measure to interrupt temporarily the water service, but also the rushed execution of costly desalination plants of doubtful usefulness in the long run, especially taking into account the deficiencies of the distribution networks that make water so costly to the consumer, water whose flavour only makes to increase the sale and consumption of bottled water. One must also recall the numerous edicts exhorting moderation of consumption, the excessive number of public works suddenly and urgently classified to be of *general public interest*, as well as the exaggerated utilization of tankers to transport water. There were even public prayers because water reserves reached a historical minimum level that were already threatening to overwhelm the capacity of the political powers. Therefore, it does not seem reasonable to forget, in so short a space of time, a chronic shortage that caused so much bitterness.

In regions with a long agricultural tradition and limited water resources, such as the Spanish Mediterranean area, any discussion about water generates passionate debate and conflicting opinions. There will be, however, unanimity in admitting that situations such as those we have lived should not be repeated again in a country such as Spain that advances firmly towards the twenty-first century.

However, the strategy to follow so that such singular events are not repeated on the occasion of the next drought is not likely to excite unanimous consensus. So, as we



*E. Cabrera and J. García-Serra (eds.), Drought Management Planning in Water Supply Systems,* 1–21.

discuss in this paper, we are conscious of the fact that the approaches presented will give rise to opposing opinions. In any case, ideas that generate discussion, independently of their wisdom or folly, have always been positive for the advancement of society. With this constructive spirit of those who want to contribute, in all modesty, to the modernization of the water management in Spain, we move on to a series of reflections which have been the result of reading, analysis and the direct contact with the reality of many urban water supplies in Spain.

## 2. Structural defects of the supplies

A diagnosis of the structural defects of the Spanish water supply systems constitutes the necessary starting point so as to be able to outline, with some chance of success, a series of actions that support and provide an impulse to its modernization. And when we deal with diagnoses, one must at once make two important reflections that avoid misunderstandings and situate the content of the reflections in this paper within the desired framework or context. They are:

- When assertions are made, it applies only to a situation where there is a standard water supply. There are well-known, notable cases that escape this type of idealized context. Therefore, many of the present reflections are not always applicable in these cases.
- When dealing with water management, it is easy to diagnose but rather more complicated to carry out reforms. There are so many variables and conditioning factors present in connection with this natural resource that to restore balance to the system, having been disturbed by any kind of action, even if it is almost imperceptible, constitutes a great amount of engineering work. Of this we are very conscious.

Taking into account the above-mentioned points, we would like to highlight the principal structural defects of the urban water supplies as the following:

### 2.1 THE PRICES POLICY

Borrell, an ex-government minister, has affirmed in his article "The Water Debate" (El País, 26/06/95) that in the current economic regime water is a void-of-cost public good, being the consumer the one who has to pay the costs of hydraulic infrastructures that make possible its transportation and subsequent use. He uses the metaphor of paying for the glass but not for the water it contains. This is theoretically correct, in the same way as any Marxist economic policy may be theoretically correct. But praxis demonstrates that reality is very different, as is nowadays unanimously recognized (OECD, 1997; Llamas, 1997; Barraqué, 1995; IWSA, 1993, etc).

According to Courey and Hall (1995), political prices are equal to squandering and bad administration. In effect, all water supply technical managers know which is the optimum level of leaks for their distribution networks. And they know perfectly well that this level has direct effect on the sale price of water. Figure 1 represents the economic value of the water volume that are lost per year due to leaks in a network, for two different water prices. Adding these curves to that for the maintenance cost of the network in function of the losses level, it is obtained an optimum efficiency. From this analysis, we can conclude:

- If the sale price of water is zero, the network efficiency tends also towards zero. In other words, the price of water determines, almost completely, the greater or lesser sealing of the distribution system.

- As the water sale price increases, the optimum efficiency of the network will be approximated, in a natural way, to the unit. Due to this fact, certain countries such as Italy (Gazzetta Ufficiale, 19/1/94) permit the increase of the selling price of water when the network efficiency increases. Such a measure is justified because it compensates with a greater income level the more elevated cost of having a more efficient network.

In other words, the more faithfully real management costs of supply (amortization, economic and technical management, energy operation costs, maintenance, rehabilitation, etc.) are reflected in the final amount that consumers have to pay for their water bill, the more the management of the supply network will tend, in a natural way, to improve its efficiency.

Few exceptions escape the affirmation that subsidies and political prices are synonymous with barely efficient management. And this is, without doubt, one of the principal Achilles' heels of the management of water in Spain.

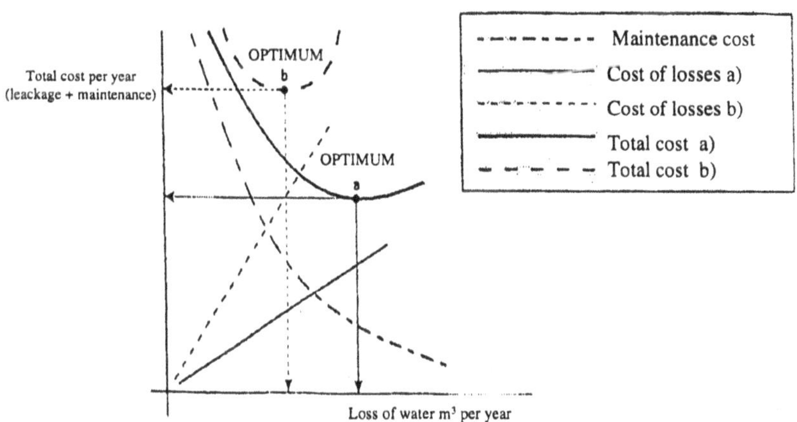


*Figure1 Optimal efficiency in a network and its relation with water price.*

## 2.2 INSUFFICIENT NETWORKS. LOSSES OF WATER QUALITY IN THE DISTRIBUTION SYSTEM

Subsidized water supply managers spend more time and effort in obtaining official financial support than in managing efficiently their distribution systems. Consequently, the diameters of the pipes, as time goes by, become too small because of the growth of cities, because of the increase in urban consumption, and because of the obsolescence of the conduits (physical reduction of the diameter through incrustations and deposits inside of pipes) or even, through the increase of water leaks. One could talk about a *consumed and amortized network* that has not been renewed and/or enlarged because of a lack of adequate planning, or because the cubic meters that are invoiced and/or the prices of water that are charged do not permit such updating or modernization of the system.

The managers responsible for the supplies that are not expanded and modernized at the same rate as the cities to be supplied, find their best defense in their fight against this marked differential growth, in the construction of household water tanks that allow the storage of water during hours of low consumption to conceal or, at least to cover up, the insufficiency of the network during peak consumption hours.

Likewise, the inadmissible temporary interruptions in the water service invite the consumer to build storage tanks that protect them in case of a temporary absence of supply. When people consider that there is going to a regular water supply cut, they start storing water in the bath tub. But, if the perspectives are such that there is a daily repetition in the cutting-off of the water supply, one must appeal to more operative solutions: the household water tanks. An example of this kind of limit situation is the water supply in Havana (Cuba). Their water supply network has become so insufficient and obsolete that the water service can only be provided once a week. Logically the household water tank should be of sufficient size to supply such a long period of time.

There are, in Spain, many water supply systems that present this problem. It is typical in tourist coastal cities that, as far as their growth is concerned, we can say that they have exploded. Such brutally fast expansion has not been accompanied, far from it, by the necessary improvement of urban infrastructures. Symptoms of this lack can be clearly seen: very low water pressure levels in the network and insufficient water supplies in the most distant points and/or highest points of the city. As they are completely insufficient, this quickly leads to complaints from consumers unless they have resorted to emergency measures (often "botched jobs"), which is the raison d'être of the promotion of household water tanks.

Household water tanks present multiple problems, forbidden and unthinkable in any developed country. We can highlight, in order of importance, the following:

- Their lack of hygiene. The pressurized pipe water is stored in an open water tank out of any sanitary control. There may be filtrations, sewer leaks, etc., that contaminate the tank or cistern. Furthermore water, like all food, has its sell-by date. It is designated as *optimum residence time in the network* and should not to go beyond two or three days since the water was treated till it arrives at its consumption outlet (normally, a tap). With household tanks or cisterns all control is lost, and the probability in tourist populations, during the low seasons, to drink water that was treated a week ago is high. Water tanks have been and will be the origin, in many water supplies, of not just a few infections. In spite of all this, and to conceal the weakness of the water supply network, their construction is promoted by the town halls.

- Really it is a bad solution. A problem arising from the lack of planning and investment in the network is shifted on to the constructor of the building that has in the execution of this construction an additional cost. This cost is passed on to the house buyer.
- Loosing network water pressure when filling an open tank is, from an energy point of view, inefficient.
- If water consumption is measured in a global way for all the consumers of a large building, an error of calculation is generated through undermeasurements (Cabrera and García-Serra; 1997).

We close this section, related to the insufficiency of the network with the consequent installation of household water tanks in our distribution networks, with three final considerations:

- Their use does not signify any water saving.
- The temporary interruption of the water service needs to be supported by water tanks. This consequence is typical of not developed countries where these practices are carried out regularly.
- This invites the user to consume bottled water at 50,000 pesetas/m$^3$. In Italy this percentage has reached 44% of the population (Greco, 1997). In Spain, without official data in this regard, the percentage may well be superior.

## 2.3 UNKNOWLEDGEMENT OF THE SYSTEM

The correct management of urban supplies in drought periods demands full knowledge of possibilities and limitations. However there are numerous water supplies that:

- Do not audit regularly, from a water supply perspective, their networks. Unaccounted for water, the totality of leaks and uncontrolled consumption, should be known with the greatest possible precision in each one of their potential sections.
- Do not have enough detailed information about the real state of the networks. The diameters of the pipelines, existing obsolescence, the state of the different materials, etc., are not known in due precision. In other words, no reliable data bases exist in this regard.
- Evolution of time consumption is not quite known. Nor are these facts compared with the real possibilities of the distribution network.
- The possibilities of operation improvement that the system may offer through regulation and control have not been studied.

- And, finally, a data base has not been built which determines the value of Performance Indicators of the supply, by which one could know precisely the general state of the system.

There is no doubt that to manage correctly an urban water supply, above all in drought periods, demands to know the strong and weak points of the system.

## 2.4 LACK OF PROFESSIONALISM IN THE MANAGEMENT OF WATER SUPPLIES

This section, in a certain manner, is a synthesis of what has preceded. And one has to hastily say that, although we refer to the lack of training of the technical personnel in charge of the management of the water supply, one also has to think above all in the responsible political authorities that take decisions with scarcely any knowledge of the matter. Consequence of this, in a drought period, the most serious situation of emergency that this country is likely to face, there is not any global plan to overcome this problem. A drought management plan requires:

- Knowledge, at all times, of the available water resources and the time consumption forecast. This allows to identify the seriousness of the situation and therefore to determine, at any moment, the most adequate strategy.
- The ability to carry out precise and in detail water measurements, in the injection phase as well as in the consumption. The possibility of measuring the consumptions in short periods of time is the starting point for any modern strategy of drought management in urban networks. The network must be well endowed with flow meters, not only in the injection points, but also in the main distribution pipelines.
- The use of a well calibrated mathematical model for the network. This demands detailed knowledge of the network. The importance of the model is significant, inasmuch as the modern control techniques to monitoring networks require a perfectly updated mathematical model.
- The distribution network must have a very high efficiency. In fact, the only advantage that one can gain from the temporary interruption of the service is that, during the time in which the water supply is cut, there are no losses due to leaks. Everything else is merely an inconvenience. There is a lost of reliability and safety in the network, - for example problems with trapped air -, as well as in the quality of the water that is supplied to the users. And this besides the fact that it causes severe nuisances to the consumers.

To be able to confront, without traumas, a deep drought does not permit improvisations in the water supply. What is needed is to outline in advance a modernization plan that only qualified professionals can carry out with guarantees. And this needs time to be developed. From this point of view, certainly, this period of abundant rains would be a convenient time to plan a better future for the water supply to our populations. To incur in the same defects, having already entered the twenty-first century, there would be no justification or excuses possible in our country.

## 3. Reasons that explain the preceding situation

Two fundamental causes explain, to our mind, the situation that has been explained above. On the one hand, there is a culture of water use. On the other hand, there is the structure of the Administration of the water. Subsequently, both factors are analyzed.

### 3.1 THE CULTURE OF THE WATER USE IN SPAIN

As it has happened in many other countries, throughout the twentieth century the phrase that best summarizes the water policy existent in Spain is: permanent increase in the offer of water supplies. That is to say, consumption has always been increasing, based on increasing the total number of hectares irrigated and hardly any rationalization of use. While other countries, such as in the case of Israel (Bruins, 1993), have obtained as time goes by, decreases in the consumption, consequence of an important effort in the rationalization of use, in Spain the consumption has increased in an uninterrupted and uncontrolled way.

The well-documented research by Pérez Díaz et al. (1995) analyzes how water is consumed in Spain. Their conclusion is worrying: *We have a rather imprecise knowledge of the real situation about water consumption in Spain and about its recent evolution. In the irrigation sector, good data does not exist neither about the total number of hectares being irrigated nor about the consumption per hectare according to the crops being cultivated. For urban uses we already have data consumptions from a survey on the sector carried out by the Spanish Water Supply and Sewerage Association*[1]. We, therefore, know very imprecisely how irrigation water is consumed. And as far as urban use is concerned, the subject of this paper, the only moderately reliable data come from surveys promoted by the Spanish Water Supply and Sewerage Association (AEAS[2], 1994) have not been checked. Each water supplier sent a reply to the survey according to their own criteria. It is, furthermore, scarcely representative as far as the medium and small size municipalities are, very often, outside the structure of the A.E.A.S.

So, there is a culture of hardly any control over use, as much on the part of the Administration as on the part of the user. And if one adds the fact that water prices are strongly subsidized and, therefore, fenced off from any economic policy based on real costs, one can understand the relatively low predisposition of the Spanish user to rationalize on water consumption. The consumer is normally only sensitive to saving water in times of drought, when there is no longer time to carry out such a rational way of saving. With respect to water saving, improvisation does not work.

The consumer who, during periods of generous rainfall, feels comfortable with this policy of use, must be educated so that management will be sustainable in periodes when there is a shortage of resources. But given the deeply rooted nature of this culture in our country, a change of mentality needs the involvement of the media. Sociologists are requested to provide answers and to indicate, as a previous step to the necessary political action, the way to modify a mind-set which has become so consolidated. The water saving campaigns in the middle of a drought only are effective if, during the years

[1]In Spanish, this is "Asociación Española de Abastecimiento y Saneamiento de Aguas" (AEAS).
[2]See footnote 1.

before it, one has been working on the modernization of the supplies. Otherwise, they only deserve one qualifier: demagogic.

Certainly a new culture regarding the use of the water is making decisive headway in culturally more advanced countries. Fortunately, Spain has not escaped from this winds of change (Martínez Gil, 1997; Llamas, 1997) and the international reports about the management of water in Spain follow similar paths (Barraqué, 1995; OECD, 1997). But nearly everything is still to be done (Cabrera, 1997). One might consider and compare it to a similar movement towards change which was sparked off by the energy crisis in 1973 and that fired a series of energy saving and reutilization policies. This similarity is emphasized in the paper "The economic impact of water conservation: Case studies in Ontario" by Economic Research Limited (1995). And among many other advantages of these saving policies, they emphasize the fact that it can generate more employment and of greater stability than the traditional expansive policies. There is no doubt that the twenty-first century is going to be presided over by a change of culture towards a more efficient use of water. It is the best way to be able to overcome, in an effective and modern way, future droughts.

### 3.2 THE STRUCTURE OF THE WATER ADMINISTRATION IN SPAIN

There is a need, without any doubt, for some kind of dynamic force to impel the work required to modernize in Spain all the water supplies. This dynamic force must defeat a secular inertia and it is obvious that only the Administration itself can act in such a way. But not with the current structure which, when envisaged from the perspective of urban water supplies, it has come to be inefficient and insufficient. The reason for this lack of operability is the consequence of the confused organizational structure currently existing in the administration. At present, the following institutions have more or less direct responsibility for our water supplies:

a) Environment ministry
b) Ministry of Industry
c) Ministry of Health and Consumption
d) Hydrographic Federations
e) Public Works Councils
f) Health Councils
g) Industry Councils and their Price commissions.
h) Provincial Delegations
i) Town Halls, as the ultimate responsible authority

An organizational structure, which is so complex as well as so diffuse, does not facilitate coordinated decisions and, at the same time, dilutes, almost completely, any kind of responsibility. One has to outline the way to follow and plan it in time, establishing some mechanisms of control that verify the advances that are being accomplished as well as the drawbacks that, following-up the organized plan, could appear.

The administrative powers responsible for water supplies should urgently establish a series of technical and legal regulations to clarify the current situation. The objective of these technical regulations would be to determine which are the minimum levels of quality within which the water supplies should be managed. Among other questions, the regulations should stipulate the degree of reliability of the water sources and the water quality, not only from its source point but also at the consumption points. They should also stipulate the minimum water network efficiency; the monitoring and automatization required to control the water supplies in each case; the preventive and corrective maintenance to be carried out for each element of the network, their rehabilitation or their renovation. Finally, they need to stipulate the guidelines and order of magnitude of Performance Indicators that make possible an economically balanced development of water supplies. All this, in short, to favor water saving and to reach the necessary quality standards of a modern country.

There is also a need for legal regulations that clarify the increasingly intensive and confusing privatization process to which water supplies are being submitted. Regulations should be established to stipulate how such privatizations should be carried out; at the same time they have to state definitions for any outstanding and existing contracts, and to clarify the responsibilities of each party to the contract. All this should be carried out from the perspective of a clear defense of the consumer or, what is the same, that all kinds of commercial relationships are based and presided over by the ultimate objective of carrying out an impeccable management of a modern and reliable system of water supply at the least possible cost.

With the current structure of the management of water in Spain, referred previously, there seems to be no administrative organism that could assume all these responsibilities. Perhaps, because of tradition and greater protagonism, it should be the Environment Ministry concerns, without doubt, with the collaboration and support of the town halls who are ultimately responsible for the water supplies. It also seems logical that there should exist an institution, at the national level or autonomous regions level, entrusted with auditing and effecting control over the state of the different water supplies. It seems clear that, with the current diffuse designation of responsibilities, nobody feels that they have the ability, the responsibility, the authority, or the necessary motivation to assume such a challenge.

Analyzing recently the management of water (Cabrera, 1997), within a much broader context than the framework of water supplies which is our preoccupation in this paper, we proposed an administrative structure that grouped under the same umbrella the control of resources as well as that of consumption and use, which would end the current divorce that exists in Spain in this regard. That idea is not original, far from it, and since being a common sense answer, it is in line with the tested and efficient model used in the modern state of Israel. Its structure has been described by Tamir (1993). Sincerely, we believe that it will be difficult for the water supplies in Spain to get being modern and periodically controlled and audited, so as to be able to face with full guarantees a drought, if previously no administrative structure has been set up to make it possible.

**4. Drought management in urban water supply systems**

Referring to temporary water service interruption Lund and Reed (1995) assert, in a recent paper, the following: *Many urban water supplies in less developed countries lacking effective customer metering and other means of curtailing water consumption, ration water through rotation of service outage, allowing each sector of a city, say, only several hours of water service each day. Although doubtless inconvenient and economically inefficient by almost any standard, and having public health risks, it is a practical approach for desperate and relatively uncontrolled conditions.*

It does not seem necessary to insist on the obvious inconveniences that consumers have to suffer from the temporary interruption of the water service. As far as sanitary risks are concerned, they are a consequence of the shutting-down and the starting-up of the system. When the service is interrupted, the pipelines at the high points of the network are depressurized as consumers continue demanding water. The existence of leaks also favours such circumstances. This makes possible, in a set of pipelines in the system, either the air entrance, or the entry of untreated water coming, for example, from the subsoil and through the cracks where previously water leaked out (the greatest potential danger lies in the water from subsoil originated from leaks in the sewer system). The breakdowns, which are caused mostly by the trapped air previously admitted, as well as turbidity as a consequence of the reversal flow when starting up the system, increase notably (González, 1995), causing quite a lot of sanitary problems as well as affecting the management of the water network. Inconveniences and unhygienic conditions are, in drought periods, the high prices that the consumer has to pay as a consequence of a barely professional management and, in any case, inappropriate of a modern country at the doors of the third millennium.

In other words, the temporary interruptions of the water service submit the network to operational conditions which are far more unfavorable than under normal circumstances. Breakdowns increase and the reliability of the system is deteriorated. This practice, without doubt, must in our country be eradicated completely. And the only way that leads to this is the improvement on a day to day basis of the system and taking advantage, above all, of periodes such as the current one, of abundant resources, and which provide sufficient time so as to be able to obtain important results.

We can think that only habit and custom, that has familiarized us with and made us accept this situation, prevents us from becoming aware of the serious defects that underlie this practice. It is inconceivable in other urban services (electricity, gas, telephone, etc). Only two *advantages* justify this practice:

- The possibility of water saving on leaks in low efficient networks.
- It avoids the important extra management effort that more coherent methods of rationing water demand. To these more coherent methods we will refer in the next section.

It seems evident that there is a need to ration water in periods of drought, as a consequence of the less quantity of available water resources. It is a customary practice in any country in the world. But this can be obtained, in addition to interrupting temporarily the service, with alternatives that are more logical and modern. Lund and Reed (1995) envisage, in addition to temporary interruption of the water service, five other alternatives:

- Rationing by fixing a given consumption level.
- Rationing by percentile reduction of the consumption.
- Rationing based on an increase in the water price.
- Rationing through the control of specific uses.
- Rationing by saving credits.

Each procedure presents advantages and inconveniences, therefore one should choose the alternative which is best adapted to each case. One can even choose the combination of two of these procedures. A short discussion of each one of these alternatives follows. More details appear in the original reference.

## 4.1 RATIONING BY FIXING A GIVEN CONSUMPTION LEVEL

This consists in defining a maximum consumption quota for each consumer taking into account their specific needs, fundamentally the number of family members. For industrial consumers, the usual practice is to consider the number of employees, the occupied area, and, of course, the manufacturing process. Previous consumption is also taken into account. Once established their consumption limits, when consumers exceed the preset quota, the economic penalty, if one wants to obtain satisfactory results, must be significant.

This procedure presents some inconveniences because of the rigidity of fixing a single quota without taking into account particular cases. Its drawback is its potential inflexibility. Consequently consumers that, because of their needs and/or habits, demand in normal periods a quantity of water superior to that by other consumers, are clearly prejudiced. Lund and Reed illustrate this drawback, by referring to the management of drought in the municipality of East Bay, California. Detached single family housing, with a small garden and facing the sun (towards the East) have a habitual average consumption of 2.24 $m^3$/day, while those which face towards the damper West consume around 0.90 $m^3$/day. In these conditions to ration family consumption to 1 $m^3$/day impacts strongly on the first group of housing, while the adopted restrictive measures do not suppose any inconvenience for the other group of houses.

Before carrying out such a policy in any population, an analysis must be made that enables the outlining of the most just and impartial strategy possible. As there may be many circumstances that influence the situation, the problem of weighting them adequately is not a simple task.

## 4.2 RATIONING BY PERCENTUAL REDUCTION OF THE CONSUMPTION

This is the option most often selected, in a drought period, by many municipalities in the area of San Francisco. This strategy is based on limiting the month average consumption at a given percentage which depends on the quantity of water one wishes to save. Usual values set around 75%.

The disadvantage it presents is that it provides incentives for water consumption in normal hydrological periods. On the other hand, and because a lack of experience, one does not know the restriction to be applied to consumers who have been recently connected. However, it is a rational criterion. The abuses can be penalized with an adequate policy of water rates, based on increasing prices of the cubic meter, applicable to progressive consumption blocks.

### 4.3 RATIONING BASED ON AN INCREASE IN THE WATER PRICE

Increasing the water price of a second consumption block, without modifying the costs of the consumer basic needs, permits the water management companies, in drought periodes, to maintain their income level. It is a right decision, because in this situation a maximum management effort is done. Other methodologies are accompanied by an important loss of economic resources, a frequent complaint of those who manage urban water supplies. As they cover with incomes a good part of the fixed costs, imbalances may appear between incomes and expenses if the water consumption is reduced.

It is a highly effective decision, if prices are notably increased. The reduction of the consumption is given by the elasticity of demand defined as:

$$elasticity = \frac{percentage\ \mathrm{var}iation\ of\ the\ water\ consumption}{percentage\ \mathrm{var}iation\ of\ the\ water\ price}$$

Considerable reductions in water consumption only appear as a result of important increases in water price, since water is absolutely essential to the consumers and so they do not easily dispense of it. Consequently, it presents low elasticity. Its value depends on the economic level of the consumers and on previous water prices. Its value normally moves in the interval between -0.2 ÷ -0.5 (AWWA, 1992). Therefore, supposing an elasticity value of -0.2, doubling the water price is equivalent to reducing the consumption by 20 %. If located in the other end of the interval, this effect could be obtained with a much more moderate increase in the price (40 %). Therefore, it is of great interest for a water supplier to know the demand elasticity. If one intends to apply this methodology, the elasticity must be determined in advance. In this way, the price can be adjusted according to the objectives one is pursuing.

### 4.4 RATIONING THROUGH THE CONTROL OF SPECIFIC USES

The prohibition of certain water uses (car washing, irrigation, street cleaning, water supply to fountains, filling of swimming pools, etc.) is a measure usually to be adopted in drought periods.

This method is useful to draw the user's attention and, therefore, only serves to be applied during a short time. It presents a double advantage: it is easy to put into practice and makes the user aware about the need of saving water. Consequently, it is useful as a first measure. If the drought goes on, this method must be reinforced with complementary restrictions.

### 4.5 RATIONING BY SAVING CREDITS

Certain rationing strategies can be made more flexible by permitting consumers to accumulate saving credits by low water consumption in normal periods. Thereafter, they will be able to spend these credits in drought periods. This allows those who have a moderate water consumption, not to be so affected by the restrictions in drought periods. The system, logically, is activated when the drought is envisaged, being this moment the starting point for the plan.

This method is not appropriate for short duration droughts because a limited time does not permit to establish and to communicate the credit system that is going to be adopted. Nor does the consumer have time to be mentally prepared for such a system. The consumer understands that it is a transient situation, of a rapid normalization, and that, consequently, it does not offer the consumer the possibility of benefiting from the saving and effort that is asked for. On the other hand, this procedure enhances the water conservation, because the consumer try to get saving credits in periods of normal supply.

### 4.6 SOME FINAL CONSIDERATIONS ABOUT THE RATIONING POLICY

The five strategies we have just discussed have a point in common: they are useful in high efficiency water supplies, with a considerable technology level and a detailed knowledge of consumptions. If these conditions are not given simultaneously, there remains no other solution: the service must be temporarily interrupted.

It is clear that an adequate combination of the five strategies described above enlarges the possibilities of the process. With their application, or through any other procedure properly studied, the drought management can be adapted in a very effective way to the needs and idiosyncrasy of each case.

On the other hand, if the supply is properly managed, there is no doubt that to act within this framework is beneficial for the different parts involved. The water distribution company acquires prestige earning the confidence of the consumer. The citizens tolerate the drought difficulties in a much less traumatic way and, what is more important, without any type of sanitary risk. The water saving is always superior to that provided by the temporary interruption of the service, since at no time the water is lost in leaks. All the water is used in an effective way. The only losers are the mineral water traders.

## 5. Technological level of the urban water supplies and their relationship to drought management

Only rationing through the control of specific water uses does not require in its implantation a high technology level. However, as it has already been discussed, such a procedure is not so useful in the long term. Nevertheless, with the remaining alternatives we have been discussing, or any other combination in which they might appear, what is really required is a water network with a high hydraulic efficiency and a detailed and exhaustive control of consumptions. We, briefly, review these questions.

## 5.1 ACTIONS TO IMPROVE THE HYDRAULIC EFFICIENCY OF A NETWORK

The hydraulic efficiency of a network is the rate between the water measured by the household flow meters and the water registered by the flow meters installed at the supply points to the network. The difference between both quantities is designated as *not registered water*, and it is the sum of the network leaks and of the uncontrolled water consumptions, mainly official consumptions without flow meters, public fountains consumption, parks and gardens irrigation, illegal connections, undercounting by household flow meters, etc. The hydraulic efficiency equals to one when all the water injected to the network is registered by consumption flow meters.

The calculation of the network efficiency must combine the continuous injected flow measurements with the measurement, generally bimonthly, of household flow meters. Furthermore, one must take into account the impossibility of measuring simultaneously the consumption of all consumers, because readings of household flow meters are taken according to preset itineraries on which the workmen spend the two months between two successive billings.

Supposing a network properly equipped with flow meters, the improvement of its efficiency has to pass through a set of investments, actuations and strategies that demand a significant level of technology and professionalism. We highlight the most relevant ones:

- Periodic and systematic campaigns of leak detection throughout the whole distribution network and, most especially, in those areas where the pipelines have a considerable age.
- A programmed preventive and corrective maintenance plan, that takes into account the mean life of the different elements of the network, for their inspection, rehabilitation or replacement.
- A monitoring (permanent flow rate and pressure measurements in the most strategic points of the network) to detect any breakdown or important failure in the shortest time period.
- A network sectorization permits to determine by areas, and independently, the efficiency for every one of them. By this way, one can get to know the areas in which one must carry out a more intensive strategy of leak detection. The network sectorization will only be fully effective when the flow rates are registered along the lines that feed it. This allows to find the relationship between consumption in that sector and entry supplies, evaluating the corresponding partial efficiency.
- A command and adequate control of the different valves that regulate the distribution network so that the pressure levels get their optimum values. Neither too low, inasmuch as they would not give a service of quality to the consumers, nor too high which can cause potential existing leaks to increase their water losses. To keep the network pressures within an optimum operation range demands sufficient pipeline diameters so that not to originate excessive head losses at peak consumption hours. This, furthermore, is the only solution to avoid household water tanks. The water

must always be stored at the network reservoirs, where the managers responsible for the water supply can guarantee its sanitary quality.

- To use a calibrated mathematical model of the network. It constitutes an excellent tool in the technique of locating leaks, either in a direct way, or through the so-called inverse method.
- To control, periodically, the reliability of the different measurement devices. As time goes by, the flow meters tend to register inferior volumes to those that pass through the pipe, a phenomenon known as under-register.
- All consumptions must be measured, including public areas (fountains, garden irrigation, public buildings, etc). Otherwise, and even when accepting that there were no leaks, it is impossible to obtain high efficiencies. It is, furthermore, a way to detect potentially illegal connections.

This set of action plans can not be improvised in any way. Furthermore, it has an important cost and, therefore, it cannot be carried out unless it is reflected in the water price. There are many water supplies in this country that do not keep a good part of the preceding points.

## 5.2 THE IMPORTANCE OF HAVING A COMPLETE AND DETAILED DATA BASE

Evidence showing that one can not manage what is unknown remarks the importance of having complete databases about the water supply to be managed. From the drought management perspective, knowledge of consumptions is possibly the most important fact. The reviewed strategies of droughts management are based on this. But complete and modern management of a water supply requires the knowledge of other data besides the water consumptions on a day-to-day basis. And it is the actual trend to organize them into two blocks:

a) Data and characteristic parameters of a water supply. With greater frequency, databases are structured within a general context which has been called the "Geographical Information System" (GIS). They should define perfectly the structure and constitution of the water supply as well as the most representative values of their daily management. From an exclusively technical perspective (not in an administrative sense), the principal sections to be included would be:

- Supply sources and production plants.
- Complete and updated cartography of the distribution network, as well as physical details of its constitution, indicating diameters, materials, conservation states, repairs, etc.
- Detail of the regulation and control devices, with indication of their state.
- Exploitation and maintenance operations in the whole system.
- Structure of the consumers. Consumption data and other details. Remember that to apply certain rationing strategies it is also necessary to know the number of persons behind each consumer and their connection point in the network. For example, it is

very important to know rapidly the relationship between a given breakdown and the affected consumers.

- •Details about the operation of electromechanical installations to elaborate a preventive maintenance program.
- •Evolution of the significant variables of a water supply system: Injected and circulating flow rates, reservoir levels, pressures, etc.

b) Performance Indicators of a water supply. They are not in themselves specific magnitudes but rather meaningful relationships between them. The best known, perhaps, is the hydraulic efficiency of a network. Logically, a good part of the relationships that provide for Performance Indicators can be obtained from the data structure which the Geographic Information System provides.

Performance Indicators inform about the state of a water supply system and, consequently, they are fundamental to make successful decisions. Literature details reference optimum values for the Performance Indicators, which allow to know the strong and weak points of the system. It is clear that, for a good drought management of water supplies, one must take into account mainly those which in such circumstances present greater relevancy.

Performance Indicators are structured into different sections depending on their assigned objectives. The following five ones configure quite a logical structure:

- Performance Indicators about the supply structure.
- Performance Indicators about the staff and their productivity.
- Performance Indicators about maintenance and operation.
- Performance Indicators about economic management.
- Performance Indicators about the quality service.

It is not the object of this paper to discuss about the configuration and contents of these databases. There exists in the technical literature information about that topic: Hirner (1997), Happy (1997), Cabrera and García-Serra (1997), Janssens et al. (1994), Parsons (1997), Buenfil and Piña (1996), SGWA (1997), Cubillo (1997), Larsson et al. (1997), etc. The interested reader might like to refer to them.

The objective of this short description is to raise the awareness of the political agents with responsibility for our water supplies, as well as the managers and technicians in charge of these services, of the importance that the Performance Indicators have in the modernization of a water supply. In fact the Performance Indicators could be an excellent and comfortable mechanism for the Administration to control the state of a public or private water supply system.

## 6. Drought management planning for urban water supplies

A modern supply must supply water, without any interruption, in quality and quantity and at the pressure required by the consumers. This is the day-to-day objective. But, in a drought period, it is essential to carry out a greater control of the available resources and, to do this, one must design a clearly defined strategy. It is what is known as a drought management plan for the water supply. According to the American Water Works Association (AWWA, 1992), a plan can be structured in six clearly defined steps:

- Implication of the consumers and the public agents.
- Defining the basic goals and objectives of the drought management plan.
- Assessing the supply and demand conditions accordingly to the preceding objectives. This is equivalent to deciding what rationing methodology, or combination of procedures described in section 4 of this paper, is adopted.
- Monitoring the intensity and severity of the drought through specific drought indicators.
- Identifying and assessing drought mitigation measures.
- Developing a drought index and management strategy.

These steps are described and fully developed in AWWA (1992) and also in Cabrera and García -Serra (1997). In Spain, to our knowledge, only the Canal "Isabel II" has developed a Drought Management Plan. In other industrialized regions, as for example the west of the USA, the whole water supplies have one of them. There are a lot of references about how different droughts have been managed (for example, Gilbert et al., 1990; AWWA, 1992; Boulos et al., 1997). Having a well-defined plan means not to improvise. Previous planning is used so as to know in each moment what to do depending on the seriousness of the situation. In any case, it is supposed a water supply system with some reasonable levels of Performance Indicators.

Being the objective of this paper the drought management in urban water supplies, it seems reasonable to choose from the various Performance Indicators those of greatest effect on water saving. They could be grouped into a synthesis paragraph entitled Performance Indicators related to drought management.

It is important to highlight that rationalizing the water use means investment and technology. But contrary to industrial rationalization that in most sectors implies labour losses, the promotion of water saving and conservation creates stable employment and it is, in the medium term, economically profitable. There exist in this regard a lot of papers. We can emphasize from among them those of the Environmental Agency Demand Management Center (1996), Rocky Mountain Institute (1994) and the National Regulatory Research Institute (1994). The European Union, in a report by the General Directorate V (DG V, 1995), has detected efficient water management as one of the most promising sources of employment for the twenty-first century. Being unemployment the principal problem now existing in Spain, it does not seem irrelevant to consider such a circumstance.

One must plan, and to plan in time. To forget now the droughts, and above all in their potential incidence on the most sacred uses of water, urban water supply, is a serious irresponsibility. In Spain cities as Bilbao have suffered from drought, which in fact was unimaginable (Eizaguirre and Silveiro, $1991_a$ and $1991_b$), and the same could be said for countries such as England, whose restrictions in their water supplies, during 1995, have been important. In line with these problems, we conclude with the following thought (AWWA, 1992):

During the last decade practically all the regions in the United States have suffered droughts and their consequences on urban water supplies. The lesson to be learnt is that even areas which are relatively rich in this resource are exposed to the risk of severe droughts. Besides, the vulnerability of the water supplies in relation to droughts grows in time due to the increasing water demand. The permanent possibility of a climatic change adds, furthermore, uncertainty and additional risk if in the future the droughts reappear with greater frequency and intensity, as some people foresee.

## 7. Conclusion

We can draw the following conclusions:

- Only in periods of hydrological prosperity, and only with time and the necessary planning can the adaptation and modernization of a water supply be carried out so as to be able to manage efficiently the next drought.
- It is essential to have a modern and reliable water supply system. Control over its particular characteristics can be carried out through Performance Indicators.
- The current state of the water supplies in Spain is not the most adequate to face up to extreme events. To improve such a situation some political actions have to be done. The strategy to follow could be based on:
  - Setting-up an administrative organization in order to monitor and control more efficiently our water supplies.
  - Establishing technical and legal regulations for urban water supplies; at present they are non-existent at a national level.
  - Providing training and technological support for the public water supplies. Not to do this is the same as pushing them towards privatization, by placing municipal management of water supplies in a position of technological and professional inferiority with respect to the private companies. Denmark, for example, provides this through the "Danish Supply Foundation".
  - Setting a greater control over the water supply operations, so in systems with public as with private management.
  - Persuading the consumers to assume the economic sacrifices that these actions require, in the same way that the people understood the need of fulfilling the economic convergence criteria of Maastrich. It is the only way to get a more sustainable management of water.

The current hydrological circumstances, the political and economical situation of this country, and the will to transform and modernize the Spanish Society make the actual moment the best one. If we do not act in this direction episodes such as those happened in the period 1991-95 will happen once more. It does not seem reasonable.

## 8. References

ALEGRE, H. (1997). *A General Framework of Performance Indicators in the Scope of Water Supply.* Proceedings of the IWSA Workshop on Performance Indicators for Transmission and Distribution Systems. Lisbon, Portugal May 5-7, 1997. Ed LNEC

AMERICAN WATER WORKS ASSOCIATION.(AWWA) (1992). *Drought Management Planning.* Ed. AWWA. Denver. Colorado. USA.

ASOCIACIÓN ESPAÑOLA DE ABASTECIMIENTOS DE AGUA Y SANEAMIENTO (AEAS) (1994). *El Suministro de Agua Potable y Saneamiento en España. 1.994.* IV Encuesta Nacional de Abastecimiento Saneamiento y Depuración. Ed. AEAS. C/ Orense nº 4. 28.020 Madrid.

BARRAQUÉ, B. (1995). *Les politiques de l'eau en Europe* . Cap. V. p.p. 85-102. L'Espagne. Editions La Déconverte. 9bis, rue Abel Hovelaque, 75013 Paris.

BOULOS, P.; MAU, R.; RINGEL, D.J.; GLASER H.T. (1997). *California's Approach Managing Water Supplies During Droughts.* Actas del Curso Internacional de la U.I.M.P. Gestión de Sequías en Abastecimientos Urbanos. Valencia. 9 al 13 de Diciembre de 1997.

BRUINS, H.J. (1993). *Drought Risk and Water Management in Israel. Planning for the future.* Drougth Assesment, Management and Planning: Theory and Case Studies. Ed. D.A. Wilhite. Kluwer Academic. Publishers. pp. 133-156.

BUENFIL, L.M.; PIÑA, R. (1996). *Sistema de evaluación de eficiencia en empresas de agua. Sec. A. Manual de usuario.* Instituto Mexicano de Tecnología del Agua. IMTA. Cuernavaca. Mexico.

CABRERA, E. (1997). *La gestión del agua en España. Necesidad de planificar un futuro mejor.* Ponencias de XXIII Reunión de Estudios Regionales. Revista Valenciana de Estudios Autonomicos. Generalitat Valenciana. pp 187-211.

CABRERA, E.; GARCÍA-SERRA. J.(1997). *Problemática de los abastecimientos urbanos. Necesidad de su modernización.* Ed. Grupo Mecánica de Fluidos. Universidad Politécnica. Valencia.

COUREY, P.J.; HALL, M.J.(1995). Rehabilitation and Leakage. A Joint Approach. *Journal Aqua*, vol 44. Nº 4. pp 196-201.

CUBILLO, F. (1997). *Information Systems: A Key Factor to Manage Performance Indicators.* Proceedings of the IWSA Workshop on Performance Indicators for Transmission and Distribution Systems. Lisbon, Portugal, May 5-7, 1997. Ed. LNEC.

DG V. EUROPEAN COMMISSION. (1995). *Local Development and Employement Initiatives.* Internal Document. March, 1995. SEC 564/95.

ECONOMIC RESEARCH LIMITED. (1995). *The economic impact of water conservation: cases studies in Ontario.* Burlington. Ontario. October 1995.

EIZAGUIRRE, J.M.; SILVEIRO, A. (1991a). Experiencias de una sequía. 1ª parte: Descripción del problema. *Revista Tecnología del Agua.* October 1991. pp 93-103.

EIZAGUIRRE, J.M.; SILVEIRO, A. (1991b). Experiencias de una sequía. 2ª parte: Actuaciones y Resultados. *Revista Tecnología del Agua.* November 1991. pp 33-87.

ENVIRONMENT AGENCY. DEMAND MANAGEMENT CENTRE (1996). *Water Conservation Planning. USA Case Studies Project.* Ed. Amy Vickers & Associates. Inc. Amherst. Massachusetts. USA. June 1996.

GAZZETTA UFFICIALE (1994). *Disposizioni in materia di risorse idriche.* Boletín Oficial del Estado Italiano de 19.1.1994. pp 5-20. Legge 5 gennaio, n 36.

GILBERT, J. et al. (1990). Reducing Water Demands during Drought Years. *Journal of the American Water Works Asssociation*. May 1990. pp 34-39.

GONZÁLEZ, A. (1995). *Impacto de una sequía en un abastecimiento urbano.* Actas del Seminario de Iberdrola sobre los problemas del agua. Valencia, November 1995. pp 307-334.

GRECO, C. (1997). Il sistema tariffario dei servizi idrici fino al 1996. *L'Aqua. Rivista bimestrale dell'Associazione Idrotechnica Italiana.* Gennaio-Febbraio 1997, pp 74–77.

HIRNER, W. (1997). *Technical, Operational and Economic Performance Indicators of Water Utilities. Part 2.* Proceedings of the IWSA Workshop on Performance Indicators for Transmission and Distribution Systems. Lisbon, Portugal, May 5-7, 1997. Ed LNEC.

INTERNATIONAL WATER SUPPLY ASSOCIATION (IWSA) (1993). *10 Theses for a Drinking Water Tariff Policy.* Internal document.

JANSSENS, J.G. et al. (1994). *Development of a framework for the assessment of operation and maintenance performance of urban water supply and sanitation.* Proceedings of the IWSA Specialised Conference on The Quality of the Service. Amsterdam. September. 1994.

LARSSON, M.; STAHRE, P.; ADAMSSON, J. (1997). *Performance Indicators. The Swedish Experience"* Proceedings of the IWSA Workshop on Performance Indicators for Transmission and Distribution Systems. Lisbon, Portugal, May 5-7, 1997. Ed LNEC.

LUND, J.R.; REED, R.U. (1995). Drought Water Rationing and Transferable Rations. *Journal of Water Resources Planning and Management.* ASCE. Vol 121. N° 6. November December 1995. pp 429-437.

LLAMAS, R. (1997). Declaración y financiación de obras hidráulicas de interés general. Mercado del agua. Aguas subterráneas. Planificación hidrológica. *Revista Ingeniería del Agua*. September 1997, vol. 4. N° 3.

MARTÍNEZ-GIL, F.J. (1997). *La nueva cultura del agua en España.* Ed. Bakeaz. Avda. Zuboroa, 43-B. 48012 Bilbao.

ORGANISATION FOR ECONOMIC COOPERATION AND DEVELOPMENT (OECD) (1997). *Environmental performance reviews: Spain.* OECD. Publications 2, rue André-Pascal, 75775 Paris Cedex 16.

PÉREZ-DIAZ, V.; MEZO, J.; ALVAREZ-MIRANDA, B. (1995). *Política y economía del agua en España.* Edita Circulo de empresarios. Serrano 1, 4°. 28001 Madrid. Marzo 1995.

PARSONS, D.P. (1997). *Managing Benefits of Mains Rehabilitation Through Structured Surveys.* Proceedings of the IWSA Workshop on Performance Indicators for Transmission and Distribution Systems. Lisbon, Portugal, May 5-7, 1997. Ed. LNEC.

SGWA. (1997). *Water Statistics 1.995.* Swiss Gas and Water Association. April, 1997.

TAMIR, O. (1973). *Administrative and legal aspects of water use in Israel.* Water in Israel. Part A. Selected Article. Ed. Ministery of Agriculture. Water Commission. Hakirya. Tel-Aviv. March 1973.

THE WATER PROGRAM ROCKY MOUNTAIN INSTITUTE (1994). *Water efficiency. A resource for utility managers, communitary planners and other decision makers*. Rocky Mountain Institute. Colorado. USA.

THE NATIONAL REGULATORY RESEARCH INSTITUTE (1994). *Revenue effects of water conservation and conservation pricing: issues and practices.* The Ohio State University. 1.080 Courmack Road. Columbus. Ohio. USA.

# GEOGRAPHICAL INFORMATION SYSTEMS (GIS) APPLIED TO WATER SUPPLY SYSTEMS (WSS)

IGLESIAS REY, PEDRO
IZQUIERDO SEBASTIÁN, JOAQUÍN
LÓPEZ PATIÑO, GONZALO
MARTÍNEZ SOLANO, JAVIER
*Fluid Mechanics Group*
*Universidad Politécnica de Valencia*
*Camino de Vera s/n*
*P. O. Box 22012. - 46071 VALENCIA (SPAIN)*

**Abstract.** In this paper, a general overview of a Geographical Information System is presented. First, a definition of some basic concepts about what a GIS is will be given, paying special attention to the topics related to the Georreferenced Databases. Afterwards, some general applications of the GIS technology to the Management of Water Supply Systems will be presented. Finally, a discussion on the potentiality of the GIS technology is developed.

## 1. Introduction

The management of a Water Supply System (WSS) involves the use of a huge amount of information, either to know its hydraulic performance or to efficiently administer the existing resources. This information, even though diverse in nature, can be gathered into three main groups depending on its nature and the usage it will be given. These groups are:

- Physical features of the network elements
- Economic information of the WSS
- Spatial information about the location of both economic and physical data

Traditionally, this information has been saved in different formats. The information about network elements (diameters, lengths, starting dates, suppliers, etc.) was saved in work plots or small inventory databases. The economic



*E. Cabrera and J. García-Serra (eds.), Drought Management Planning in Water Supply Systems,* 22–51.

information was the most jealously protected within the system, where consumer data (demand, address, registering date and other data necessary for a correct economic management of the system) were recorded. Finally, spatial information was usually scattered in various topographic maps where the isolines of the supplied geographic area, the location of the mains and the distribution pipe network layout appeared most of the times as handmade material needing updating. In any case, connection among the three information systems hardly existed.

However, nowadays the need of linking spatial, economic and physical information together is more frequent. This is possible thanks to the implementation of a proper Geographical Information System (GIS). This system allows not only to link geographic or spatial data with another alphanumeric data, but also to update data in a simple way, by means of an appropriate graphical interface.

In the present chapter, an introduction of the GIS concept will be given, briefly reviewing the elements it consists of and different solutions for the information storage.

Afterwards, various applications of the GIS within the limits of the WSS will be presented. In this point the so-called Leakage Management System, one of the GIS applications, will be addressed. In such a system, a policy of GIS usage for the reduction of the uncontrolled flows and the improvement of the volumetric efficiency of the network is proposed. This proposal is included on the framework of *Drought Management Planning in Water Supply Systems*, which is the aim of this book.

Finally, some proposals of future trends of the GIS technology applied to the Water Supply Management are developed.

## 2. Georeferenced Databases

### 2.1 DEFINITION OF A GEOGRAPHICAL INFORMATION SYSTEM

There are several definitions of GIS. In any case, even though some are more comprehensive, all of them coincide in highlighting two GIS basic features:

- A GIS is basically a database that must have the same tools than ordinary databases.
- A GIS stores and links spatial data (position or location) together with thematic data (alphanumeric attributes).

Accordingly, one of the simplest definitions that can be given for GIS is *a system for the geographical information capture, storage and analysis* (Hernández Rodríguez, 1995). This is, without any doubt, a definition that shows the origin of GISs as the integration of CAD software with digital cartography management utilities and linked with database manager software.

The difference among the different GIS commercial solutions comes, for one thing, from the type of spatial data they manage, and for the other thing, from the way these data are stored and related with attributes.

It is also possible to understand a GIS as a management philosophy. In this case, it would be a way of making decisions within an organization based on information that is managed in a centralized way but ordered depending on its geographical location.

A more global definition should identify three interconnected components within the GIS concept, as shown schematically in Figure 1 below (Parsons, 1997):

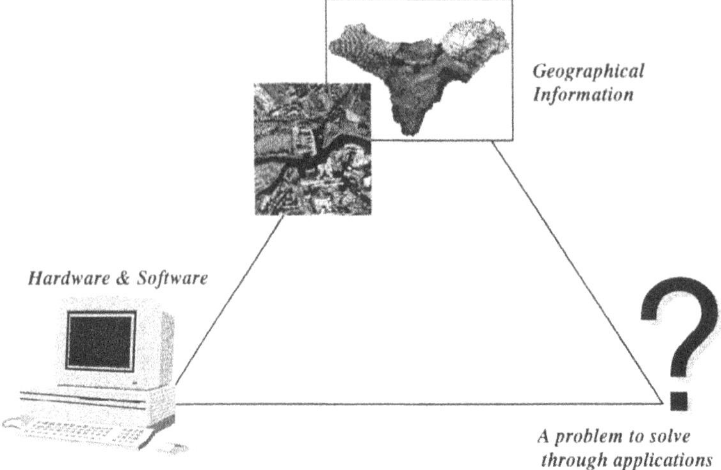


**Figure 1. Schematic Representation of GIS components**

- A set of data ordered depending on their geographical location
- Some equipment and software to manage and link these data
- A specific problem or target that is intended to solve from the data spatially distributed with the aid of the available tools

### 2.2 MODELS OF SPATIAL DATA

The spatial data models used by GISs can be basically classified into two elemental models: *raster model* and *vector model.*

#### *2.2.1 Raster model*

Raster models of spatial data representation emulate the real world by making use of a regular grid. In a raster model, both contour and interior of objects are represented. A grid or a mesh whose cells share shape and size is usually applied. In this way, a kind of matrix in which cells contain values corresponding to certain variable magnitudes is obtained. Stored magnitudes may be both qualitative (as is the case of land uses or type of water demand) and quantitative (for example, elevation or demand in every point of a network).

Several data structures have been described to store permanently this information. One would be based on exhaustive enumeration. It consists basically of gathering in an individualized and sequential way the contents of each cell. Obviously, when the plot is too big and/or the cell size too small, the amount of information turns to be overwhelming, what implies the main inconvenient of this kind of structure. As an alternative to optimize the resources, a row enumeration is proposed in which the value of the variable and the first and the last column in which the value is taken are given. This structure is especially suitable in systems with clear spatial correlation, since many contiguous cells have the same value. Occasionally, only the variable value and the last cell taking it are stored. The column where the next set of cells sharing a different value starts is understood. Sometimes, the value and the number of cells sharing it are stored.

No matter the data structure chosen to store the information contained in a plot, the main limitation of the raster model is that a different plot must be recorded for each variable. This means that it is necessary to store a *layer* of information in the way described above for every variable. In Figure 2, a scheme of how the information from a plot is stored in a raster model is shown. The different cells of the superimposed grid are given a value depending on the land use in the original plot. In this case, a color code was associated to each land use in the plot. For a given cell the land use taking the biggest percentage within it has been represented through an alphanumeric code (a letter).

This kind of data model is specially suitable for geographical variables in which few regular geometric shapes are found and the exact shape of the areas with same value should be described by using many polygons and segments. Thus, its main application is terrain description (Digital Elevation Models, Aspect Maps, Slope Maps, etc.).

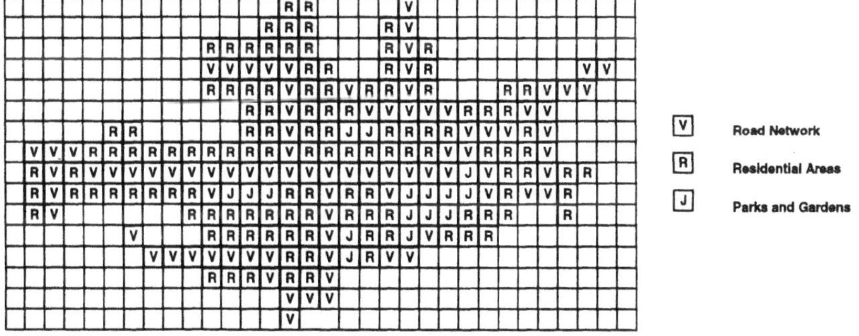


**Figure 2. Scheme of the raster representation of the information contained in a map**

### *2.2.2 Vector model*

Another way of representing spatial data is based on some reduction of the areas where variables take a fixed value to simple geometrical objects (such as points, lines or polygons). In its simplest form, spatial data are stored as sets of points or vertexes and instructions on how these elements are connected. Another alternative is that data suitable to be represented by lines (the case of the layout of a water distribution network) be stored by means of mathematical functions, for instance a combination of polynomials. Two main differences between vector model and raster model can be drawn from the above descriptions. The first one is that in the case of a vector model what is explicitly stored is the boundary or limits of the map geographical objects but not their inner contents. The second difference is that only one value of the considered variable is stored in each object for a vector model. For example, for a specific main consisting of several pipes or segments with homogeneous characteristics, only one value of this characteristic will be recorded, while, as mentioned above, in the raster model one value of the characteristic will be recorded for each cell where the main goes through.

In Figure 3 a schematization of the vector mode representation of the information of the map shown in Figure 2 can be observed. Streets are represented by lines of a certain color and width, while areas covered by vegetation (parks and gardens) or buildings are represented by polygons with different fillings.

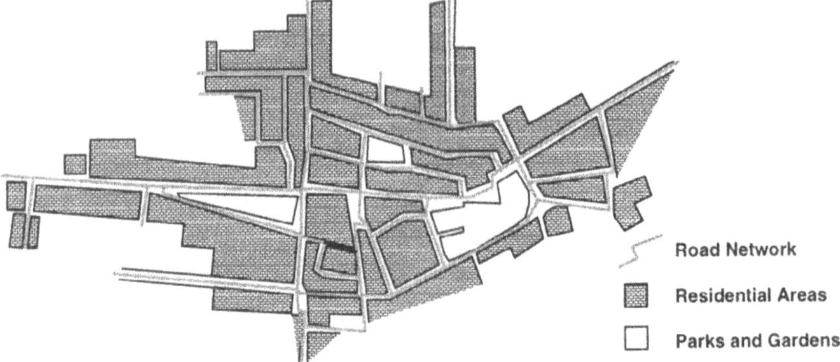


**Figure 3. Schematic representation in vector mode of the information contained in a map**

## 2.3 STRUCTURES OF GEOGRAPHICAL DATABASES

One of the characteristics of Georeferenced Databases is the capability to hide their structure and contents to the final user, in a way that some safety rules can be set. Even though it is not necessary for a GIS user to understand in detail how data are ordered and stored in the memory of the computer, some knowledge does help to understand how the system works and which the advantages and limitations of the database manager system are.

The main characteristic of any data storage system is that it allows both quick access to data and cross references. There are several ways to achieve this, from simple *file lists,* trough *sequentially ordered files* and *indexed files,* to more complex structures used by the database manager. As the number of records in the database increases, its structure plays a more and more relevant role.

There are four main types of database structure of common use in the GISs. These are *hierarchical, network, relational and object oriented* databases. The first two types were very widespread during the seventies and eighties while relational databases dominate nowadays. The object-oriented structure is still under development. It will be commercially available with the advent of more powerful computers. Currently, the object-oriented approach applied to relational databases is becoming more and more popular, mainly because it adds the new *object taste* to the well-known relational structure.

### *2.3.1 Hierarchical Structure*

This structure provides quick and effective ways of accessing data when those are stored with a one-to-many relation and also when a clear relation father-son among data can be established. Some authors name the hierarchical structure as *tree structure* or *branched structure*, since the model is formed with a number of links connecting every record in a tree. The root is the first level of the hierarchy. The root is linked with one or more records of lower level in the hierarchy, and these with one or more records of an even lower level. As a consequence, each element is connected to only one from a higher level, called *father*, and one or more of a lower level, called *sons.*

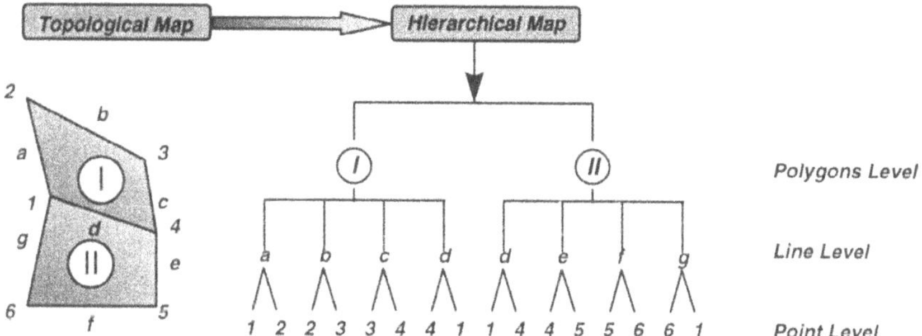


**Figure 4. Hierarchical structure of a Database**

Figure 4 shows an example of a hierarchical database. As can be observed, there are neither links between elements in the same level nor diagonal links. This is the main inconvenience of this structure, since once the model has been established, it is not possible to create new links between elements. On the other hand, this structure allows high-speed access in huge databases, together with easy update and inclusion of new data in the lowest level.

### *2.3.2 Network Structure.*

In most of the GIS applications, the only one-to-many relation offered by the hierarchical structure leads to a great amount of redundant information (as shown in

Figure 4). However, the many-to-many links are necessary, since diagonal connections between elements in different levels are allowed. This kind of structure is called *network structure* (Figure 5). Pointers do the link between entities and no links between same level elements are permitted.

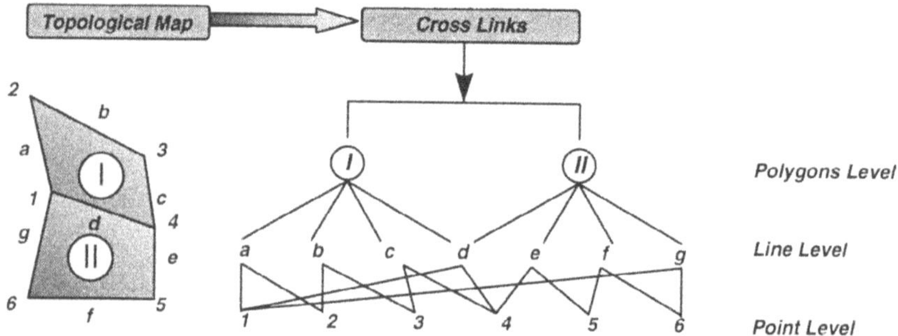


**Figure 5. Network structure of a database**

Network data systems are very useful when the relations or links can be specified in advanced. By using them, data redundancy is avoided and quick queries through the existing links are allowed. The main disadvantage lies on the fact that the database increases its volume with the inclusion of the mentioned pointers, what in complex systems may become a substantial part of the database. In the same way, the biggest problem is that only predefined queries are allowed, as also occurs in the hierarchical structure. In order to get a really flexible as well as expandable database relational or object-oriented approaches are needed.

### *2.3.3 Relational structure*

Hierarchical and network structures are known as navigational structures. This comes from the fact that pointers are used as links: the application *navigates* through the database thanks to those pointers. The only paths for search are those previously defined.

In contrast with what happens with navigation by pointers, the concept of two-dimensional interconnected tables is introduced in relational databases. Every table has a number of rows or records, and every record has a set of predefined attributes or fields. Tables are linked through a set of keys, a primary and some secondary keys. In this case, it is the user who creates the necessary links at will either when creating the database during the data input phase or later during the daily work. During a particularly complex query the database manager can create temporary links and tables. Without any doubt, the concept of *query* is the cornerstone of the relational structure of the database. In the first step of the design of tables, these are separated in such a way that data redundancy is minimized.

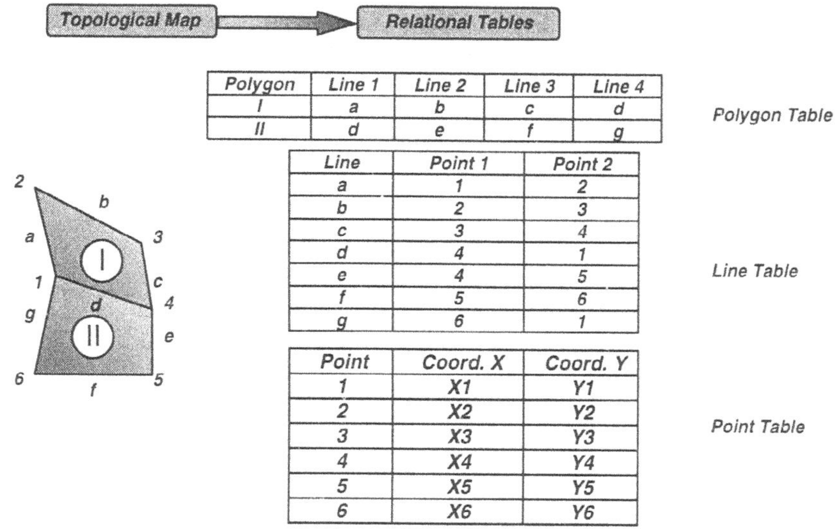


| Polygon | Line 1 | Line 2 | Line 3 | Line 4 |
|---|---|---|---|---|
| I | a | b | c | d |
| II | d | e | f | g |

| Line | Point 1 | Point 2 |
|---|---|---|
| a | 1 | 2 |
| b | 2 | 3 |
| c | 3 | 4 |
| d | 4 | 1 |
| e | 4 | 5 |
| f | 5 | 6 |
| g | 6 | 1 |

| Point | Coord. X | Coord. Y |
|---|---|---|
| 1 | X1 | Y1 |
| 2 | X2 | Y2 |
| 3 | X3 | Y3 |
| 4 | X4 | Y4 |
| 5 | X5 | Y5 |
| 6 | X6 | Y6 |

**Figure 6. Relational structure of a database**

Figure 6 shows the same two adjacent areas of the example used to describe the hierarchical and network structures and the resulting set formed by four tables. In order to get information from the database, the user must built a query by using the key elements as selection criteria. For example, the user may form a query as *'Select all points belonging to both polygons'*.

### *2.3.4 Object Oriented Structure*

Currently, the generalized trend in the development of information technologies is the *Object Oriented* approach (OO). Object Oriented systems are more modular and easier to maintain and describe than traditional procedure systems. Consequently, a number of specialized programming languages allow the implementation of object orientation in different fields easily. The term *Object Oriented* is commonplace in modern software and, hence, in database aspects.

Frequently, object orientation is used without a clear definition of what it actually implies. One of the definitions of object orientation is that a given entity, independently of its structure and complexity, can be represented in a precise way as an object. In contrast, in other data structures an entity must be separated into its basic geographical constituent parts. This definition considers the fact that the database system will store and handle objects as they are and not as a collection of their parts. An object is a software element having state, behavior and identity, and that has been endowed with its own functions and operations.

## 3. Applications of GIS to Hydraulic Network Management

GIS, as a georeferenced data management system, works by integrating these data with the functions of the different water supply management subsystems.

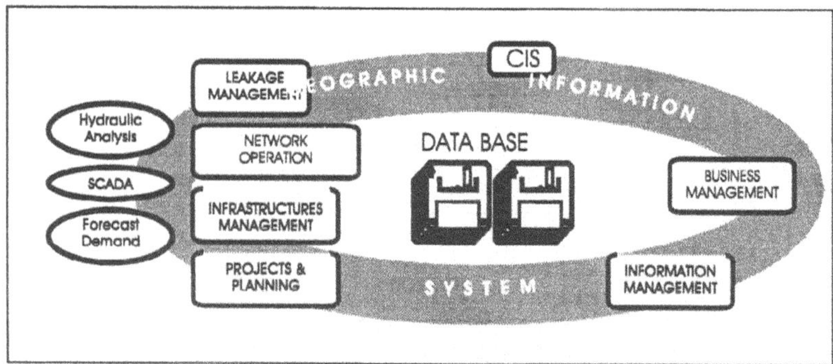


**Figure 7. Integrated Model of Information Systems**

Different areas, with specific tasks dealing with the information of the georeferenced database, can be found in the global water supply management system:

*Customer Information System (CIS).* All the aspects concerning customers are controlled in this area: contracting, registering and dropping out, billing, reclamation, failures, ...

*Updating information system.* The information is introduced, managed and updated within the database in this area.

*Planning and project.* GIS is used to plan new implantation actions on the WSS. From the action is planned until it is executed and verified, a specific monitoring is performed to ensure that GIS is suitably updated with the new data.

*Network operation.* Daily network operations, maneuvers, pressure controls, hydraulic analyses, demand forecasts, ... are managed by using information on the GIS and running applications from it.

*Infrastructure management.* A well defined and updated inventory of the elements of the network is fundamental for the optimal management of the WSS. Maintenance and reliability strategies need accurate knowledge of the elements and available resources.

*Leakage management.* Control and monitoring of failures within the network causing loses of water are one of the main advantages in using GIS for water distribution networks management.

*Financial management.* WSS, owned by public or private companies, must be subject to financial control. Economic and financial information of any WSS is fundamental for its management and no WSS lacks of a more or less effective information system. The possibility to deal with georeferenced information gives additional efficiency to the service.

In many cases all these systems share the information. Every system has its own application and characteristics. GIS controls the access of any application to the information.

It is obvious that many of them are not relevant for drought management, but indirectly all them must be considered, because the whole ensemble is important to achieve a smooth WSS integral management.

We are going to focus in the GIS applications directly participating on drought management, but prior to the advent of the shortage situation in any case.

## 3.1 PLANNING, DESIGN AND PROJECT

The GIS possibility to connect to other similar systems together with its huge storage capabilities allow establishing close control over the network from its conception to its construction. This control ensures good quality data, which is absolutely necessary for efficient WSS management.

### *3.1.1 Planning and design*

Planning a network consists of designing the network layout and sizing all the elements taking part in the new implantation zone.

Planning a new network begins with need detection and location.

The information management system must provide the urban background cartography of the new zone, known or estimated from city urban plans. Once urban background topology is known, optimal layout of the network is addressed by using the GIS. The system evaluates different alternatives, using the available information on the database and according to the established criteria, and selects the better.

Knowledge of water demand on the zone is necessary for pipe dimension. Since it is a new implantation zone, water consumption is estimated based on water demand or historical data of similar zones. Then, the GIS demand forecast module is executed to prepare a prediction of the expected maximum water consumption in the zone during the pipeline lifetime.

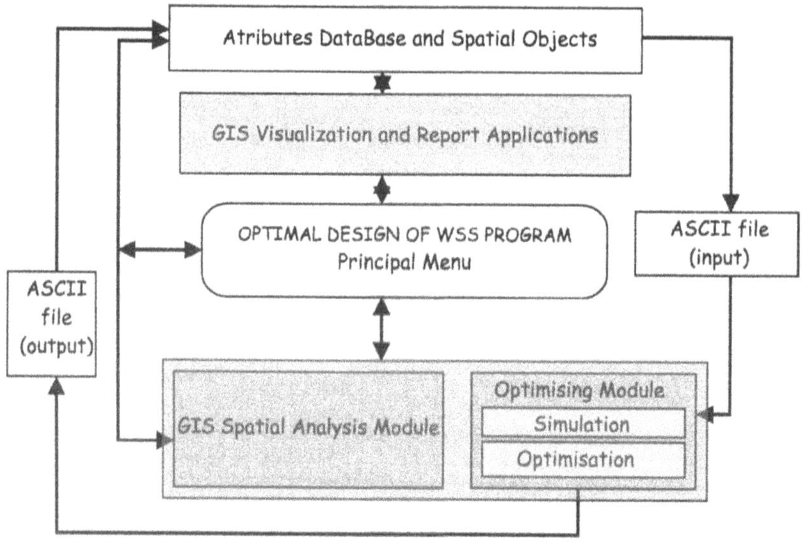


**Figure 8. Application of GIS to network optimal design**

Once water demand has been estimated, pipe diameters are calculated by using optimization techniques, maybe integrated into the GIS.

If no optimization technique is integrated into the GIS, sizing is checked by means of a network hydraulic analysis program. Then, a simulation process with the new data is run from the system. The network analysis shows if the layout and the pipe diameters are adequate to guarantee the service. In the case of a negative answer they will have to be modified. In fact, design techniques developed with GIS (Taher et al., 1996) run automatically a trial an error method.

Generated information, dimensions and topology, must not be consolidated on the database since it is still in the network design phase. However, availability of this information is important to ensure the possibility of successive updating. New data are stored in a non-consolidated database, taking part in the global database but separately from real data for basic query operations.

Data are consolidated in the main database as the design makes its way from project to executed project and, finally, to customer registration.

The database updating process can be handled more easily and in a much more reliable way if, since the network conception, the user is aware of the need of gathering GIS information.

### *3.1.2 Projects*

During the project phase both implantation of new execution zones and rehabilitation of some of the existing may be considered. The problems are similar, only distinguished by the credibility of the cartography data support. For a new implantation network the cartography is an estimation of the reality. On the other hand, when rehabilitation is the issue urban background is already consolidated in the database.

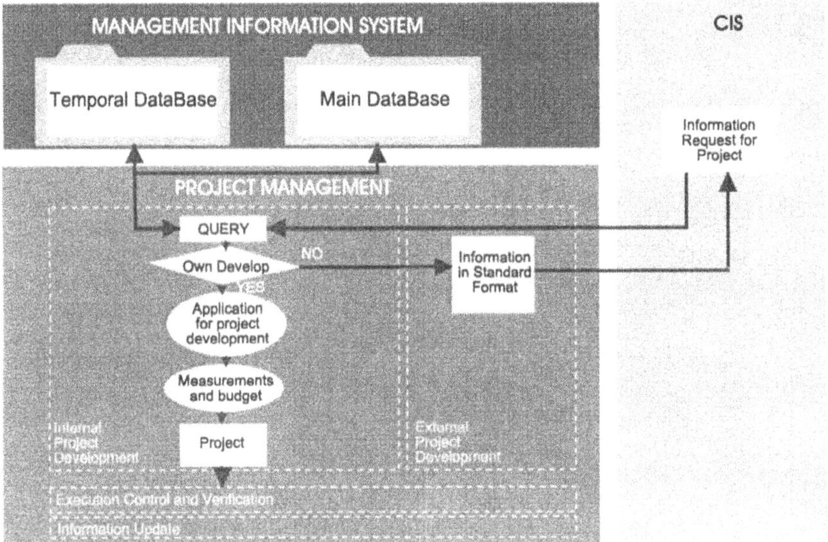


**Figure 9. Project management. Project development**

A WSS company can develop the project or not. Nevertheless, one fact is sure: execution will not be carried out by the WSS.

During the project phase consideration must be made that the WSS company will be in charge of the network maintenance. If the project exhibits flaws or materials are not suitable for the working conditions, failures would happen frequently and, as a consequence, network operation costs will increase and efficiency decrease.

During the project phase, the GIS will be used to: get urban background information (from urban plans of the zone); detect interactions with other urban infrastructures and services; select elements from commercial databases; provide measures and budgets for the works automatically; update the new information on the database and check if the project verify the standards.

Establishing a standard on the project format will make it easier the updating of data in the GIS.

The project may address elements on the main and secondary network or hose connections. In the first case, the standard ensures updating of graphic and thematic information. In the second one, the standard also includes the urban background of the parcel involving the building so that it may be consolidated in the database.

Project drawings in the standard format are introduced in a non-consolidated database. Later, control and surveillance of the project execution will allow the validation of existing and modified data on the GIS, if produced during execution.

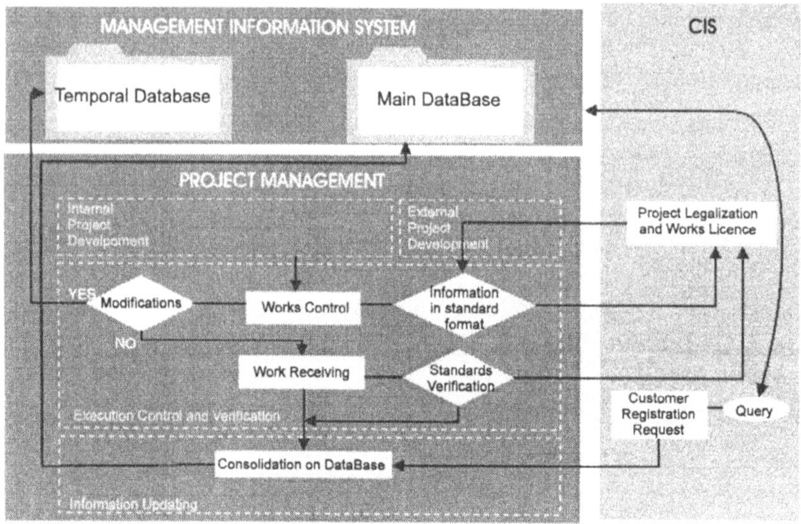


**Figure 10. Project management. Execution control and verification**

For the works to be legalized, updated information of the parcel, with all the modifications, is required in the standard format. This information is sent to the *management system* and, in the case it receive the approval of the works supervisor, it is consolidated.

When a customer is registered, the *customer information system* checks if his hose connection exists on the database. If the information is right, graphical data is completed with customer thematic one, thus improving the database.

Users of the *Project and planning system* deal with only a small part of the whole information. They have access permission to these data exclusively. The access to other information not concerning this system is not allowed, not only as a security measure but also as a way to keep data reliability.

Access to the demand forecast application, used to estimate future water demands, and to the hydraulic analysis module, to check network sizing from pipe commercial databases, has also to be free.

Access to non-consolidated information on the database is exclusive of this *project and planning system.* No other system on the GIS may have access permission to these data.

## 3.2 NETWORK OPERATION AND EXPLOITATION

Maneuvers performed with the network elements are included on what is called *network operation*. Maneuvers are devised to achieve higher efficiency on the management of water and economic resources.

### *3.2.1 Remote Control and Measures*

Real time field measurements are fundamental for network control: the decision making on how the WSS is going to be operated is made based on these measures.

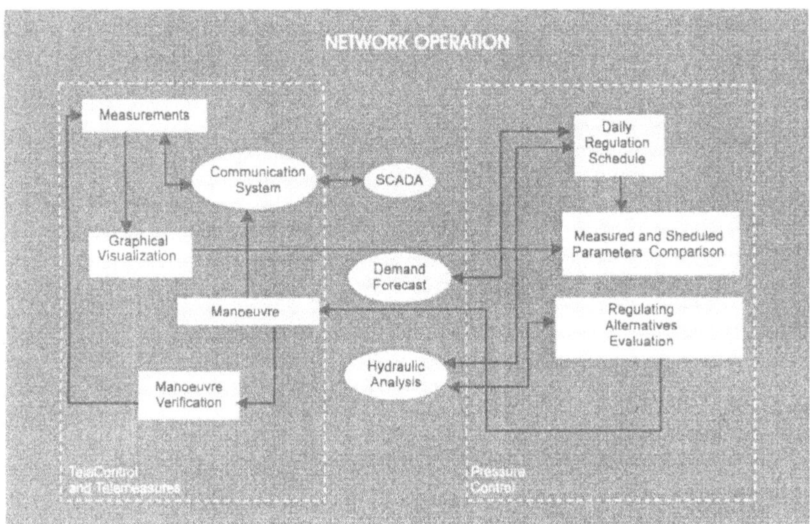


**Figure 11. Telecontrol**

Knowledge of the network behavior is borrowed from SCADA, older than GIS, which managed to work successfully for years in an autonomous way.

Nevertheless, GIS advent multiplied their possibilities. Information acquired with SCADA is transferred to GIS to be used in their applications.

It does not mean that SCADA stops working by its own. Not all the information managed by SCADA, which uses a great memory capacity, is useful for GIS.

Filter communications are used to transfer only useful data: daily maximum, minimum and average values, or instantaneous values at one-hour intervals, are enough to keep historical data.

For example, the forecast demand application uses all registered data by SCADA to calculate daily forecast, but only average values are used to perform mid or long term predictions.

In the same way, maneuvers generated by GIS applications can be transferred to SCADA to be carried out. The difficulty in this issue has been the conversion of GIS to SCADA data format.

Besides technical possibilities, measures registered by SCADA are used to compare real time pressure values with hydraulic analysis results, and flowrate values with demand forecasting.

Thus, leakage and failures within the WSS may be detected, and sent to the infrastructure management system, as seen bellow.

### 3.2.2 *Pressure control*

Pressure measurements are taken from instantaneous data registered by SCADA to establish a criterion for the state of regulating elements at any time.

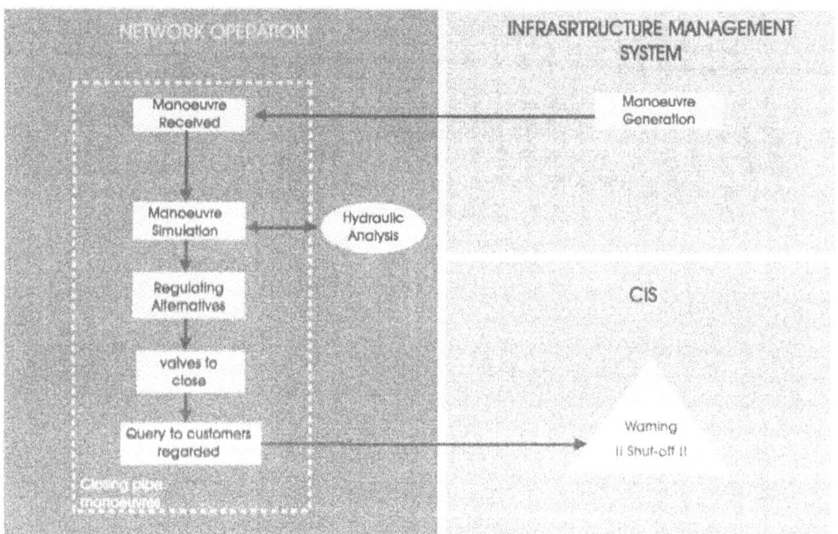


**Figure 12. Maneuvers management**

*Operation system* must ensure that network pressure falls within certain range, suitable enough to guarantee the service without decreasing WSS efficiency.

The module of demand forecast estimates water consumption at every hour on every day. Pressure evolution is simulated by using the hydraulic analysis module, and an a priori optimal regulation strategy for the day is established to maintain pressures within an acceptable range.

Controlling the pressure levels on the network, helped by modern network direct pumping, implies lower production costs, thus improving WSS management.

### *3.2.3 Closing pipes and maneuvers management*

The *Network operation system* continuously generates maneuvers on the network. Besides typical maneuvers due to WSS regulation, others, not generated by this system, may be necessary.

The *Infrastructure management system* performs maintenance operations. Any operation generates a maneuver surveyed by the network operation system.

When a maintenance operation is generated in response to a failure on a pipe, the first step to take is known as *closing pipe management.* The network operation system determines the valves to be closed to isolate the pipe.

In response to the closing pipe analysis, the hose connections affected by the shut off are spotted on the database. Then, CIS notifications are addressed to the affected customers indicating the service interruption and estimated time.

When a scheduled maintenance operation or rehabilitation is the issue, there is time enough to analyze some different alternatives and chose the smoothest. CIS also notifies affected customers.

### *3.2.4 Mathematical model update*

The Mathematical model of the network is used for hydraulic analysis in most of the systems. Model calibration is fundamental for reliability of analysis results.

The steady expansion of the network and the changes that pipe roughness experiments along the time force maintenance and temporary update of the model.

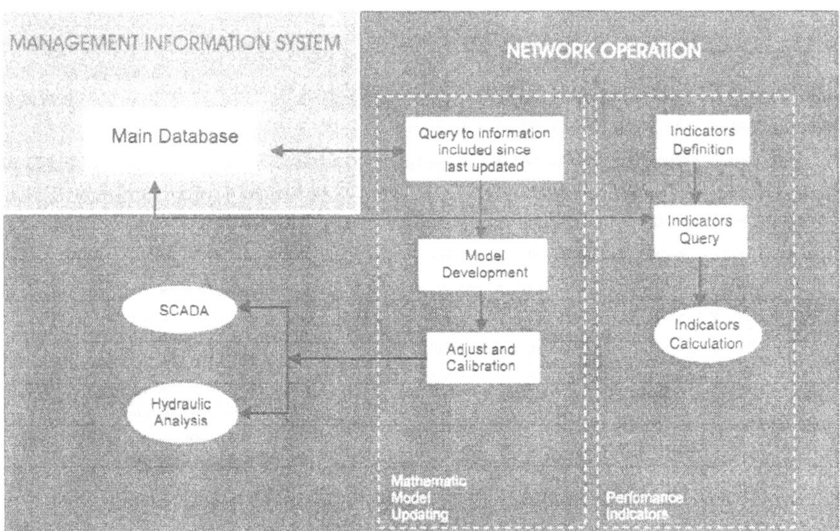


**Figure 13. Mathematical model management and performance indicators**

Being the aim of another chapter, the mathematical model is not addressed here. Nevertheless, it is obvious that the graphic information management possibilities offered by GIS and the applications promoted by this information are really useful to automate the model development and the updating process.

### *3.2.5 Performance indicators (PI)*

Performance indicators, as suggested by their name, are parameters that enable to assess the WSS management behavior.

Availability of management and processing information tools to work with big amounts of information, which GIS integrates in its applications, is needed for the identification of performance indicators.

For example, the number of incidences on the network, water service pressures and its daily variation, water quality parameters, leakage levels, are parameters stored on the GIS database or easily calculated with basic query operations to the database.

GIS frequent operations made by means of PI are:

- Infrastructure database queries
- Register of historical data of incidences
- Graphical linking of incidences and infrastructures
- PI linking to customers, network elements, network zones,...
- PI calculation

## 3.3 INFRASTRUCTURE MANAGEMENT

The network is an extremely dynamic object, subject to daily changes on elements and equipment looking for the best efficiency.

The infrastructure management system controls corrective and preventive maintenance and network reliability.

### *3.3.1 Corrective maintenance*

Corrective maintenance operations are linked to *network operation system* and *CIS* because they generate work orders.

A work order from the *network operation system* emerges from the fact that, as mentioned, a difference between measures registered by the remote sensing system and results of the mathematical simulation has been detected.

A work order from *CIS* is generated as response to a customer notice.

Any work order triggers a whole mechanism on the information system.

Failure must be located. When a user complaint is the consequence of a service deficiency, this is easily located. The database is then queried and a map, either on paper or on a portable PC screen, including the network elements of the WSS for the zone, is generated and, if available, other infrastructure information is included, to achieve a faster solution.

When work orders come from the *network operation system*, precise location of the failure is difficult. In most cases a warning is transferred to the *leakage management system* for leakage detection and, once located, it is returned to the maintenance system. As long as location is known the method is similar to the previous one.

One of the actions GIS runs when a work order is produced, is a warning generated to alert that information on the database may not be updated as a consequence of the repair. Working crew, once the repair has been executed, report all the modifications to be updated in the database. When updating is verified, the warning is canceled. Thus, information accuracy is maintained on the database.

Besides, the history of failures is kept if maintenance operations are registered. Historical data are treated statistically to be compared with those of other zones and prevent failures or, as see later, to review prevention maintenance programs.

Communication capabilities among GIS systems allow the minimization of the time elapsed between the user notice and the failure repair.

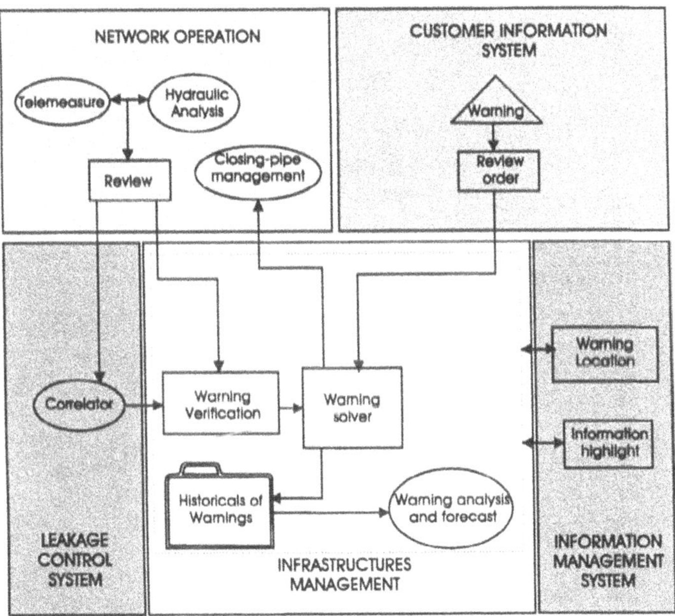


**Figure 14. Corrective maintenance**

Thus, better quality is achieved, and, if the failure is due to water leakage, the shortest the repairing time, the lower the water lose and the better the WSS efficiency.

### 3.3.2 *Preventive maintenance*

Preventive maintenance techniques are used to avoid failures before they appear. Rehabilitation programs together with preventive maintenance improve WSS efficiency, not only on water resources but also on service management.

Maintenance programs are developed from the information given by the network supplier. The supplier establishes frequency of reviews at first stage. As time goes by, maintenance programs are reviewed based on failure historical information. For this task, a statistical application is used.

Once maintenance program has been established and reviewed, the system generates work orders automatically.

Work tasks are introduced in the network and GIS studies optimal routes to minimize time and resources and availability of personnel.

While a work order is generated, information is marked on the database, thus warning that data is going to be modified.

Programming preventive maintenance works helps to anticipate operations. The system sends all the expected short term works to the *network operation system* for study, maneuvers are simulated and a regulating operation is then proposed.

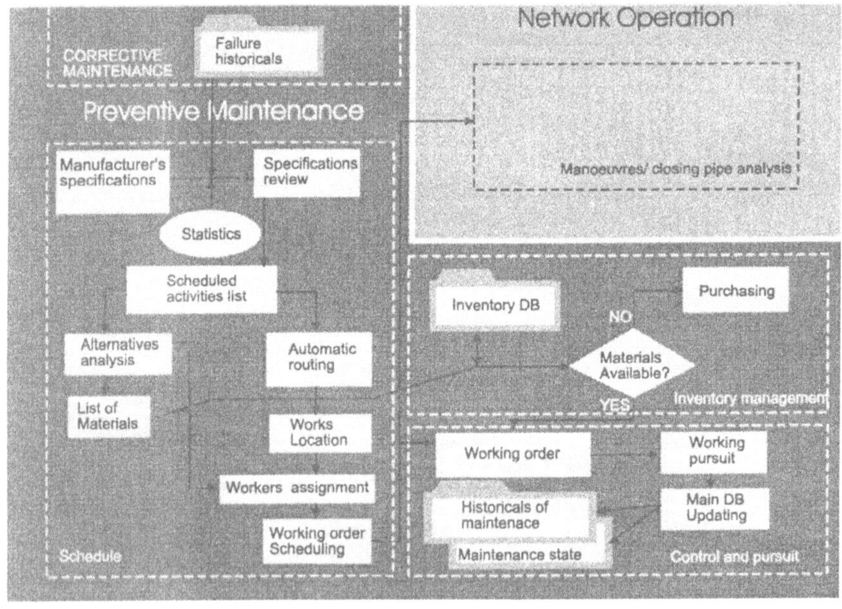


**Figure 15. Preventive maintenance**

The stock database is queried to check availability to perform the repair. Following a negative response, the system itself emits the buying order.

All these tasks reduce the execution time of the maintenance operation.

### 3.3.3 Rehabilitation

Rehabilitation may be considered as a drastic preventive maintenance implying element replacement. So, financial resources are determinant in order to establish rehabilitation programs.

To assess a rehabilitation plan, some items must be taken into account: available financial resources, statistical information on repair history and non-replacement costs, such as water leakage, energetic loses and quality of service.

Information items the GIS contributes to rehabilitation plans are brand, model and installation date of any element of the network, working conditions, maintenance history (number of repairs and last review date), failure history and, by using graphical related information, incidence on the environment. Availability on human and financial resources, from the *business information system*, is also needed.

All this information is managed by a specific application that generates and optimizes the annual rehabilitation schedule.

Once the schedule is designed, the rehabilitation work may be either substantial enough to need a project or not. In the first case it is sent to the *project system* and in the second to the *infrastructure system.*

If the rehabilitation work is sent to the *project system* the procedure has already been described. When it is sent to the infrastructure system the following actions must be performed:

- Working location query
- Generation of a warning on information as possibly modified
- Study of alternative evaluation for works
- Linking with *operation system* to run a hydraulic analysis of the closing maneuver and generating regulation alternatives
- Stock database query and generation of buying order, if any
- Work order generation for rehabilitation
- Work supervision and control and validation of modified data
- Information updating of GIS. Release of the warning

## 3.4 LEAKAGE MANAGEMENT

Next, a strategy for facilitating the location and repair of anomalies in a network (leakage) based on the facilities of a GIS is proposed.

As a way to approximate this strategy, let us go closer to the leakage or uncontrolled flow concept in a WSS. Usually, leakage in a network element represents a flow loss through a fault on it (say an orifice, a groove, a joint, a crack, ...). From a hydraulic point of view, leakage can be represented by valves at node whose manometric pressure are set to zero (see Figure 16). The value of the coefficient $k$ of such a valve gives an idea of the size of the considered fault. The bigger the crack, the lower the value of $k$. Furthermore, this leakage representation considers its dependence on the pressure in the network. As commented below, this is one of the reasons for performing network diagnoses during off-peak demand hours (that is, at night).

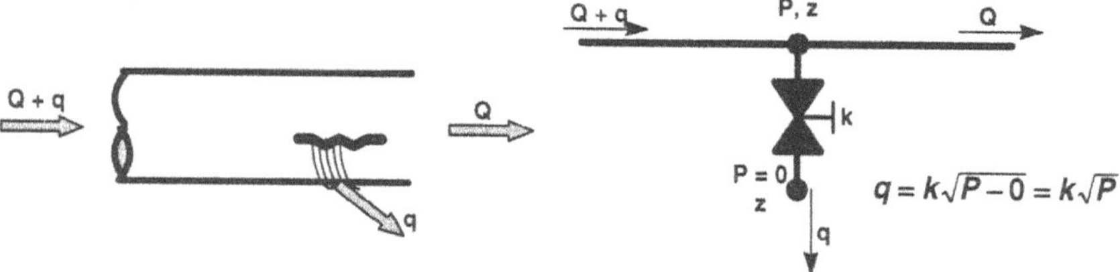


**Figure 16. Leakage model in a Water Distribution Network**

Once familiarized with the concept of leakage and its behavior in a WSS, a strategy for detecting leakage based on the GIS capabilities can be described. This strategy could be split up in four phases:

- Characterization of the controlled demands in the network. Analysis and prediction of water demands
- Sectorization and zonal flow balance in order to determine the volumetric efficiency of the network. Determination of leakage *risk maps*
- Detection of demand variations and pressure patterns, especially during the lowest demand periods (off-peak hours). Contrast of theoretical performance results with measures made on the network
- Reduction of the area with high leakage probability and finer detection of leakage with the help of auscultation techniques

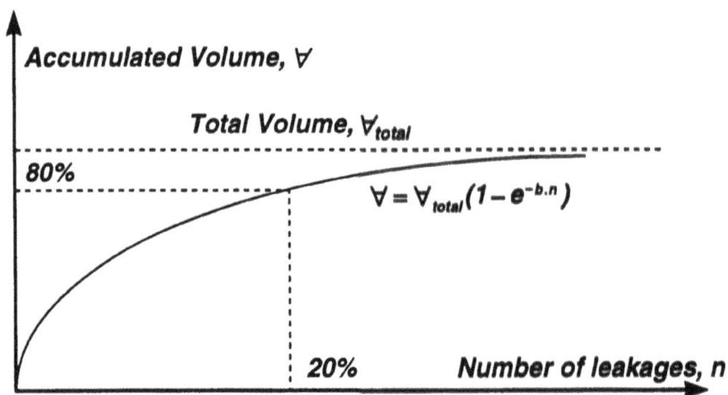


**Figure 17. Typical distribution of water losses due to leakage in a WSS**

Before presenting the steps to be carried out in each phase, it is advisable to point out that the target is not to eradicate leakage, but to find those points that represent the bigger lost water volumes. In this sense, an illustrative fact may be that approximately 20 % of leakage in the network causes 80 % of lost water volume, as shown in Figure 17. This implies that leakage detection may be worth the investment or not depending on its location.

GIS may assist with the demand pattern prediction.

Knowledge of the demand distribution on the network is an important aspect (maybe the first to be taken into account) in the analysis of its volumetric efficiency. In the same way, to start an optimal strategy for control and tracking of uncontrolled flow rates it is vital to have a suitable estimation of the demand evolution in the network.

This estimation can be obtained with two different techniques:

- By extrapolating existing data
- With economic and demographic models

In the first case, only a database containing historical records of registered water demands along a given period of time should be enough. For the second method, GIS capabilities may be crucial. GIS allows distinguishing, for example, areas with different demographic and economic characteristics, which notably simplify the use of demographic models for predicting future water duties, even easily distinguishing among areas with different trends. In this sense, basic data that should be available are:

a) Geographical location of customers

b) Customer identification

c) Quantifying of the average water consumption depending on the type of customer

This information is relatively easy to get if a GIS has been implemented in the WSS where the study is being developed. In fact, demand estimation through a combination of GIS and demographic models has already been used successfully in other fields within economic activities, such as shop location depending on the profile of the would-be consumer or the estimation of the stock size for each product.

For determining the average water duty per inhabitant or hose connection (occasionally both duties must be superposed), some estimation based on socio-economic criteria (income level of the customer, number of inhabitants per house, etc.) and technical criteria (such as pressure level in an area or the diameter of the installed water meter) must be performed. So, by combining water duty estimation and demographic trends in areas with different growth it is possible to get satisfactory results in water demand estimation.

Another aspect the use of a GIS may be of great help is the distinction of the demand type depending on the land use. One must start from a geographical distribution of the different types of demand (ordered by areas with majority demand or by superposition of layers with percentages of each one):

- Domestic
- Industrial

- Leisure
- Agriculture (including gardening)

For the classification of land uses either techniques based on the statistical analysis of historical records or analysis of aerial images collected by teledetection techniques can be used. In general, it is advisable to combine both techniques to limit the inherent uncertainty in the estimation of water demands with these methods.

From these analyses, customer, hose connection, street or zone demands can be estimated. As a consequence, the value of the volumetric efficiency of the network comparing these estimations with the injected flow rate can be obtained. It is also possible to estimate daily variations of the demands (what is known as demand pattern). This will allow distinguishing between the performance of the network during the peak and the off-peak hours.

### *3.4.1 Concept of Sectorization*

Sectorization is the division of the whole network into several smaller subnetworks. Each subnetwork, called *sector*, will constitute a distribution unit, limited and homogeneous enough as to make the management of the data to get and analyze as quick and reliable as possible. The delimitation of every sector needs careful preparatory work. All the elements that ensure the physical isolation must be checked and eventually repaired, mainly the cut-off valves. All this process may be made considerably easy with the help of the topological capabilities of a GIS.

When the boundary of the sector is perfectly delimited, if possible with a single water inflow point, the analysis of the sector performance in order to detect leakage may be performed always observing the phases listed next:

- Continuous observation at regular time steps of the injected and demanded volumes, especially at night (off-peaks hours) when the expected demands are lower
- Leakage location within an area smaller than the observed sector
- Leakage exact location by using the classic auscultation techniques

So, sectorization is a strategic option that reduces the area to inspect for anomaly detection and placing (points of breakage, leakage, pressure deficiency) and also improves the management of the global operation of the network through the implementation of a control system, optimizing pressure in each sector.

*Viability Study of a sectorization.* Prior to the sectorization of a network, it is necessary to carry out a pilot experience in a delimited area of the network. Such an experience will provide enough information to perform:

a) Analysis of the control element behavior: flowmeter, data logger, throttling control valve and pressure gauge
b) Comparison analysis between invoiced and measured flowrates
c) Possibility of running real leakage simulations (artificially creating points of leakage and checking if they are easy to find)

d) Analysis of the efficiency of every sector

e) Analysis of the global results of the experience

It is advisable to implant the pilot experience simultaneously with the leakage study, since it will allow identifying basic sectorization criteria leading to the sectorisation of the remaining of the network, thus evaluating, by extrapolation, the approximate number of resulting sectors. For this purpose, the results of provoked leakage may be contrasted with those of a mathematical model of the sector. This mathematical model, assisted by the GIS, should include artificial leakage simulation.

When devising a network sectorization two general characteristics of the big WSSs should be taken into account:

a) Big WSSs usually are functionally desegregated into pressure levels.

b) Big WSSs are densely meshed, mainly in those parts of the network that supply the most densely populated areas of the town.

It is hence indispensable to ensure that changes due to sectorization are not excessively drastic, and the level of service that the sectorized network will provide has, at least, the same quality that the level existing previously. This compels to perform theoretical checking of the performance of the sectorized network, through simulations with a mathematical model.

In addition to the previous technical viability analysis, a short and medium term economical analysis should be done. This analysis will allow choosing the most profitable sectorization. In this sense, costs and benefits should be estimated. The costs are related to:

- necessary control devices
- civil work
- network adaptation

and the expected benefits come from:

- analysis of volumetric and energetic losses
- savings duc to the new pressure regulation
- savings due to the reduction of the probability of breakage or leakage events
- savings related to the reduction of detection and location time for leakage.

*Basic criteria used for sector definition.* The main criteria to be borne in mind in order to sectorize a network are the following.

1. An initial sectorization of the network has to be taken into account: its division into pressure levels. Work should start from this first iteration. This will help in designing a leakage detection strategy, since those sectors with higher average pressure will also present higher leakage probability.

2. Another argument to be observed is the delimitation of sectors depending on the age of the pipes. Most of the times, there exist polygons that have been created in relatively short time intervals and, except for some replacements made during maintenance tasks, the installations on them have same age. A strong correlation between the risk of breakage and the age of a pipe has been demonstrated.
3. In order for the whole network operation to be suitably operative and effective, a number of sectors between 1 and 10 per 100Ha is recommended, depending upon the network density. In areas of population high density, sectors should be between 10 and 15 Ha wide. In terms of pipe length, an approximate range between 5 and 10 km might be established.
4. Provision of a single point of inflow for each sector could lead to some localized problem and occasionally it will be necessary to foresee a double feeding for the sector. For this kind of operations, topological analysis capabilities most of the GIS include may be of great help for the location of feeding points of each sector.
5. Uniformity of pressure at inner points in each sector, that is, minimization as much as possible of the spatial variations of pressure within each sector is highly desirable.

The main problems to face during the phase of sector definition are:

a) Selection of a suitable boundary of the sector (size of the sector)
b) Appearance of low and even negative pressures in some points, what would imply absence of service on these points
c) The sectorization may cause that in some pipes (especially in the inflow pipes of a sector), the water transport velocity exceed the maximum advisable value, around 1.5 meters per second.
d) Too high pressures during the off-peak hours (night flow)

Different solutions can be adopted at a first moment to solve these incidences, at a simulation level. They involve redundancy (two feeding points for the sectors and installation of new pipes), iteration (trial of other boundaries) and installation of pressure reducing valves in order to control high pressures.

*Leakage control and sectorization of a network.* All over the water distribution network and, consequently, over its zones and sectors, the presence of uncontrolled water demands has to be taken into account. These uncontrolled flows can arise from unaccounted flows, errors on water meters and, as mentioned above, faults and leakage.

Two lines of proceeding to reduce water losses are possible:

- To act within the network infrastructure itself, mainly by systematic gathering and replacement of elements such as joints, hose connections, older pipes and specially water meters in domiciles.

- To reduce the time between leakage formation and the moment it is repaired.

The combination of the sectorization and the rest of the modeling, monitoring and maintenance techniques allow to increase the effectiveness and the efficiency of the methods for reducing the lost volume.

### *3.4.2 Inverse problem*

A typical analysis in a WSS consists of determining node pressure and pipe flow rates, the data being magnitude and location of water duties. Such an approach is known as direct problem. Of particular interest is the simulation of faults and leakage. In this case, the magnitude and location of the fault are known, usually in one or more nodes, and pressure in nodes and flowrates in pipes are calculated.

The inverse approach is known as inverse problem. Then, from known heads at certain nodes and flowrates through the lines leakage location and/or magnitude are determined.

In the inverse problem the system characteristics and the useful demands in its nodes are known, while the existing faults, leakage location and magnitude are unknown.

For simplicity in many of the solution methods based in mathematical models both useful demands and likely leakage are supposed to be located at nodes. However, even though neither in the case of the demands nor in the case of leakage it happens to be that way, the approach is accepted to be a good approximation.

In order to solve an inverse problem, water demands at nodes again, and a set of values of the measurable variables (unknown in the direct problem) such as node pressures and line flowrates are needed.

Once the corresponding mathematical model of the hydraulic system to be analyzed and evaluated has been performed, and being the model ready to use in the sense of calibration and experimental validation, it may be used to identify leakage (among some other uses). Sectorization may help in both model calibration and validation and in the determination of its behavior as far as leakage concerns.

The setting of the inverse problem is as follows:

- The network topology is perfectly known.
- The useful water duty distribution is also known in the nodes of the network model.
- The value of the pressure in some points (nodes) of the network is known.

The unknowns are the values of the coefficient loss, $\boldsymbol{k}_j$, of the valves simulating leakage in the model.

The solution of the inverse problem is based on the same program used for the solution of the direct method, but using a node method. Using this method, some attempts have been carried out, getting satisfactory results.

The results are about:

- Pressure on the remaining nodes in the network
- Circulating flowrates in every line
- Nodes exhibiting leakage (those with $k_j > 0$)
- Leakage flowrate in those nodes

### *3.4.3 Fine detection of leakage trough auscultation*

For the obtained results being reliable, they must be first contrasted with measures made in some points of the network. The final objective is not a fine location of leakage (for this purpose a traditional detection method should be used), but the identification of the area where leakage is likely to occur.

As a result of the demand analysis, a distribution of the different water uses and an estimation of the evolution of water duties in the short and medium term can be obtained. Classifying the demand types in different areas may be very useful in order to estimate the daily variation of such a demand. Usually water billing during long time periods (one or two months) is the known data and frequently deficiencies in water meters or procedures may occur (demand estimation or different measurement periods depending on the zone). This can drive the modeler to distinguish among domestic, industrial and irrigation uses. In this way, the network behavior may be reshaped for those instants of especial interest, the off-peak hours, when the average pressure increases and the network volumetric efficiency decay is more evident.

On the other hand, if during the sectorization process sectors with more or less homogeneous pressure levels have been considered, probably the first step to solve the problem for the sectors with higher pressure levels has been taken. Another step should consist of grouping sectors by the age of their pipes. This parameter has also great influence in the water volume losses. Two leakage risk maps may be created, one based on the pressure level and another based on the pipe age within the sector. By overlapping both maps, it is possible to detect areas where the combination of both effects leads to a great break risk on the pipe and hence, risk of leakage. This procedure is shown schematically in Figure 18.

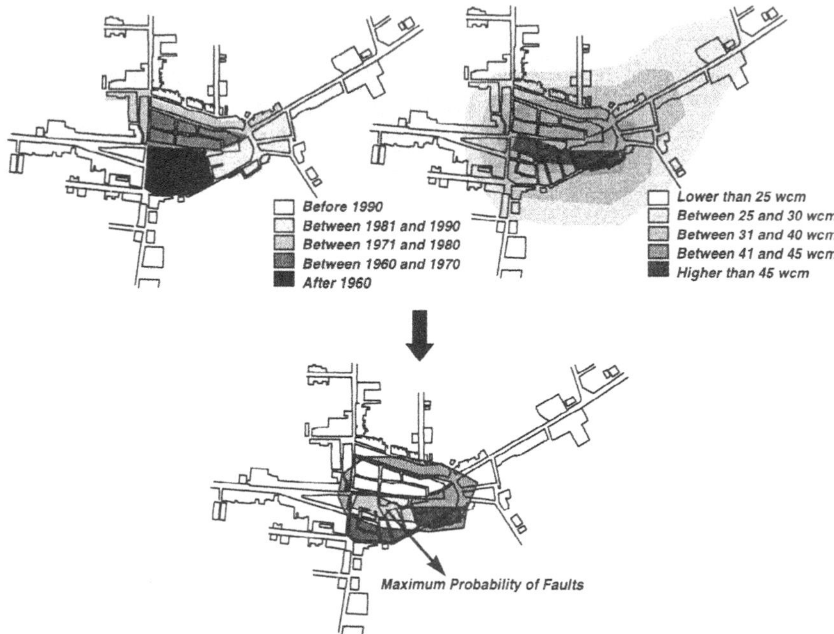


**Figure 18. Schematic determination of risk maps through superposition of GIS thematic maps**

In any way the objective is the reduction of the area with high leakage probability. This will turn the detection of leakage simpler and cheaper by using the traditional auscultation methods. Nevertheless, avoiding them is not plausible.

## 4. Future trends in GIS development

After what has been exposed in the present chapter, it is necessary to declare that GISs are currently in a basic stage in Spain. Except for some cases, most of the WSS that have faced the challenge of implementing a GIS are still developing preliminary phases, such as data introduction or design of applications to manage them. In this sense, the fulfillment of the initial phases of GIS implementation, at least in those systems whose size justifies the investment (Cubillo, 1997), can be identified as a first short-term objective regarding the application of this technology. From this point on, more ambitious targets could be planned, even though a review to the trends of the different water supply companies may pinpoint some characteristics GIS should have in a near future.

So, a constant feature in all the WSS trying to incorporate the use of GIS is the trend to unify technical data with other administrative data in a single Information System. Some people even point to sharing data with other utility (gas, electricity or sewage) companies, in such a way that a huge corporate database should be developed that allows tasks as works management or predictive maintenance.

Another aspect, in which some authors (Martín, 1997 y Zaragoza, 1997) coincide, is the need to converge to open platforms in graphical object oriented environments which make it easier the use of these systems, independently of the user skillfulness. This will have positive implications on the user motivation and on the reliability of the works performed.

A tendency towards decentralization of the Information System can also be observed. This decentralization will allow the work tracking and file updating with the help of portable computers and Internet/Intranet connections. The future in this case points to Client/Server-like architectures.

Finally, other trends that may be enumerated are:

- Joint modeling of spatial and temporal variations in a single information system
- Implementation of teledetection and real time control systems for networks
- Incorporation of knowledge based systems and expert systems to support decision-making
- Three-dimensional organization of the spatial information, in order to allow accessibility studies

It can be concluded that GIS perspectives regarding WSS are really wide. In the case we are involved, interesting consequences pointing to more efficient use of the resources and improvement of the service quality may be reasonably expected.

As a conclusion, it could be said that technical solutions when water shortage has been produced are always desperate and in this sense GIS do not bring magic solutions. It is possible to look for emergency solution, but the smartest strategy will be able to avoid those limit situations through a suitable management of the existing resources.

## 5. References

Bernhardsen, T. (1992). *Geographic Information Systems.* Viak IT & Norwegian Mapping Authority. Arendal (Norway), 1992.

Cabrera, E. et al. (1996). *Ingeniería Hidráulica aplicada a los Sistemas de Distribución de Agua.* Grupo Mecánica de Fluidos. Valencia (Spain), 1996.

Cubillo, F. et al. (1997). *Guía para la implantación de Sistemas de Información en la gestión de redes de suministro de agua.* CENTA, Seville (Spain), September 1997.

Cubillo, F. (1997). *Características Específicas de los Abastecimientos de Agua.* Curso de Sistemas de Información Geográfica aplicados a Redes Hidráulicas. Máster en Gestión y Uso Eficiente del Agua. Valencia (Spain), November 1997.

Hernández Rodríguez, F. (1995). *Modelización de Información Espacial mediante Tecnología orientada a Objetos.* PhD Thesis. Seville, March 1995.

Iglesias Rey, P., López Patiño, G, y Martínez Solano, J. (1997). *Simulación, Interpretación y Presentación de Resultados en Sistemas de Distribución de Agua Potable.* Curso de Sistemas de

Información Geográfica aplicados a Redes Hidráulicas. Máster en Gestión y Uso Eficiente del Agua. Valencia (Spain), November 1997.

Martín Navarro, A. (1997). *Aplicación de los SIG a la Gestión Integral de los Sistemas de Distribución de Agua.* Curso de Diseño, Análisis, Operación y Mantenimiento de Redes Hidráulicas a Presión. Máster en Gestión y Uso Eficiente del Agua. Valencia (Spain), June 1997.

Parsons, E. (1997). *The Essential Guide to GIS. A hands-on GIS Workshop.* Internet Seminar, Kingston Center for GIS. Kingston (United Kingdom), July 1997. (http://www.future-geomatics.com/esguide/start.html).

Prodanovic, D. (1997). *Data Bases for Urban Infraestructures.* Curso de Sistemas de Información Geográfica aplicados a Redes Hidráulicas. Máster en Gestión y Uso Eficiente del Agua. Valencia (Spain), November 1997.

Taher, Saud A., Labadie, John W. (1996) *Optimal design of water-distribution networks with GIS.* J. Water Resources Planning and Mangement **4**, 301-311

Zaragoza, J. (1997). *Gestión de Redes utilizando el SIG. El SIG en Aguas de Alicante.* Curso de Sistemas de Información Geográfica aplicados a Redes Hidráulicas. Máster en Gestión y Uso Eficiente del Agua. Valencia (Spain), November 1997.

# THE MODELLING OF WATER DISTRIBUTION SYSTEMS

FUERTES, V., GARCÍA - SERRA, J., PÉREZ, R.
*Fluid Mechanics Group.*
*Universidad Politécnica de Valencia. Spain*

## 1. Introduction

The decision making process when deciding which strategy to adopt must be based, in any activity, on good information. Broadly speaking, we can say that to elaborate a model of a network of water distribution consists in organizing adequately the available information about the water network, increasing the information if necessary and maintaining it updated, with the objective of feeding a hydraulic calculation program that permit us, as our principal task, to simulate the behaviour of the physical system which it represents as accurately as possible.

From this point of view, the model of a network is no more than a support tool to the decision making based on the existing information about the network.

The development of computers and of hydraulic calculation software, improvements in knowledge of the system through SCADA (Supervisory Control and Data Acquisition System), GIS (Geographical Information System) etc., and the ever greater technological modernization of the sector have influenced in a decisive way in the fact that a great many water supplies have decided to build a model of the water network that they manage.

The demand for continuous improvement of the technical management of the system has converted modelling into a need for water supplies, since thanks to simulation it is possible to improve the planning and technical management of the system. The Department responsible for taking decisions about investments to be carried out in infrastructures has a powerful support tool to make the best decision. Likewise, it is possible to improve the operating conditions of the network introducing timely correction measures. It is no less important the help that the model can lend when planning actions and remedies to take when faced with breakdowns, supply interruptions, breakages in pipes, etc.. that otherwise would only be guided by the intuition of the operators.

Therefore, nowadays, it is fundamental to count on a model of the network for any water supply that wants to modernize its management. A detailed knowledge of the way the system operates can only be approached with the aid of models. The intuition and experience of the personnel in charge of the management of the network is still fundamental, but it is not sufficient. It is necessary to have adequate tools that permit us, not only to carry out analyses of a qualitative type on the repercussions of measures and action plans to be adopted, but also of a quantitative type.



*E. Cabrera and J. García-Serra (eds.), Drought Management Planning in Water Supply Systems,* 52–88.

It is important to distinguish in this introduction between what is hydraulic calculation software and a model. The first is no more than a program or a package of programs capable of solving a system of equations that one has to define. For this it is necessary to have a series of data with which to build such a system. These data (topology of the system, lengths of the pipelines, diameters, levels of the tanks, consumption, etc..) are really the model of the network. It is no use having a hydraulic calculation program that it is capable of giving results of the existing pressures with very exact precision (for example, mmca) if the set of data which one feeds into it has an error factor of, for example, up to 20%.

A fundamental criterion in the elaboration of the model consists in the adequate gathering and classification of all the necessary information for its subsequent processing based on the algorithms included in the model. This information will be duly structured and will constitute a real database, complete and coherent, so that it can be consulted and used by additional models or even for different objectives.

The purpose of this approach is to take advantage to the maximum of the most important part of any model, the supporting information, on which one depends not only for the validity of the results obtained and for the model in itself in fact, but also the diagnosis, improvement and safety of the water supply system.

A graphic representation of the importance and also of the cost of obtaining such information would be a pyramid. This pyramid would have a wide base, corresponding to the invariant information (basically cartographic, sustained in a Geographical Information System, GIS ). The following stratum would correspond to the characteristics of the infrastructure of the system and after that there would be a stratum corresponding to the consumers database, consumption and billings. The penultimate level would include all the field measurements realized and the registered values and telemeter measurements of the variables of the control of the system, more concretely speaking, pressures, volume of water flow, levels in tanks, valve positions and the rest of the parameters. Finally, at the apex one would be find the package of programs that would process the previous information, in that the programs would be checking, selecting and completing the information, to be used in relation to the developed models for gauging as well as for identifying and solving the different design problems, analysis and development of water supply systems.

In a nutshell, while one ascends the pyramid the investment cost and the length of usefulness of the parts diminishes rapidly. The apex can be improved without requiring a great economic effort, although some technological effort, but it probably demands for its effective application a change in the lower levels, that must be updated continuously.

One must take into account two phases clearly differentiated throughout the life of a model: the **construction** and the **applications**. The first of these means building a model from the available data, that can always be improved and enlarged. We can say that this phase never finishes, since it will be necessary to improve and to update the model continuously Also, the calibration of the model to increase the reliability of the results should be realized periodically. Until a few years ago it was customary to entrust an external Company (Engineering Consultants) the construction of the model of the network. This inevitably lead, in many cases, to a lack of knowledge of the model on the part of user, which provoked on quite a few occasions situations where the model remained dormant and would not be continually updated. It is becoming more and more frequent within Water Companies to find Departments that are developing models, advised in many cases by engineers with experience in this field. From our point of view, this fact is fundamental for achieving an exact knowledge of the model on the part of the user, so that they will know exactly how it has been built and the limitations of the model, being able to undertake in an adequate way the **phase of application**. The objectives that are fixed for the application of the model will also be an important conditioning factor in the construction of the model. It will define the type of model, the degree of required minimum accuracy, the degree of detail, etc..

As an example of the ever greater use of models, we have collected some statistical data about the utilization and application of these models from the reference literature (*Cesario, 1992*).

In 1992 a questionnaire was sent to 915 United States Companies to which 338 (37%) answered, of those who had responded, 290 (86%) said they were using a model of their network. Most of these Companies were using the model at least once a month, being 9% those which were using it daily. The model was used fundamentally by the Engineering and Planning Departments.

Half of the answers indicated the use of static calculation models, and the other half of the models over extended periods (simulations of 24 operation hours of the network). In most cases (86%) the models consisted of less than 2000 nodes and consumption was assigned to nodes being based on the records of the water meters of the consumers. 80% of the consulted Companies were managing water supplies for more than 20,000 inhabitants.

83% of the surveys received answered affirmatively to the question when asked if they had plans to continue improving the model of the network, which gives the idea that in most cases they saw it as an important tool.

Concerning the applications of the model, there were used largely, and in order of importance, for the following tasks:

- Long term planning
- Study of operation conditions in the event of fire
- Enlargement analysis
- Studies on operations of the system
- Action plans in emergency situations
- Energy cost optimization
- Operator training
- Etc.

Throughout this paper, we shall be discussing the two aforementioned phases, construction and development of the model. The fundamental objective is to make known the possibilities that they have and the general standards to follow in building a model.

## 2. Generalities. Types of models

From a purely mathematical point of view, the behavioural analysis problem of a network has nowadays been solved, thanks to the use of powerful computers and calculation programs.

The mathematical modelling of behaviour of the different elements of a distribution network consists of establishing a relationship between the variables that intervene in the problem (pressures and volume of water flow) and the parameters of each element (for example, length, roughness and interior diameter in the case of a pipeline). With the relationship fixed for each one of the components of the system, a system of equations is built that once solved gives a solution as the value of the variables of the network for a given situation. To define the state of the network that is to be processed, it is necessary to fix a series of parameter values of the system (for example, consumption at the nodes, position of the valves, pumps in operation), that distinguish for the same network different operational situations. Expressed simply in Figure 1 is a summary of what has been explicated above.

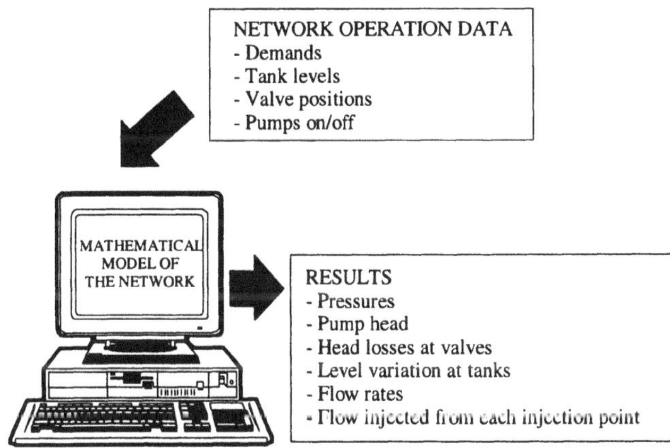


*Figure 1. Mathematical model of the network*

In reality a model is somewhat more complex, since a great quantity of data is required to model not only the elements of the system (pipelines, valves, etc..) but also the form in which the consumption is distributed, regulation and operation of the system, etc..

In Figure 2, extracted from Cesario, 1992, we can see a basic scheme for a model. As we can observe, the hydraulic calculation program is only a tool within the framework that a model makes up.

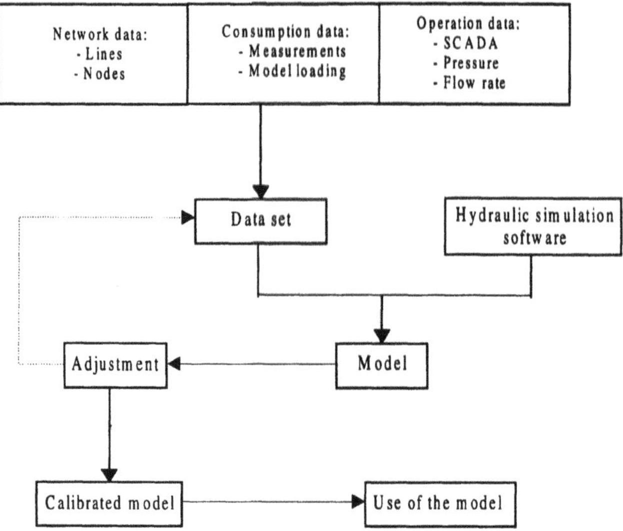


*Figure 2. Modelization scheme.*

## 2.1 SOME IMPORTANT CONSIDERATIONS.

It is fundamental when approaching the construction of the model of a network to carry out some basic thinking and take into account some aspects of the modelling process.

- According to the use that one is going to give to the model (development), it will be necessary to construct one type of model or another (strategic or detailed, static or dynamic, etc.).

- It is not possible to develop an excessively reliable model of the system without having an exhaustive knowledge of the network and if there is a lack of data about its operation, consumption values, elements of which it disposes, etc. One must be careful because on occasions one can have data but these are little unreliable or they are not updated. The quality of the model depends fundamentally on the data of that one has readily available.

- Although one may know the system, there is a whole load of "uncertain" parameters in the network, so that it becomes impossible to achieve a model that will be 100% faithful or reliable in as much as the real operating conditions of system. Comparing the pressure values, water volume flow, levels, etc.. provided by the model with field measurements will always show a difference when one is trying to analyze a concrete and particular operation situation of the system.
    - As time goes by, the roughness of the pipelines will have been increased, without being able to specify exactly that value until it has already happened.
    - In some instances there will be diameter reductions as a consequence of postprecipitation of, for example, calcium carbonate.

- It is materially impossible to know in each instant the spatial distribution of consumption in the network. As a starting point, no network has a 100% efficiency (ratio between registered and injected volumes), there exist errors in the meters, uncontrolled consumption, leaks, etc.. that we will not know in any scientifically certain way where to locate them. Even "controlled" consumption varies continually, therefore upon analyzing any situation we are realizing an estimate of the loads on the system.
- There are always mistakes in the data. For example, in the measurements of pressure, of flow rates or of levels, this will depend on the precision of the apparatus used. There exist mistakes in the characterization of some elements (for example the characteristic curve of a pump that has been modified as time goes due to use). There may also exist mistakes in the level of a node on which are effected measurements to prove the effectiveness of the model which gives rise to a bad reference point to contrast the pressure results given by the model with measurements carried out in reality.

- A monitoring process is always necessary and subsequent "calibration" of the model. This consists in modifying values of the uncertain parameters to achieve results from the model that will be as similar as possible to the measurements taken in the network. The monitoring as well as the calibration should be carried out periodically. The constant update of the model is basic to maintain the model operative.

- A model must be evaluated taking into account its capacity to carry out specific tasks. Therefore, depending on the use that is going to be given to it we will judge if it is or not sufficiently reliable. Seen as a tool that will help us to take decisions, can supervisors using them rely on them for the information that models provide?

- One has to know the limitations of a model, to take into account different aspects of the operation of the network. Above all in what affects to the modelling of consumption (the way in which loads are distributed in the network), there is always the chance of the existence of important "lagoons".

## 2.2 TYPES OF MODELS ATTENDING TO THE VARIABLE TIME.

In the analysis of the operation of a system, one can include or not the variable time, as a result of which we will find ourselves with different types of models:

- Static models: It tries to determine the values of the variables of the system for a given situation, without taking into account variation with time of the parameters of the system (consumption, levels, etc..). It is a photograph of the operation of the network for a given set of values of the parameters.
- Dynamic models: In this case the variable time is taken into account, taking into consideration the temporary variation of the parameters of the system, what means a temporary variation of the values of the variables to calculate. The analysis is, in most cases, discreet, since it is necessary to solve the equations of the system for each time instant. With the obtained results from such analysis it is possible to calculate the next time step. Within these models one can accomplish

other kinds of classification, attending to the way in which the simulation is carried out and to the equations that intervene:

- Inertial: One takes into consideration the inertia of the water fluid in movement, that is to say, the energy put at stake with changes in speed. These models are necessary when we simulate sharp speed changes in the system, because of "rapid" manoeuvres in the regulation of elements (the starting and stopping of pumps, rapid valve closures, etc..). At the same time it is possible to accomplish other classifications for inertial models, according to whether we take or not into account the elasticity of the walls of the pipeline and the fluid. In the first case we are dealing with elastic models (hydraulic ram or blow) applicable in very abrupt and sharp manoeuvres in the system, even though generally these disturbances will be absorbed quickly in the case of distribution networks. In the case of not taking into account factors of elasticity (rigid pipelines and incompressible fluid), we will be using rigid models (mass oscillation), that give sufficient accuracy if the effected manoeuvres are not excessively rapid. Inertial models take into account in the equations system a greater number of terms than static analysis models. The difficulty and time employed in the resolution is notably greater. Nevertheless, it is important to indicate that the most general of all models, the elastic model, encompasses the rest of the models. In fact, some authors (Koelle, 1989) advocate the use of a valid formulation for any type of model.
- Non inertial: The dynamic characteristics are conferred to the model by having some changing boundary conditions with time. However, the equations used are static. These models, which are called quasi-static or extended period sinulation, calculate a succession of static simulations in those which take into account the variation that is produced from the previous calculation instant of the regulation elements, demands, pumps in operation, tank levels, etc.. These methods are used generally by the existing simulation packages on the market that analyze the behaviour of a network over a period of time, generally 24 hours.

The utilization of one or other type of model will be conditioned by the type of analysis that we want to effect. To analyze a concrete operation situation a static model will be used. For example, when one wishes to determine the diameter of a pipeline that should be installed to supply an area that has been incorporated into system. In such a case the operation of the network is analyzed for the peak consumption situation, fixing the flow for this case.

When what is intended is to analyze the response of the system faced with an abrupt and rapid manoeuvre of a regulation element one should use a transient regime model (elastic model or rigid model). In most cases this means analyzing the pressure levels that are reached in a transient to test if such manoeuvres can become dangerous for the integrity of the system or to design safety and protection systems. In these cases one tends to start from a static situation, generating the disturbance and seeing how the variables of the system evolve over the time. In the case of networks these disturbances are absorbed rapidly, therefore in most cases only a few seconds are simulated. Some aspects of the modelling of elements and of outline conditions have not been totally resolved. Such is the case, for example, of the analysis of losses of load in transition, the dynamic modelling of regulation elements (for example, retention valves), the movement of air within a pipeline, the possible formation, development, movement and collapse of bubbles of steam water as a consequence of a decrease in the pressure below the steam tension , etc.

Nevertheless, the values of the first "peaks" of pressure in the system as a consequence of the effected manoeuvre are approaching real conditions, therefore the usefulness of these types of models is undeniable. Their use, nevertheless, is limited to specialized analysis not habitual analysis. It is a form of model that is not used every day.

If what is wished is to analyze the behaviour of the network over a period of time but with "slow" modifications of the parameters of the network, one should a model over an extended period. In these cases the parameters that are seen to be modified in time are the consumption of the consumers, that do not suffer generally such sharp variations that make necessary the use of an inertial model. As a consequence of these variations, the regulation elements of the network will act (for example a pump will start when the water level in a tank reaches a minimum value) giving rise to a transient situation. After a few seconds the effects of this transient situation will have been absorbed, therefore the starting of the pump is included in the model but not the immediate consequences of this. We can say that the element is incorporated into the system from the moment it is put under way. The analysis of the short time interval in which is produced this transitional situation would have to be analyzed by an inertial model.

In Figure 3, taken from Cabrera et al. (1995), we have expressed in graphic form what we have previously explicated. It represents the evolution of the pressure and of the flow in the mid-point of a pipeline of a network fed from two tanks that has suffered a sudden break at one of its end points. The continuous lines are the results of the elastic model and the dotted lines are the results of the quasi-static model. The latter is not capable of predicting the pressure fluctuations and gives as a result a sharp flow variation. Nevertheless, after some 20 seconds, the disturbance has been absorbed, and the results converge.

It is not object of this work to analyze in detail the different forms of linking the different stationary states to be analyzed in a model over an extended period. But to show the kind of working philosophy we are using we can analyze the modelling of the level of a water tank as an example of a simple case. For this we suppose that for a given instant $t$ the value of all the parameters of the network is known (consumption, position of the valves, pumps in operation, etc..). The water level in a tank of the same network is Z(t).

The static analysis of the operation of the same will give as a result the values of the flow though the lines and the pressures at the nodes. Let us suppose that q(t) is the value of the volume of water flowing out of the aforementioned tank. If what one is intending to analyze is the functioning or operation of the network a posterior $\Delta t$ (instant $t + \Delta t$) the value of the level of the tank for this new analysis will be:

$$z(t + \Delta t) = z(t) - \Delta t \, q(t)/A_d$$

being $A_d$ the section of the tank. Evidently, the rest of the parameters will also adopt the corresponding values related to the analyzed instant.

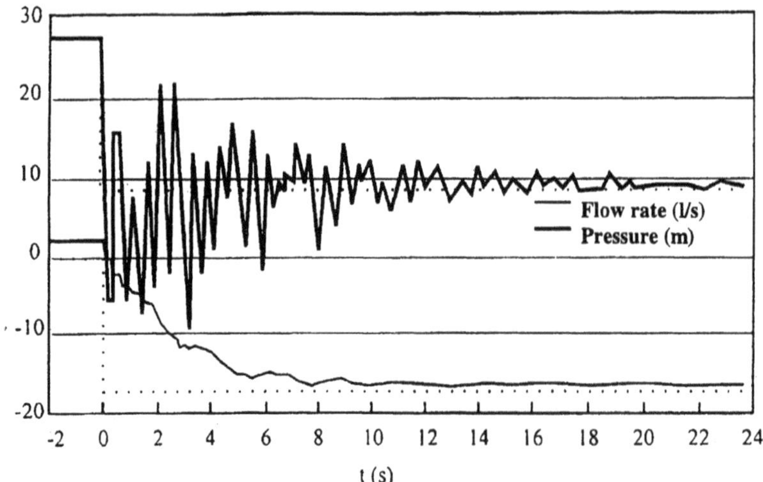


*Figure 3. Evolution of pressure and flow in the middle point o fa pipe.*

Evidently, as the water level in the tank is not modified until a subsequent $\Delta t$, the value of the water volume q(t) will not be modified over the period of time $\Delta t$. Depending on the case being analyzed, this can represent a more or less important error with respect to the real situation. Evidently, the value of the water volume will vary in a constant way from the value q(t) until the value $q(t + \Delta t)$, therefore the level will not in reality be $z(t + \Delta t)$. In the case of emptying a tank, the final value $z(t + \Delta t)$ will be less than the real one, since in practice, upon reducing the value of the level, the volume of outflowing water will be less than the initial (instant t). If the chosen time period $\Delta t$ is very small, the error will be negligible. It is therefore important to choose a value of adequate $\Delta t$ or to adjust in some way, for greater $\Delta t$, these effects. In this particular case, there are techniques that without modifying this value give more correct results.

Similarly, some computer packages take into account the possibility of modifying the state of the regulation elements between two calculation instants as a consequence of having activated an order (for example to start a pump if the level is inferior to a given value).

## 2.3 TYPES OF MODELS IN FUNCTION OF THEIR UTILIZATION.

Firstly, we can carry out a classification of models according to the size of the model or degree of detail that is to be included in the model. This is closely related to the utilization that is to be made of the model, as well as the availability of data to construct the model.

We can find:

- Detailed models: They are also called `cartographic' models, they include each and every one of the elements of which the system is comprised, either described explicitly (the detail covers each and every one of the pipelines, joints and outlets), or implicitly (groups of pipelines are described and in some instances those pipelines with a small diameter are eliminated and various outlet points for consumption are grouped in nodes).

- Strategic models: In these models, one only includes what could be denominated the main arterial network of the system, being grouped together the consumption points at nodes of the network, from which the distribution pipelines would depart.

- Macromodels: The information is even more condensed, so that through a few parameters one could come to know the operation of the system (key points of the network such as tanks, pumping stations, principal regulation elements, etc..). The scheme of the model is removed in this case from the physical distribution of the elements to condense the information, so that it works in the manner of a "black box" that receives information in real time from the system (SCADA), processes it and sends the timely orders to the regulation elements. Evidently an important degree of simplification is required for it to be able to act in real time.

Nowadays, there are no real problems in processing detailed models that include in an explicit way all the elements of the system. Using these models, it is only necessary to condense the information to feed the calculation program, since they are very close to the structured information in the Databases and the GIS. Nevertheless, the fact that one reaches such a degree of detail does not provide any additional valid information of a hydraulic type. It is impossible to contrast in the same instant the calculation results with the real measurements in an outlet, since the model is "loaded" statistically as far as consumption is concerned (not all the outlet points consume instantly at the top of their consumption range) and the measurement can be effected in the moment in which peak consumption is being produced.

It is more normal to use simplified detail models, in which are eliminated or grouped the pipelines with inferior diameters and the consumption of some 70-100 consumers are condensed at nodes.

If one lacks sufficient information, one is unable to generate a detailed model. Let us imagine a network in which one knows the consumption by sectors, but one does not know how it is distributed within each sector. The model will be able to function up to the "entry" points to the sectors, so that the degree of detail will become limited.

In reality, the difference between strategic models and detail models is not perfectly defined. For example, the size of the water supply, as well as the quality and quantity of available data influences. Similarly, it makes no sense to develop a detailed model to study, for example, the transfer of water between the principal tanks of the network or the optimum scheme for starting and stopping mechanisms of groups of pumping stations. The greater the degree of detail, the more complex the calibration or adjustment of the model, since the number of parameters is increased considerably and the statistics begin to play against the model designer.

Without being excessively rigorous one could say that detailed models are used to study the behaviour of specific areas of a network. For example, with design objectives and/or rehabilitation, these models can be used for the analysis of fire and breakage situations or localized breakdowns, sectorization, modifications of the topology or diameters that could increase the pressure levels in a given zone, the impact of a consumption increase or an enlargement of the network on the pressures of the surrounding nodes, etc.. On many occasions only a static model will be necessary to analyze these situations.

To begin a study of the levels of quality of water (quality models) in which is analyzed the temporary and spatial evolution of a pollutant or other substance in the water (for example residual free chlorine), detailed models are also necessary over an extended period (dynamic simulation) in order to determine substance evolution over. The use of static models for this type of analysis restricts the analysis of parameters denominated "conservative" (there is no reaction between the aforementioned parameter and the environment) such as, for example, fluorine, origin of one or other source, etc..

Strategic models are used as much as static models as in extended periods. For example, to design enlargements in the arterial network or to analyze the filling or emptying of water tanks, the regulation of the system, to carry out an energy study of the network, etc.

In Cesario (1992) another classification of models is realized in function of the objective that the models pursue. Hydraulic models (of planning, operations and of training) are distinguished from models of quality. Nevertheless, to implement a model of quality, it is necessary in the first place to have a hydraulic model.

- Planning models: They are used for pipeline and installation design, development of operation strategies, determination of the needs of the system and, as a rule, for studies into the improvement in the system in the middle to long term. They do not require as a rule a high level of detail and the degree of accuracy is not so important as in other types.

- Operation models: They are used to study specific problems, current or future problems, of the system. Depending on the application they can represent all the system or a part of the same, but with a greater degree of detail.

- Training models: They are used to train personnel in charge of acting on the regulation systems. In this way it is possible to simulate operations in the system and to compare them with the methods that are used normally. They can be used to train the operators in a manner of working with new installations such as, for example, a new pumping station. It also serves to train new operators or even with an educational end, so as to make known to students the operation of a system of water distribution. All this requires user friendly software that allows for the easy carrying out of operations that normally are carried out at a water supply, even to have additional facilities such as management and treatment costs, easy visualization of the results, etc.. Generally, the use of these models on the part of the operators meets with a certain resistance on the part of these operators, accustomed to work in a totally different manner based more on intuition, experience and, in many cases, routine, where the orders tend to come "from above" (superiors). A considerable effort is still needed to incorporate this kind of technique for applications of this type.

- Quality models: They are used to study the flow and the distribution of certain substances within the network, as we have mentioned previously. In the case of non-conservative parameters (for example residual free chlorine), this requires a hydraulic model over an extended period with a significant degree of calibration that, at the same time, will have to be recalibrated in terms of the parameters of quality, which increases the level of complexity of the problem.

**3. Necessary data. Modelling of elements.**

A distribution network model demands a great quantity of data. The existence or not of the same, as well as the reliability that these databases may have will condition in an important way the accuracy of the results of the model and its usefulness.

Unfortunately, on most occasions, we find incomplete information on the system or information which has not been updated, and with a lack of historical records on many aspects of the operation of the network.

It is normal to find water supplie systems that lack flowmeters at their entry point, without having to make reference to flowmeters installed in the interior of the network. In some instances, there are not even statistics to calculate customer consumption because of the lack of water meters.

Even if we have flowmeters, on occasions these are not read with a regular frequency, therefore the data that they provide only refer to the mean daily consumption at the most. To find systems that have data of the volume of water entering the network at each injection point during each hour of the day is not common. As a result of this arises the difficulty of constructing a model over an extended period of 24 hours, when not even the evolution of the consumption of the network during that period is known.

An essential first measure, therefore, when beginning the modelling of a water supply is to gather the existing data, to analyze them and to design an information structure to solve the problem of any lack of existing data.

Nowadays, it is necessary to set up structured information in Geographical Information Systems (GIS) so that this can be condensed to feed the hydraulic software. Similarly, the information about the system SCADA and any other available source will be used to feed the data model.

From all the available information about each and every element of the system the model uses only a part of this information. The basic information that the model should have is the following:

* Information related to pipelines:

- Number of the initial and final nodes
- Diameter
- Length
- Roughness
- Minor losses (equivalent length or loss coefficient)

* Information related to valves:

- Number of the initial and final nodes
- Diameter
- Type (isolation, control, pressure reducing, pressure sustaining, flow control, check, etc.)

- Loss coefficient in open position (or flow coefficient).
- Loss coefficient for different degrees of opening (or flow coefficients).
- Upstream setting for pressure sustaining valves, dowmnstream setting for pressure reducing valves and flow seeting for flow control valves.
- Minor loss coefficient in accessory elements.
- Regulation procedures in the case of control valves. They can be connected to a pressure in a node, to a level in a tank (for example, float valves for the filling of water tanks), at a certain time of day (totally opened or closed or partially closed).

* Information related to pumping equipment:
  - Number of the initial and final nodes
  - Characteristic curve of the pump
  - Minor loss coefficient in accessory elements.
  - Nominal rotation speed.
  - Starting and stopping procedures for the pumps or for the modification of their rotation speed (for example, in function of levels or pressures).

* Information related to the nodes:
  - Number of the nodes
  - Elevation of the nodes
  - Type of node (connection, demand, injection, tank or reservoir)
  - Node <u>of step</u>: Pipeline intersection.
    - Demand node: Demand is fixed. One should assign a level of demand and/or the evolution of the demand over the period to be analyzed.
    - Injection node: The pressure is fixed. One should assign the evolution of the injection pressure over a period of time.
    - Tank or Reservoir: The water level is fixed in a static simulation and variation in the level of water is permitted in an extended period simulation. The necessary data in the latter case are: Bottom elevation, overflow level, control levels (generally highest and lowest) and area.
    - There are demand nodes in which the flow rate is not fixed as a parameter, but instead a relationship between the pressure and the demand exists. This is, for example, the case of a node that functions as an outlet into the atmosphere through, for example, a fire hydrant or a sprinkler in a fire-protection system.

Evidently, the information related to how all the elements are interconnected is fundamental. This information is the one which permits us to define the topology of the network.

Some more advanced calculation models require additional information. We can look at some examples:

- Efficiency curve of the pumps and of the electrical motors to calculate energy consumed over a period of time.

- Operation costs (for example, of production, energy, etc.) to analyze different strategies of operation and to compare their costs.
- Databases related to commercial diameters (nominal and interior) and their cost to effect an optimum design taking into account factors of an economic type.
- Location of the nodes in the network to be able to draw a plan.
- The demand allocation to the different demand nodes of the model (load allocation) can be realized automatically from a data base if the data base information is related to the nodes to which consumption must be allocated.
- Reaction coefficients of the different substances to analyze in water quality models in relation to the water, the pipeline walls, as well as the tanks.
- Initial values of the quality parameters of the water at the beginning of a simulation.
- Concentrations and evolution of the quality parameters at the injection nodes.

To process a static model or a model over an extended period, it is necessary to know the different load states (or evolution of the same), the state of the regulation elements and the pumping stations (the operation procedures over an extended period), the water levels in the tanks for a static situation and the initial levels and operation procedures over an extended period and the parameters and initial values associated with the water quality models.

The results obtained will be:

- Pressures (or its evolution) at the nodes.
- Flow rate (or its evolution) in the water tanks and at the injection nodes.
- Water level variation in the tanks.
- Flow rate (or its evolution) in the elements of the system (pipelines, valves and pumps).
- In the case of extended periods simulation, the start-up and shut-down of the pumps in relation to some given procedures and operation of the regulation elements.
- In the case of quality models, evolution over time of a given substance at each point of the network.
- Head losse in the pipelines, valves and accessory elements, velocity in the pipelines, pump head, etc..

A fundamental part of the databases is all the data referring to consumption. With them one should be able "to load" the model to process different operational situations. In the following section we will deal with the distribution of consumption. Now we want to demonstrate how consumption is modelled in most models.

In practice, the consumption of the network is affected by pressures and by the regulation elements which the consumers have. It is evident that the pressures will influence decisively the flow through leaks (therefore the consumption of the network), the flow of water that is extracted through fire hydrants, or even the consumption of the consumers. In the case of a consumer who is supplied directly from the network, an increase in the pressures will result in an increase in the flow (we can think, for example, of someone filling a bath tub) and also, to a lesser degree, of the consumed volume. If the consumers have intermediary regulation elements (tanks, pumps, pressure reducing valves, etc..), the relationship pressure-flow is no longer direct, but will depend on a great number of factors (power and characteristics of the pumps, characteristics of the service pipe and of the float valve of the tank, etc.). There are few studies which have been carried out on the relationship between water pressures and flow rates or pressures and volumes of consumed water, but there is no doubt that the relationship exists. One only has to observe how a decrease in pressure causes a reduction in the flow rate, above all at night. The problem is how to quantify this fact.

In reality, to correctly model the consumption of the consumers one would have to take into account every real boundary condition of the network. To arrive at the final limit point, one would have to model down to the last water tap of every consumer and to know how often these taps are opened or closed over the period of the simulation of water consumption.

Similarly, in this way, we have to analyze, for example, the facilities installed for fire protection. Concretely, in the case of sprinklers, we have to model each and every one of them and the existing pressures in the network are determined, the flow in pipelines and the flow emitted by each open sprinkler.

It is clear that this is not viable in the case of a distribution network. Because of this one has to recur to the study of the volume consumed by the consumers and to modulate the same over time to determine the consumption at any given moment in function of the type of consumer: domestic, industrial, commercial, recreational, hotels, etc.. Within each one of them can exist, at the same time, different typologies. Once the model is loaded and the simulation realized one analyzes if the existing pressure at the service pipes is sufficient to feed the required consumption. Generally one does not take into account the possible variation of the flow with the pressure. Nevertheless, since we are trying to consider how to distribute a measured consumption at a given hour of the day between the different nodes of the system, the model will actually remain loaded. The difference between the injected water volume and the consumption assigned to the consumers will be distributed between the nodes of the network.

This can give cause for satisfactory results when one is considering the analysis of normal operation situations, assuming the errors derived from the difficulty of distributing the water volume injected between the nodes. However, if one is considering analyzing a situation in which there exists a decrease or notable increase of the pressure in the system, to use this methodology without any other considerations can give rise to results which do not accord with reality. It will be difficult, therefore, in these cases, to realize a true estimate of the real consumption of the consumers.

We can think, for example, of a case such as fire simulation. Upon opening the hydrants, the pressure in the nearby nodes may be affected enormously. If, for example, it has fallen to 15 mca, when what is normal is to have 30 mca, the consumption of the affected consumers can be really diminished. Generally, in these cases, there is a tendency to process the model and to analyze the results to test if the water volume supplied to the hydrants is sufficient allowing for a mean value of consumption for the consumers (not a peak value), maintaining a minimal pressure from between 10 and 15 mca.

## 4. Elaboration of a model

An important factor to take into account, as has already been commented on previously, is who is going to elaborate the model. There are several alternatives. The adoption of one or another will depend on the application that is going to be given to the model, and of the availability of resources, above all human, that the water supply company has. Basically there exist three alternatives:

- Development by personnel of the water supply company.
- Mixed development with internal and external personnel.
- Totally external development

In any case, knowledge of the system, in principle, belongs to the personnel of the water company. Their participation will consequently be indispensable at the hour of providing data. Nevertheless, we think that the personnel who will in charge of the operation of the model must participate actively in its elaboration, therefore it seems to us more adequate the first and second alternative, the latter if there are no members of personnel with sufficient experience in the elaboration of models.
The elaboration of the model is realized covering a series of stages:

- Information gathering.
- Skeletonization. Simplification of the real network pipelines, according to the use and the available information.
- Analysis and allocation of recorded consumption. This means "to load" the model with demands recorded at the consumption points.
- Analysis and allocation of the non-recorded consumption. To study and to distribute the non-recorded consumption: leaks, meter errors, illegal connections, etc.

Following these stages we have a first model, without validation, from the network. The following stages are centred on the alteration and adjustment of the parameters of the network, so that the model reproduces it with a certain reliability.

- The taking of measurements, generally of pressure and flow rate. They are realized through a series of measurements at some points of the network, for the different load states in the case of a static model, and over several days with records of the values measured in the case of a dynamic model. These measures will serve as guide values for the adjustments in the following stage.
- Adjustment of the model. This is also designated the model calibration. The load states with the measurements (static model) or the dynamic simulation are reproduced with the model The pressures and flow rates measured in the network are compared with those obtained from the model and the different parameters are adjusted so that the values measured coincide with those calculated through the simulation.

At the end of this process we obtain a calibrated model, that will permit us to realize simulations of the operation of the system. The thus obtained model must be updated and recalibrated periodically to incorporate the enlargements in the network, as well as the variation in the calibration parameters over time through the operation of the network.

In various references (García-Serra, 1988, Martínez, 1982, Cesario, 1992), one can find the detailed development of each one of these phases. Nevertheless, we go on now to carry out a short review of each one of these.

## 4.1 THE COLLECTION OF INFORMATION

We have insisted throughout this paper in the importance that, for the correct functioning of the model, it is the collection of specific information of the elements that make up the distribution of the system to be modelled.

When we speak of system, we are referring not only to the pipelines, valves, pumps, and tanks, but also to the production plants and to the wells, to their operation plan, to the regulation plan of the network, and all those aspects that serve to reproduce the behaviour of the network in real time.

During the collection of information phase we will have the opportunity to check and update the sources of information of the network object of the model. In many cases, this work is sufficient justification of its elaboration. Many of the elements that form the network are as old as the network itself, and their state of conservation as well as of operation is not known. Independently of who elaborates the model, an important effort of collaboration of the personnel of the water distribution company is needed. They are the people, who work day in day out with the network, and are the ones who know where to locate the necessary data, or in the case of a lack of data how to obtain information about the network. Historical data, if it is available, about the operation of the network is of great importance.

It is also useful to refresh and to update information. Many errors are corrected in data which have been maintained over long periods of time.

The information that it is necessary to gather has been described in the previous section. Nevertheless there exists additional information that will permit us to place the value of some uncertain parameters. For example, in order to fix a value for the roughness of the pipelines it is important to know the material, the age, the conservation state, the type of water and if there exist or do not exist statistics on values of the real roughness and its variation over time. In some references estimates are given for aged pipes ( Sharp, 1988).

In (Walski, 1984) a method is proposed in order to be able to carry out Hazen-Williams Coefficient ($C_h$) measurements on the network. From the values of this coefficient, which very often used in the United States of America, it is possible to determine the roughness of pipeline material. It is not a bad idea to realize a campaign of measurements of this type because, in addition to providing us values for parameters of the system, it will allow us to test the conservation state of the network.

The physical location of the different elements of the network (including consumer service pipes) must be known. If a GIS system is available, it will contain this kind of information. In any case, it would be desirable to have a cartography of the network, at least to a scale 1:1000 in order to know the details of the network. Likewise, associated with this kind of information one should have data on the consumers: type of consumer, meter number, calibre and age of meter, records of meter readings, type of feed (direct, tank with pump, pump without tank, a raised tank, etc.).

The elevation of the different nodes of the model might be available in the GIS, or in the case of cartographic plans, with elevation curves every 0.5 m. If necessary, one would proceed to a topographic outlining of the points (in the measurements phase we will see how convenient it would be to carry this out on those points where pressure measurements will be taken).

There would be no harm in characterizing elements of the system in the event of the available information not being excessively reliable. This would lead us, for example, to test groups of pumps to determine the current characteristic curve.

Once the information has been gathered, the following step is to organize it in such a way that we can introduce it into the analysis program. The synthesis process depends on the program. Each simulation program requires some data in function of its capability.

The possibility for enlargement and update of the model must be taken into account. In subsequent enlargements or updates the simulation program can be changed or, during the life of the model, it can be coupled to an Information System. Because of this, we should never lose contact with the origin of the data.

To organize and handle the information, it is convenient to create different data bases. From these, the data will be extracted that each simulation program requires.

Attending to the form in which we have gathered the information we can create the following databases:

- a database for each type of element
- a database for the topology of the network (GIS)
- a database of historical records of production data.
- a database containing historical records of the operational mode of the regulation elements in the network
- a database of records of measurements taken on the network.
- a database of records of consumption (billings)

Programming different applications on each database is going to permit us to automate, up to a certain point, the elaboration process of the model.

## 4.2 SKELETONIZATION OF THE NETWORK.

To work with a model that collects absolutely all the elements of the distribution system, such as to be found physically in the field, may result in some difficulties in some cases. It may be necessary to filter the data that we have collected so that it is converted into useful information to be introduced into the model.

The skeleton of the network consists, in some form, in this treatment of the information, and it can be summarized in a simplification of the interconnecting pipelines, and a schematic design of the remaining elements of the network . This simplification will allow us to visualize better the results in the case of having an analytical program that contemplates this possibility.

The simplification of the interconnecting pipelines is realized by eliminating those which are smallest, and substituting the ramifications of the network by demand nodes. The diameter after which the pipelines are going to be eliminated depends on the type of model and on the size of the network.

In strategic and planning models, the mains with greatest water carrying capacity are considered. In quality models, it will be necessary to envisage distribution pipelines with smaller diameters, since it is where the water suffers greatest deterioration in its quality.

In small networks, pipelines of 80 mm and 100 mm can have an important carrying capacity, while in a large network these diameters are not in the model since they have little transportation capacity (they serve only as distribution) as compared to other pipelines of greater diameter (400 mm, 600 mm, ...).

Within a same network there could exist consumption areas in which small diameter pipelines may have an important transportation capacity. This happens, for example, in the oldest areas of the city. One must take this into account so as not to eliminate systematically all the small diameter pipelines.

There are cases in which, to maintain the connection between some pipelines of greater diameter, it is necessary to include some smaller pipelines. One must also take into account that a pipeline, that in normal operational conditions of network does not have an important carrying capacity, can become important if, because of a breakage or breakdown, we find ourselves obliged to close down some important pipeline. Taking into account that in many cases the model is used to observe the behaviour of the network when confronted with these kinds of situations, it is important to include these pipelines.

In any case, it is always better to add an extra pipeline instead of eliminating one which could come to have some importance in the model.

In addition to these considerations, other possible simplifications that can be realized are:

- Elimination of branched zones and service pipes. The known demand in branched zones is accumulated at a node that represents the point of connection with the principal network.
- Nodes which are close to each other are unified especially those which, due to the short length of the pipelines that join them, will be observed to have the same pressure value.
- Associating pipelines serially or in parallel representing in this case only one line for a set of pipelines.

It is very useful for the update and maintenance of the model, to have a record of any of the simplifications that have been carried out. Thus, we will create a new database of simplifications. Neither must one eliminate from the existing databases those elements that are not going to form a direct part of the model since they could do so in a subsequent enlargement.

The schematic design of the pumping stations implies reproducing the behaviour of these pumping stations without having to include in detail all the elements that they are comprised of.

Hydraulically, the pumping stations can be modelled, for a static model, as injection points of a flow rate whose value corresponds with the one which the pump provides for the water volume that is being injected at that moment. If the model is dynamic, such simplification is more difficult to realize since a priori we do not know the flow rate that the pump is going to provide for us in different consumption situations in the case of there existing compensation tanks or other supply points.

The last step in this phase consists of numbering the nodes and the lines that have resulted from the simplifications. The numbering criterion that is followed can become very useful for the subsequent location of elements in the model.

In the following figures an example of a skeleton is represented. In the first (Figure 4) all the pipelines of the system are reflected, while in Figure 5 only those which form part of the calculation model (the skeleton).

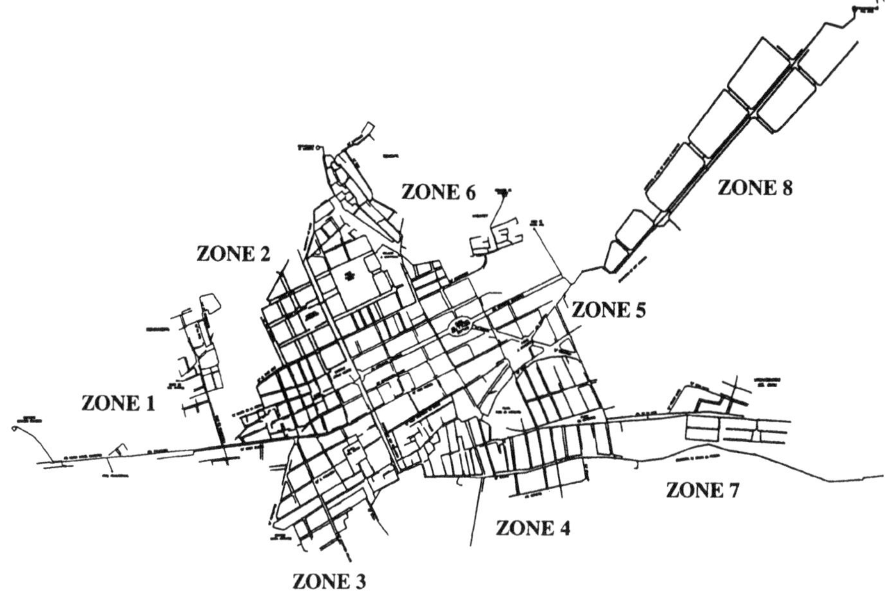


*Figure 4. Real Network*

*Figure 5. Network Skeleton.*

## 4.3 STUDY OF CONSUMPTION AND ALLOCATION OF LOADS

The distribution network is designed to satisfy the water demand at the consumption points of the same. To load the model means to allocate the demand, that is supposed to exist in the network, at the nodes of the model. We will call "the load state of the network" to the consumption situation that is produced in the network at any given instant.

In order to the study the load state of the model, one should distinguish between static models and dynamic models. In a static model, the network is analyzed for a single, unique state of load on the network. This aforementioned load state must be sufficiently meaningful or significant for the use that is made of the model, generally in the planning of the network. One analyzes: the situation when consumption is at its peak, when pressures on the network (and consequently in the model) are lowest, being detected thus possible deficiencies in the water supply service; and the situation of lowest consumption ( nocturnal consumption) when the pressures are greater and breakages in the pipelines can be produced.

In a dynamic model, one generally considers an hourly modulation of the consumption, being able to vary this from one day in the week to another (working day/holiday) or between different times of the year.

In the consumption allocation, one must also distinguish between registered and not registered consumption. We consider registered consumption as those controlled as to their value as well as their location. Fundamentally, this is the consumption that is billed to the customer. One must pay attention to billed values which are not consumed. Such is the case of water supplies that bille minimum values of consumption that may have been produced or not.

Those that are not registered are those that remain out of control. Mainly they are considered as such as errors in meter readings (a total lack of readings because the meter has stopped or underregister), the failure to carry out a reading of the meter (resulting in an estimated billing), leaks in the network, illegal outlets, uncontrolled municipal services, water consumption by firemen during fires, etc.

There are cases of distribution networks that do not have consumption data about their consumers because they lack meters. In such a case, it will be necessary to carry out measurements in the system in order to discover the estimate loads on the model.

### Registered consumption

With the reading of the volume consumed in the period of billing (for a month or more than a month) one can obtain a mean consumption for said period. This mean consumption is corrected through a series of coefficients to adapt them to the load state that is simulated (peak-hour, minimum-hour or a temporary modulation of the same).

The process of assigning loads consists of going from this registered consumption of each consumer to the demand at the node of the model. The method used will depend on how the water supply service has structured its information .

Below we go on to describe some of these techniques of load allocation in the model:

**Allocation influence areas**. The process consists in relating each consumer to a node in the model, and to assign their consumption to the said node. One always seeks to relate the consumer to a nearby node, or to the node that can be considered to supply the consumer. For this, to each node is assigned an area of influence. Only the consumers database, in which their consumption is registered, and the other database related to consumption at the nodes of the model are handled.

**Allocation by sectors**. The process consists in adding up the consumption of all the consumers that are within a sector, whose perimeter can be defined by the loops of the model. In this manner, the total consumption for the sector is obtained that then is distributed between the nodes that form part of the lines of the model that enclose the sector. The distribution may be equal to all the nodes or there may be established some different criterion taking into account, for example, the diameter of the pipeline that supplies the sector starting from each perimetric node of the same. In this case one handles: the consumers database, in which in addition to the registered data is included the consumption sector to which the consumer belongs; the sectors database, in which are included the nodes that belong to each sector, and the consumption database on the nodes where the total demand is calculated on the nodes. To automate the process related databases are used that permit, through the programming language (SQL), to design applications that relate and operate with the data that they are comprised of.

**Allocation by consumption units for each street.** The process is begun by identifying the distribution pipelines and calculating the meters of the pipelines that run through each street. A pipeline database is created with this information. With the consumer data the consumption for each street is deduced and, operating with the previous databases, the consumption unit for each street. As we know the ml. of each pipeline that runs through the street, multiplying this value by the consumption unit, pipeline consumption is obtained. Once we have the real pipeline consumption of the network, we use the database for the skeleton framework to transfer the consumption of water through the real pipeline of the network to the consumption of water through the lines of the model. The line consumption is distributed between the extreme nodes and is added to the consumption assigned directly to the nodes (if there is any) to obtain demand. In Figure 6 a plan can be seen that relates the different data bases of this process.

With these procedures we assign a mean demand on the nodes of the model, that corresponds to an average load state within the network. To realize static simulations we should adapt the average load state in the network to that which is produced at the time of the simulation. In dynamic simulation we should moreover reproduce a modulation of the demand at nodes throughout the simulation.
In static simulations, to obtain at every moment real load states in the network, we multiply the mean demand at the nodes by a coefficient that relates consumption at any given moment to the mean consumption.

The demand at any given moment depends on a series of factors: type of consumption (commercial, domestic, industrial, public), standard of living of the consumer, consumer habits, etc. All this means that the consumption of one or other sector or even at the nodes of the same sector, the momentary demand varies and as such the correcting coefficient of demand.

With a dynamic model, the process of load allocation includes also the characterization at the nodes of the temporary modulation of demand. This is obtained taking into account the temporary modulations of the consumers whose consumption is assigned to the node. The temporary modulation of each type of consumer (domestic, commercial, industrial, etc..) can be different. Even for the same group we will have different modulation curves. Thus, for example, in the case of domestic consumption, the volume consumed as well as the modulation can vary with the standard of living of the consumer, their consumption habits, the age of the consumers, etc. The type of household installation also influences, since it is matter of representing, not so much the manner of consumption of the consumer, but the way in which the consumer extracts a volume of water from the network. A household installation with a greater storage capacity gives cause for a different modulation curve from an installation in which the water supply is direct.

As a result of this, we can affirm that the standard modulation curve varies from one network to another. It is convenient to take measurements which characterize the tendencies of the demand curve of domestic consumption of the network that we are going to model.

The standard modulation curves of each type of consumption can likewise vary according to whether the day is a working day or a public holiday and with the change in the seasons, which must be taken into account when trying to characterizing the consumption curves. The case of tourist areas is a clear example.

To obtain the modulation curve at a node from the modulation of consumer consumption (Llopis, 1996) we started with the Allocation of the mean demand at the node. We obtain the percentage from the mean demand at the node that corresponds to the consumption of each consumer. On the other hand, we assign to each consumer a standard modulation curve. Weighing the modulation curve of each consumer we obtain the modulation curve of demand at the node.

It is convenient to characterize in an individualized way the large consumers, due to the incidence that an error in the consumption allocation can have on the results of the model.

**Non-registered consumption.**

The only reliable flow rate data that we can have is, as a rule, the value of the instantaneous flow rate injected into the system. It is convenient to also have the water volume of filled or emptied water tanks, in order to establish the continuity equation and to know the water volume that one has to distribute between the nodes. On occasions one can have flow data from some measuring equipment installed in the network, that without doubt will provide valuable information. There will always be a difference between the value of the water volume to be distributed and the water volume distributed amongst the consumers based on recorded consumption, being greater the former. It is necessary to distribute this difference between the nodes of the model "to load" the model with what is actually being "consumed by the network".

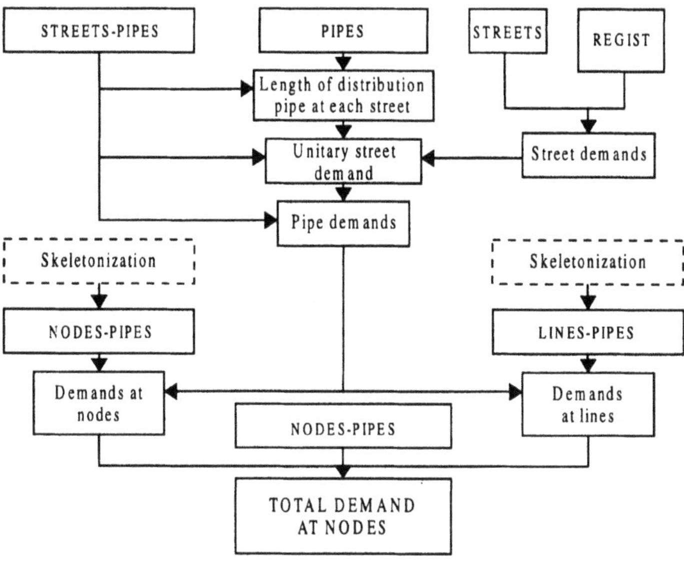


*Figure 6. Demand allocation scheme.*

In the immense majority of the occasions the meter tends to mark less, therefore there will always be some volume of water which is not recorded to distribute. Now, if one does not have available statistical studies on non-recorded water or inactive meters, it is not possible to dissociate this factor of consumption from the rest of the unrecorded consumption.

A leak in the network can be considered to be similar to a half-opened valve letting out water towards the exterior. The flow rate that the valve allows to pass depends on the difference in pressures between its extremes. Being opened towards the exterior, the downstream pressure is constant and equal to the pressure in the atmosphere, therefore the given flow rate of the leak depends on the pressure levels in the network. When the pressures are greater (as a rule in a situation when there is a low consumption hour) the leaks are also greater, while when the pressures in the network are minimal (peak consumption time) the level of leaks in the network is also at a minimum.

It is difficult to know how to assign them as demand in the model because neither their value nor their location is known. When a leak is located, it is generally because it has a certain importance and one proceeds to repair it, as a result of which it disappears.

Good part of the nocturnal consumption corresponds, in many cases, to leaks in the network. However one can not carry out an assignment of leaks in the network using this value because, as we have already said, the level of leaks is greater at night due to the pressures in the network being greater. In any event it is possible to make a global estimate taking into account the fact that the water volume of leaks will vary, in an approximate way, with the square root of the existing pressure.

**The loading of the model**

In function of the available data the loading of the model will be realized in one way or another. One must never lose sight of the fact that we work with a great quantity of estimated parameters, and some of them are even unknown, therefore the uncertainty is increased. These parameters will be object of modification in the adjustment process, in order to " square" the values measured in the network with the ones calculated for each situation.

On the other hand one must take into account the distribution of the non-recorded water volume which will influence in a decisive way the case of systems with low efficiency (the ratio between registered and injected water volumes), their importance is reduced as the efficiency of the network is increased.

In the case of a static model, the simplest way to load the model, knowing the flow rate injected into the network for a given situation, consists of multiplying the value of the mean consumption at the nodes by a coefficient, the relationship between the flow rate injected at that moment and the sum of the aforementioned average consumption. This supposes the assumption that the modulation in consumption is the same for all the nodes, and that the non-recorded consumption is distributed proportionally to the recorded volume of water. In the adjustment phase, the value of this coefficient in an individualized way will be modified for each node. Nevertheless, it would not be logical to think that there will be very important variations.

The way we load the model can be complicated even more if one takes into account the lack of uniformity in the modulation of the consumption of the different nodes. It is undeniable that the modulation coefficient of each one of them as well as the distribution of the non-recorded consumption are uncertain parameters subject to calibration, more the second than the first. In reality one does not know, at every moment, how much of the injected flow is consumed by the consumers and how much is lost in leaks or, simply is not read by the meters.

In the case of a dynamic simulation the problem is similar, with the added drawback of the fact that one has "to square" thereafter the results of the model for, as a rule, 24 hours. The premise is maintained that, knowing the flow rate injected into the system in each calculation interval, as well as the flow rate sent to tanks, one should be able to distribute the rest. The modulation coefficients, as well as the distribution of the non-recorded water volume are uncertain parameters of the problem, that it will be necessary to adjust thereafter.

## 4.4 MEASUREMENTS IN THE NETWORK

Having arrived at this point we have a model of the network without any kind of validation. The following measurement phases and calibration of the model are closely linked, and they are carried out with the intention to arrive at a stage where the model that we have constructed reproduces faithfully the behaviour of the distribution system that it represents.

One must take into account the values of the parameters that we have assigned to the model are those of design (roughness, diameters, located losses, characteristic curves of the pumps and valves,...) or they have been obtained by simplification, something which adds to uncertainty (demands at the nodes,...).

Therefore one has to contrast these values to prove that in fact they are correct, and if they are not so, one must correct them so that, at least, *they produce the same effect* as the real parameters of the network.

In the reference literature (López et al., 1995; García-Serra, 1988; Martínez, 1982; Cesario, 1992 or Walski, 1984) there are more exhaustive description of the measurement process to be effected on the network. Nevertheless, in what follows, we attempt to give an overview of the process and to focus on some important matters to be taken into account.

The measurements to be taken are:

- Pressures at nodes in the system.
- Flow rates in pipes.
- Injected flow rates.
- Temporary evolution of the water level in the tanks.
- Substance concentration (for example residual free Chlorine)
- Operation of the regulation elements.

Some considerations about measurements:

- We will take advantage of the available measurement infrastructure in the water supply, but it will also be necessary to increase the number of measurement points. It is convenient to have meters with recorders (dataloggers), to determine the temporary evolution of the variables (in the case of dynamic simulation it is indispensable).
- Generally one has available many more measurements of pressure than of flow rate. The reason, in spite of the fact that the latter are of great usefulness, is their elevated cost. In any case, it is necessary to control all the entrances and exits to the network to, for example, other water supplies that we do not want to include in the model.
- On occasions tracers have been used (for example, fluoride) to determine "journey times" of the water through the interior of the system.
- It is very important to define clearly what state the network which we are measuring is in. Upon simulating this state we have to reproduce the operation of the system under the same conditions as when we were taking measurements, therefore it is necessary to know the boundary conditions (operation of the regulation elements, number of pumps under way, position of the isolation valves, etc..). It is convenient to contrast the characteristic curves of the pumps and of the regulation elements.
- In the case of static models, it is necessary to measure under different situations and on different days. The measurements carried out under high consumption conditions give rise to greater losses of load and, therefore, greater differences of pressure, which facilitates the subsequent adjustment. On occasions there exist forced increased consumption conditions in certain areas of the system when opening fire hydrant, which facilitates the adjustment for roughness.

- When we are dealing with dynamic models the measurements will be realized over periods of 24 hours, over several days.
- In any case the measurements should be checked, eliminating any considered as incorrect because of different motives.
- If there exists the possibility of dividing the network into controlled sectors the subsequent calibration will be simpler. For this it is necessary to have instantaneous flow rates measurements at the entry and exit points to the sectors, as well as pressure measurements in the contour and interior of the sectors.
- The measurement of pressure must be realized at points where they are not distorted by sharp fluctuations. For example, we should avoid putting meters near an inlet that supplies a pump that feeds directly into the network.
- Since the movement of water within the network is provoked by differences of piezometric levels (the sum of the pressure plus level), it is necessary to have levels with the greatest possible precision, above all those points at which we are realizing the measurements.
- All the meters that are used will have to be calibrated before and after the measurements.

The aforementioned measurements will not only permit us to have a set measurements to contrast the validity of the model, but also they can be of great interest to know in depth the operation of our system and to detect important mistakes in the data of the model.

The measurements with tendencies towards estimate values for the parameters, as for example to determine roughness or modulation coefficients, will be carried out prior to the measurements here described.

## 4.5 ADJUSTMENT AND CALIBRATION OF THE MODEL

As we have already commented, the calibration of the model involves adjusting the results of the simulation to the real values measured in the network
The principal causes of the discrepancies between values calculated by the model and the measured values can be imputed to:

- Modification of the diameter and original roughness of the pipelines as a result of salt, lime and oxide deposits, increasing the headlosses.
- Ignorance of the minor losses produced at joints, bends, t-junctions, reductions, etc. all of which are impossible to quantify. Generally they are taken into account as equivalent length pipeline.
- Headlosses at the shutoff valves that, under operational conditions, do not remain perfectly closed or open due to some manoeuvre, and whose state is difficult to know.
- Ignorance of the correct consumption distribution, at a given instant, in the network due to the randomness of demand. The mean values can be known based on billings, but their modulation over a 24 hour period is difficult to predict.
- Ignorance of the value and location of the leaks, and of errors in meters, that add uncertainty to the load Allocation to the network.

- The measurements, necessary as we will see for the adjustment phase of the model, are affected by errors in the measurement apparatus, in the reading, and in the fluctuations of the variable that is measured.
- Errors in the determination of elevation at the nodes.
- Errors in the data of the model (lengths, nominal diameters, characteristic curves of the pumps, etc..).
- Incorrect definition of the topology of the network (connection between the network elements).
- Errors in the boundary conditions (for example, setting pressures at the automatic valves).

All these indeterminacies will condition the process that we will follow for the elaboration of a mathematical model of the network in operation.

Previous to the adjustment process we should obtain a faithful simulation of the state of the network at the time when the measurements are taken. It will be sufficient to simply let the model run with the simulation program and to check that the connection between the elements is correct and that the results are logical in a model which has not been adjusted. If during the simulation we find some especially abnormal result, we will locate and correct it by locating the cause of the problem before beginning to adjust parameters.

For the calibration we will adjust the parameters that were introducing greater uncertainty to the model: roughness of the pipelines (diameter reduction in some cases) and demand at the nodes.

It is fundamental to execute the adjustment while taking into account several operation states of the network. Otherwise this might give rise to error compensation. In effect, the headlosse in a pipeline is increased with the circulating flow (dependent on the water consumption) and with roughness. With measurements realized for a single, unique load state of the network, we will not know whether to increase the consumption lowering the roughness or vice versa so as to reach a state where the headloss (and consequently the pressures) is adjusted to the measured value.

If the error detected after a first simulation is excessive (in the order of 30%) it is advisable to check the initial data (connections, partially closed valves, etc..).

Since the number of measurements is inferior to that of the adjustment parameters, it is not usual to begin modifying the individual values of these. There is a tendency to group them (for example, the roughness of all the pipelines of an area and the material of the pipelines are modified in a general manner) so that the number of unknowns is reduced.

It is usual to carry out a first adjustment through tests and corrections (trial and error). In this first phase, it is also possible to detect large inconsistencies in the model. Realized in this way, the adjustment is based very much on the experience and knowledge of the system that the group of personnel have who are in charge of carrying out such adjustments. It is also vital how the grouping of parameters has been carried out.

Nevertheless, in a first phase, it is possible to apply certain methodologies to adjust the parameters globally, based on in most of the cases, in techniques of optimization (García-Serra, 1988, 1990). For this purpose, an objective function is defined, called the discrepancy function, formed from the quadratic errors (quadratic deviations between the measured and calculated values) adequately weighted in function of the magnitude measured (flowrate, pressure, level, etc.) and of the importance of the same (arterial network, distribution pipeline, etc..). Through optimization algorithms one progressively reduces the value of this function, modifying the adjustment parameters.

Techniques have been proposed to realize the adjustment of the grouped parameters based on deterministic techniques. They are based on the enlargement of the set of equations and unknowns in the equations system that define the operation of the network, which force the values of the variables measured to be equal to or match the real values. The problem that these techniques present is that the selection of the set of parameters influences in an excessive manner the adjustment, so that the results obtained which are not logical in most cases (for example, negative roughness).

It is interesting, in any case, to realize a sensibility analysis of the discrepancy with respect to the parameters to be modified. This means determining that the parameters have a greater influence on the discrepancy, so that these will be, in principle, the candidates to be modified in the first place.

To have a network which has been divided up into sectors, in which one has been able to define perfectly the areas of the system is of great help, since this permits us to carry out the adjustment in phases, beginning with the sectors to go on thereafter to adjusting the parameters of the arterial network .

From the model obtained after the first adjustment (which we could call "gross adjustment"), it is possible to continue adjusting applying more sophisticated techniques if the results (the value of the discrepancy) are not completely satisfactory. In bibliographical references, some of the techniques proposed are gathered together, based mostly on optimization techniques. In Ormsbee (1997) the state-of-the-art is described.

On the other hand, it is necessary to distinguish between static and dynamic models. For the calibration of a static model, one uses parameters such as the consumption and the roughness. In a dynamic model, an important role is played by those parameters that experience a temporary variation: position of the valves, characteristic curves of the pumps, levels in the tanks, .... For a correct simulation, one must have these parameters at each and every moment during the simulation under control and characterized. Generally, one starts in these cases with a model in which the roughness has been adjusted previously through static simulations.

The calibration of the model is a continuous process that must be realized often if we do not want the model to become obsolete. The regularity of the calibration of the model will indicate to what degree action plans have been executed on the network. In all water supply services that have a model there must exist a maintenance plan of the same.

With respect to what must be the maximum error limit admitted in order for us to consider the adjustment successful, there is no unanimity. We can cite various criteria of several authors, all reasonable but, in many cases, not strictly in mutual agreement. It is undoubtedly the case that the final accuracy that is obtained will be intimately related to the quality of the data that is available.

In Ormsbee (1997) an error of 5% in the measurements (±1,5 mca in a system with average pressures of 30 mca) is given as a limit value. In a small system in which the headlosses will not be excessively important, this may be feasible. Undoubtedly it will be simpler to calibrate than a system of greater dimensions in which the pressures might vary more between the different points.

However, in other references, the headlosses which has been measured as an adjustment indicator is chosen between the injection points and the nodes (Walski, 1985), fixing an error limit between 10 and 20%. This means that the absolute error permitted for the most distant points in the system will be greater than for the nearest, since for these headloss is smaller.

In this respect, one can consult various references given at the end of the paper. In Walski (1995) more elaborate recommendations are drawn up, issued by the "Sewer and Water Mains Committee of the Water Authorities Association in U.K.". For static models, the following limits are given:

- Flow rate measurements:
  - Maximum error of 5% in the measured flow rates if these surpass 10% of the total demand.
  - Maximum error of 10% in the measured flow rates if these are inferior to 10% of the total demand.

- Pressure measurements:
  - 0,5 mca or 5% of the headloss in at least 85% of the nodes of those which have been measured.
  - 0,75 mca or 7,5% of the headloss in at least 95% of the nodes of those which have been measured.
  - 2 mca or 15% of the headloss in 100% of the nodes of those which have been measured.

Nevertheless, it is always convenient to give the opinion of various authors. In Cesario and Davis (1984,69), it is asserted that: "*knowledge about the particular system's operation and performance, gained through calibration, is more important than obtaining that last increment of accuracy*".

In Walski (1995,55), it is asserted that:

"*Because models are so complicated and models are used in such widely varying application, a simple, rigid rule for assesing the adequacy of calibration cannot be developed. Instead a model must be judged on its ability to perform specific tasks*".

"*The answer to the question of whether the model is accurate enough really begs the question, "How is the model going to be used?" With the answer to that question, the model user can then attempt to decide if the model is sufficiently accurate for the application.*"

"*Instead, the model must be viewed as a tool to support decision making. Can the decision mekers can confidently base their decisions on the results of the model?.*"

In this respect, the same author, in Walski (1995,57) advocates the realization of a sensible analysis of the problem: "*are the cumulative effect of approximations, simplifications, uncertainty and errors so great that the model cannot clearly distinguish between alternatives?*". It is necessary , therefore, to analyze the sensibility of the decisions from the results that the model can offer us.

In fact, the person in charge of its development is the one who must act as the "devil's advocate" and, in the last resort, to know the limits of the applicability of the model and feel comfortable with its use.

## 5. Applications of a model.

We have already commented in previous sections the most important possible uses of the models. In fact, the objective is to realize simulations of the operation of the system with the purpose of planning, operation modification, contingency analysis, etc..

In this section, we want to give an exhaustive review of these tasks without forgetting a complementary objective that, in fact, is a consequence of the simulation, which is to acquire a greater knowledge of the operation of the system.

**During the elaboration of the model** we can obtain our first fruits. We would highlight:

- Analysis of the existing information, in quality as well as in quantity. This will lead us, knowing the current situation, to be able to structure the information system better, correcting the errors that are being committed. In the design of the information system it is convenient to count on a GIS of the system and to implement a SCADA system.
- Detection of anomalies in the operation of the network (for example, low or high pressure zones, low efficiency from the pumping stations, improperly setting values or incorrect performance of the regulatory valves, etc..).
- Determination of the state of the pipelines (roughness, diameter reduction, etc..)
- Estimate of the non-registered volume of water and how it is distributed.
- Knowledge of the positions of isolation valves.
- Knowledge of the chlorine levels at various points of the network, and not only at the input points of the network and at some significant points.
- Exhaustive knowledge of the topology of the system. On occasions one does not even know the basic outline of the system.
- Knowledge of the energy efficiency of the system.

Only with the information that is obtained in the elaboration phase one could justify the construction of the model. We can say that it is a good excuse to justify the need "to carry out measurements".

**In the application phase:** here we find ourselves with a multitude of tasks that would be possible to classify in a number of ways. For example, in relation to the department of the Company that would be in charge of these tasks (Development, Planning and Engineering, Projects, etc..). Another way of classifying them would be in function of their incidence on the system: Planning in the short, middle or long term, maintenance, energy analysis, operation analysis, contingencies, fires, quality of water, restrictions, sources of supply, etc..

In any case, it is not simple as a rule to classify any given action, since in many cases these are interrelated. In any case, the fundamental objective is to simulate in order to know and have information that helps us to take decisions. It is also of interest to carry out a follow-up of the action plans carried out based on the information that the model provides, to contrast the said effects with this and the real consequences.

We list below a series of tasks with clear objectives:

- The continuous update, improvement and recalibration of the model, so that the model always reflects as reliable as possible the operation of the real system.
- Improvement of the pressure levels in the system in order to achieve adequate supply values. This means, knowing the current pressure levels and detected the existence of

values outside the normal range, acting on the network to correct them. The model will give us information about the possible solutions to adopt: substitution or pipeline rehabilitation, placement of new pipelines, installation of pumping stations, automatic valve installation (pressure sustaining or pressure reducing valves), the form of operating the system (to modify the setting values of the automatic valves, to change the start-up/stop procedures of the pumps, to open or to close valves, etc.), to install end of line tanks. Logically, the information will not only be qualitative, but also quantitative, which will permit us to design the most adequate solution.

- A study of the network in function of what might happen with increased demand as a consequence of the expansion of the network. Situations will be simulated adopting various strategies such as, for example, new pipelines (diameter design), booster pumps, re-elevations (location and characteristics of the pumping station), the closing-off of loops (calculation of the pipelines and definition of the connection points), etc..
- Design of new pipelines to feed new zones to be supplied with water.
- Possible substitution of injection points in the network due to insufficient resources or the lack of water quality (for example nitrates, salinization). In these cases one can use models of quality to analyze the flow rate to be sent from each supply point and to achieve an acceptable quality of water.
- Programming of maintenance operations on the network. To analyze the consequences of leaving out of service a given installation to carry out maintenance or repairs of the same, choosing the optimum strategy to adopt so that such incidence affects as little as possible the quality of the water supply.
- Analysis of the operation of the regulation elements of the system, or incorporation of others, in order to improve the energy cost of the installation, maintaining an adequate quality of the water supply (energy optimization).
- Analysis of the maximum flow rate that it is possible to extract from a given point of the system without affecting in excess the pressures of the surrounding nodes. This circumstance can be produced, for example, before a situation of demand from a new service pipe where it may be better and even necessary to restrict the peak flow rate.
- Analysis of fire situations. Through analysis of different scenarios, to check the flow rate that would be possible to supply to the Municipal Services of Fire Extinction through the fire hydrants connected to the public network. To reinforce the network, if necessary, modifying diameters.
- Analysis of the sufficiency or not of the regulation and reserve capacity of tanks, realizing dynamic simulations of the current or future situation (increases of demand). These studies are very interesting to justify investments in storage capacity that will increase the safety in the supply, the quality of the same and the economic saving in the case of having sufficient pumping capacity to use the most adequate electrical tariffs. The analysis of this type will be realized using a simulation model over an extended period.
- Determination of the procedural curves at the regulatory valves exits or pumping stations to achieve adequate pressure levels with minimal cost.
- Training of the current operators of the system, or of new personnel intended for these tasks. For this it is interesting to have available an automatic control system simulator whose actions are reflected in the model.

- Applications of control in real time, supplying the model the optimum operation strategy at each and every moment, based on the information that we provide it about the network.
- Analysis of the different possibilities for the creation of different sectors in the network. Establishment of the number of connection points with the arterial network, selection of the most suitable and determination of the procedural curves of these. It is interesting to begin with a pilot experience in a sector, that must be adequately monitored and in which will be realized an exhaustive follow-up of the pressure levels, balances of injected and recorded volumes of water, etc.
- Reduction and control of the losses of water modifying the pressure levels of the network. It is also possible to compare values measured by flowmeters installed in the network with the values given by the model that have been determined based on recorded volumes of water. In this way we can detect possible discrepancies, which can indicate to us in what zones it may be necessary to increase the search for non-recorded water volume.
- Establishment of action plans to stop possible breakdowns or breakages at any point in the network, simulating such contingencies, evaluating their effects and analyzing the best corrective options. In this way it is possible to discover the existence of extremely vulnerable elements, which may have a serious effect on the operation of the system in the event of failure, and to correct the situation.
- Improvement of the quality of water in the network. With adequate monitoring one can know the value of certain quality parameters of the system (for example residual free chlorine) at different points in the network and to act on the regulation system to correct undesirable situations (for example, rechlorinate at the end of line tanks or intermediate points in the network). Similarly, from these studies one can derive a periodic plan of purges from the system, in an automated way even, at end branches of the network.
- Analysis of strategic or significant points in the system for the incorporation of these points as remote stations to implant a remote control over the system.
- Analysis of consequences derived from sharp manoeuvres in the system (the closing or opening of valves, the starting-up or shutting-off pumps, etc.) and design of protection elements to limit the effects of these sharp manoeuvres. For this it is necessary to use an elastic model (transient analysis).
- In simple cases it is possible to simulate the emptying or filling of the pipelines using models in a transient analysis that include the valves for filling/emptying (air valves and purge valves) in order to choose conveniently the size and location of these elements and the form of accomplishing the manoeuvre.
- The study of strategies to adopt to achieve a reduction in the consumption of the network (including leaks) in periods of resource insufficiency (restrictions). For example, analysis of how to achieve some minimal pressure levels but sufficient in order to limit the value of the leaked volume of water. Study of the incidence on the pressures in the network of closed valves or the stopping of pumps to achieve some minimum pressure levels and to avoid that the network is emptied and enters into depression (subatmospheric pressure), with the sanitary risks that this means and the operative problems that are entailed in having "to fill" the system every day.

- Analysis of the installation of pressure reducing valves with setting pressures which are adjustable to limit the pressures in different sectors of the network according to the hour of the day. This system was installed in Murcia by EMUASA, being able to achieve the maintenance of the pressures in the network at some 5 mca during the restriction schedule, but maintaining the pressures at normal supply values the rest of the day. In this way, the Company avoided subatmospheric pressures in network at the time of introducing restrictions while, at the same time, restricting the consumption.

We could list even more tasks that one can realize through the development of a mathematical model, but we think that those which we have mentioned are the most significant.

For example, it would be interesting to demonstrate the action plan that is proposed in Vela (1996) and its application to some simple concrete cases.

**Search process and selection process of possible solutions:**

1• To detect, to identify and to classify the problem, based on the values of control variables, indicative ratios, observed symptoms, deviations with respect to the guide levels, detected faults, deficiencies in the quality of service, protests received that indicate a probable state of the system which is outside the established optimum range of operation at that moment and within that scenario.
2• To determine the causes that provoke the aforementioned situations and undesirable consequences based on the behavioural knowledge of the system and on past experience.
3• Proposition of solutions and selection of the most appropriate, through information provided by the available mathematical model or of an adjustment of the model for these particular objectives.
4• Checking of the application of the solution or solutions in the real system, through the previous installation in the model of the same solutions.
5• Technical and economic evaluation of the efficiency and profitability of the collection of srategies and solutions finally adopted. Firstly, based on the results obtained with the model and then through a pilot experience, or directly on the physical system, depending on the type, importance and transcendence of the existing problem.

The previous list does not intend to be more than a short demonstration of the different problems that can be present in a water supply. It is convenient to stress that a similar cause can originate various problems, as well as a problem can be the consequence of multiple causes. It would require as such a matrix of correlations to outline and solve the problems. In any case, it is invaluable the help that mathematical models provide, as much in the detection, identification and resolution of these problems in an optimum and global manner.

## 6. Conclusion

After what has been explicated in this paper, one is obliged to recognize that the current state of the technology, its reasonable costs, and the availability of appropriate information and data processing tools, makes the situation such one can no longer consider as a "luxury" the utilization of simulation and monitoring systems to control and improve the quality of the water supply service.

## 7. Biliography.

ALEGRE, H., COELHO, S.T. *Use of simulation models of water supply systems. Processing of input and output data.* III International Conference on Computer Applications for Water Supply and Distribution. Leicester Polythechnic. U.K. Septiembre 1987.

ALLEN, R.. *Network Analysis - The Real Story.* III International Conference. Computer Applications for Water Supply and Distribution. Leicester Polythechnic. U.K. September 1987.

BASTIDA, A., GARCÍA-SERRA,J., MARTÍNEZ, F., BELTRI, R., DE LA PLAZA, C. Modelo matemático de la red de agua potable para la ciudad de Murcia. *Revista Tecnología del agua.* Abril 1990, vol. 68, pp. 36-42,.

BHAVE,P.R. Calibrating Water Distribution Network Models. *Journal of Environmental Engineering.* February 1988, vol. 114, No 1.

CABRERA, E., GARCÍA-SERRA, J., IGLESIAS, P. Modelización de redes de distribución de agua. Desde el régimen permanente hasta el golpe de ariete. Del libro *Mejora del Rendimiento y la Fiabilidad en Sistemas de Distribución de Agua.* Proceedings from the UIMP Course. November, 1994.

CASTELLO, J.J. Postprecipitation in Distribution Systems. *Journal of the AWWA.* . November 1984, pp 46-49.

CESARIO, A.L., LEE, T.K. A Computer Method For Loading Model Network. *Journal of the AWWA.* April 1980.

CESARIO, A.L. Computer Modeling Programs: Tools For Model Operations. *Journal of the AWWA.* September 1980.

CESARIO, A.L. ,DAVIS, J.O. Calibrating Water System Models. *Journal of the AWWA.* July 1984, pp 66-69..

CESARIO, L. (1992). *Modeling, Analysis and Design of Water Distribution Systems.* Ed. AWWA. 1992.

EGGENER, C.L. AND POLKOWSKI, L. Network Models And The Impact of Modeling Assumptions. *Journal of the AWWA.* April 1976.

GARCÍA-SERRA, J. *Modelización y mejora de la red de distribución de agua de Gandía-Playa. Interconexión con la red de Gandía-Pueblo.* Final Degree Project, E.T.S.I.I. U.P.V. Valencia. 1984.

GARCÍA-SERRA, J. *Estudio y Mejora de las Técnicas de Calibración de Modelos de Redes Hidráulicas.* Tesis doctoral. Universidad Politécnica de Valencia. 1988

GARCÍA-SERRA, J., MARTÍNEZ, F., BASTIDA, A. *Calibrating Techniques of Hydraulic Network Models.* International Symposium on Water Resources Systems Application. Winnipeg (Canada). June 1990.

KOELLE, E. Hydraulic Networks. A general treatment for steady, transient and oscillatory analysis and control. cases and faults. Book: *El Agua en la Comunidad Valenciana* . Ed. Generalitat Valenciana.1989.

LANSEY,K.E. *A Procedure for Water Distribution Network Calibration considering Multiple Loading Conditions.* International Symposium on Computer Modelling. Kentucky Water Resources Research Institute. University of Kentucky. Lexington. 1988.

LÓPEZ, G., FUERTES, V., AYZA, M. Modelización matemática de una red en funcionamiento del libro *Ingeniería Hidráulica aplicada a los sistemas de distribución de agua.* Ed. U.D. Mecánica de Fluidos. U.P.V. Valencia. 1996.

LLOPIS, J.F. *Desarrollo de una base de datos relacional para la determinación de cargas de un modelo dinámico.*P.F.C. Universidad Politécnica de Valencia. 1996.

MARTÍNEZ, F. *Desarrollo de un modelo matemático para el análisis de redes hidráulicas por miniordenador con posibilidades de explotación en el campo de la gestión y control.* Tesis doctoral. E.T.S.I.I. Valencia. 1982.

MARTÍNEZ, F., GARCÍA-SERRA, J. Modelización matemática de sistemas de distribución de agua en servicio del libro del Curso impartido en la UIMP *Abastecimientos de agua urbanos. Estado actual y tendencias futuras.* Ed. Generalitat Valenciana. Valencia. 1993.

MARTÍNEZ, F., GARCÍA-SERRA, J. Mathematical modeling of water distribution systems un service del libro *Water Supply Systems. State of the Art and Future Trends.* Ed. Computational Mechanics Publications. 1993.

ORMSBEE, L. E. AND WOOD, D. J. Explicit Pipe Network Calibration. *Journal of Water Resources Planning and Management.* April 1986., vol 112, No.2. pp 166-182.

ORMSBEE, L.E., CHASE, D.V. *Hydraulic Network Calibration using Nonlinear Programming.* International Symposium on Computer Modelling. Kentucky Water Resources Research Institute. University of Kentucky. Lexington. 1988

ORMSBEE, L.E., LINGIREDDY, S. *Calibrating hydraulic network models.* Journal of the AWWA. Vol.89. Febrero 1997.

ROSSMAN, L.A. *EPANET User's Manual.* U.S.Environmental Protection Agency. Cincinnati. Ohio. U.S.A. 1993

SHARP, W.W., WALSKI, T.M. Predicting Internal Roughness in Water Mains. *Journal of the AWWA.* November 1988, pp. 34-40.

STIMSON, M.J., GWYNNE, M.R. Ventajas del uso de modelos Proceedings from the UIMP Course. 1993. *Abastecimientos de agua urbanos. Estado actual y tendencias futuras.* Ed. Generalitat Valenciana. Valencia. 1993.

VELA, A. *Asignación de cargas al modelo de una red de distribución de agua potable. Aplicación al caso de la ciudad de Valencia.* Final Degree Project. E.T.S.I.I., U.P.V., Valencia. 1988.

VELA, A., AYZA, M., VIDAL, R. Aplicación y utilización de modelos. del libro *Ingeniería Hidráulica aplicada a los sistemas de distribución de agua.* Ed. U.D. Mecánica de Fluidos. U.P.V. Valencia. 1996.

VIDAL, R., MARTÍNEZ, F., AYZA, M. Aplicaciones de los Modelos de Calidad en la Simulación de las Redes de Distribución de Agua. *Revista Ingeniería del Agua.* N°3. 1994

WALSKI, T.M. Using Water Distribution Systems Models. *Journal of the AWWA*. February 1983, pp. 58-63.

WALSKI, T.M. *Analysis of water distribution systems*. Ed. Van Nostrand Reinhohld. New York. 1984.

WALSKI, T.M. Assuring Accurate Model Calibration. *Journal of the AWWA*. December 1985, pp. 38-41..

WALSKI, T.M., GESSLER, J., SJOSTROM, J.W. *Water Distribution Systems: Simulation and Sizing*. Ed. Lewis Publishers. Chelsea, MI, USA. 1990

WALSKI, T.M. *Standards for Model Calibration*. Proc. 1995 AWWA Computer Conference. Nashville, Tenn. USA. 1995.

# IMPROVING HYDRAULIC EFFICIENCY IN A WATER DISTRIBUTION SYSTEM

EDMUNDO KOELLE
Universidade de São Paulo
Consulting Engineer
Rua Jesuino de Abreu, 354
São Paulo, SP, Brasil, 05662-010
Phone/Fax: + 55-11-842-8270
e-mail: edkoelle@mandic.com.br

## 1. Introduction

The contemporary epoch of socio-economical globalization and the latent scarcity of natural and energy resources, associated to adverse climatic conditions periods, present new challenges to engineers in design preparation, works execution, operation and management of public utilities, particularly in the treatment, transportation and distribution of water.

Present technology, undergoing constant evolution in all areas of knowledge and the communication facilities, internationalize the solutions and break down the frontiers between nations, including those where restrictions of a political nature to the exchange of products and services were evident up to a short time ago and maybe still exist today.

The breakdown of the barriers between NATIONS and the formation of economical blocs, leading to a future communion between the peoples, preserving the cultural identities, forming a HUMANITY, reducing the social differences and promoting means towards equal opportunities to all citizens, is the GOAL to be reached, in spite of several contrary opinions, which insist on branding it as the "Utopia of the 21st century"!

For the GOVERNMENT (excepting those of the more developed countries), SOCIAL investments (health, housing, education, public safety) are top priority and require high investments, usually incompatible with the tax revenue and with internal saving. This leads to the necessity of private investment in the INFRA-STRUCTURE area (energy, transportation, communications, water supply, sewerage, etc.), for which there are no PUBLIC RESOURCES available, since they must be primarily applied in social development. Because of these evident facts, a tendency exists for the CONCESSION of INFRA-STRUCTURE services to PRIVATE entrepreneurs, decentralizing the action of the STATE, which operates as a REGULATOR and as a STANDARD-SETTER of the services that are essential to the economic development.

As a consequence of this scenario for the new century, the efforts towards the increase of efficiency in the management processes and of the effectiveness in reaching



*E. Cabrera and J. García-Serra (eds.), Drought Management Planning in Water Supply Systems,* 89–102.

results at short and medium term, are extinguishing the EMPLOYMENT SOCIETY that resulted from the INDUSTRIAL REVOLUTION and transforming it into a WORK SOCIETY in which the employer/employee CONFRONTATION is being changed into a PARTNERSHIP, completely transforming the action of the LABOR UNIONS from a revindicatory into a participative one.

These introductory considerations are essential to put into focus the subject of this lecture. Indeed, the understanding of the socio-economical scenario is essential for the understanding of the limitations one may find in his environment with respect to putting into practice proposals for IMPROVEMENTS in the water supply and distribution. Any solution that is proposed will be carried out only if there are, on one side, FINANCIAL RESOURCES and, on the other, POLITICAL CONSCIOUSNESS of the PEOPLE involved in the decision. The IMPROVEMENTS will be real only if there is an effective MANAGEMENT during the development of the project, during the execution of the works or of the plan and in the operation of the system.

The techniques used in design, execution and operation of water networks aimed at cost reduction, involving the analysis of risks to obtain flexible and reliable solutions (reliability analysis) are well known (1, 2, 3) but it is not always that the results are adequate for the user, since, besides the "quality" of the services provided, it is essential to set the corresponding RATES and this involves MANAGERIAL action - an aspect that is seldom focused in courses such as this one.

The changes introduced in the WATER corporations, as a consequence of the interpretation of the scenario presented above, are more of a logistical/managerial nature, rather than a result of the introduction of new techniques to increase the efficiency of the water distribution systems. Results have been encouraging when the management units COMMUNICATE and solve the problems by integrating and relating.

It has been observed that invariably something new emerges, generated by the RELATIONSHIP between management units, showing that the global result may be greater than the sum of actions taken isolately in each area; thus, this approach cannot be forgotten in the development of the theme of this lecture.

## 2. General considerations

The updating of water supply systems involves global actions (technical and management innovations) in its various components:

i - Water source
ii - Water intake, transportation and treatment (production)
iii - Transportation and storage
iv - Distribution (primary and secondary)

The actions at the WATER SOURCE are focused towards preserving the quality of the water in the natural reservoirs and its adequate operation to guarantee the supply even during unfavorable hidrological periods, during which it is necessary to establish

PREVENTIVE OPERATIONAL RULES to minimize the risk of a collapse in the supply.

For quality maintenance, it is necessary to have specific LAWS to REGULATE the use of the adjoining land; the control and disposal of urban and industrial pollutants require rigorous INSPECTION, carried out by the government. Fortunately, in the present days, the ecological consciousness is diffused and new projects pass through the analysis and approval of the ENVIRONMENTAL IMPACTS that are going to be caused. There is a worldwide action of governments and of non-governmental bodies for the preservation of natural resources in the sustained economical development; for this, the MONITORING of the quality of water resources is essential.

The availability of water resources is the determining factor in regional development planning. The division of the territory into WATERSHEDS (with adaptations to other planning parameters such as limiting areas and distances) suggests an intelligent direction towards the POLITICAL REFORMULATION and the MANAGERIAL DESCENTRALIZATION of governments. Indeed, the division of the territory into municipalities without identity and without economical sustentation is wrong for it involves the channeling of public resources towards the maintenance of group privileges at the expense of other communities.

This regionalization option is under way and quite advanced in the State of São Paulo, Brazil, and has become known and been applied in other Brazilian states with basis in a recently approved federal law (Law 9433, January 8, 1997) concerning the management of water resources and the use of water.

With respect to DRINKING WATER PRODUCTION, updating must begin with the introduction of new treatment technologies and the AUTOMATION of the units to increase the process efficiency with the recycling of effluents. Modern equipments that reduce the SPECIFIC CONSUMPTION OF ENERGY, reducing costs, have been researched by water corporations in association with Universities.

It is in the domain of TRANSPORTATION and STORAGE that one can find the greatest margins for improvement of the efficiency of systems with the OPERATIONAL OPTIMIZING of the means for pumping, transporting and storing treated water.

The use of CALCULATION MODELS (softwares) for the operational analysis in EXTENSIVE TIME makes it possible to incorporate REDUCED RATES for energy consumption, controlling the operation during peak periods; with the use of frequency variators in the pumping units it is possible to optimize the speed of the pumps to obtain the maximum efficiency in the TRANSPORTATION/STORAGE, reducing energy consumption and the maximum power demand. As an example, in the city of São Paulo (12 million people) 30 to 35% of the operational cost is for payment of the electric power used in the pumping. Efforts to reduce that proportion imply in:

- Replacement of pumping units by others with better efficiency (gains of 1% to 5%);
- Operation with speed variators and full use of the existing storage capacity (gains of 2% to 10%);
- Using smaller power demands during "peak time" (gains of 5% to 10%).

With the adoption of the above mentioned measures, energy consumption could be reduced by about 15%, which will represent a reduction of about 5% in the operational costs of the TRANSPORTATION AND STORAGE IN THE SÃO PAULO WATER SYSTEM.

The distribution system, however, is where the greatest amount of LOSSES is to be found, since about 40% of the treated water in not BILLED in the great systems. A value of 15% of unaccounted-for water is considered adequate since there are factors inherent in the distribution process that are not "economically" solved in the greater part of the systems having high levels of control and technology of design, construction and operation.

At a recent INTERNATIONAL COURSE (1) that brought together several experts from different countries, the subject of "Improving efficiency and reliability in water distribution systems" was approached in its TECHNICAL aspects; noteworthy among the papers presented was the one presented by Alegre and Coelho (2) that dealt with the strategies for water distribution system rehabilitation. Merit indicators were proposed for the several elements of the network; these indicators, which are obtained through the collection of operational information, make it possible to diagnose deficiencies in the water distribution (pressure variations, excessive energy consumption, etc.). The analysis of the variation of the indicators will point out the need for corrective measures and the rearrangement of the network to supply the consumers at an adequate standard of total quality. The guidance presented in the paper is indicated to establish operational diagnoses of the distribution network, from which we will discuss possible corrective measures for the improvement of the hydraulic efficiency of the network. The solution will depend on a MANAGERIAL ACTION involving several areas and operating units.

### 3. Hydraulic efficiency and effectiveness

The term EFFICIENCY is usually associated to a transformation PROCESS, while EFFECTIVENESS is associated to the END RESULT. A process may be very efficient and the result not quite effective (or vice-versa), as it may be observed in our daily experience. For instance, a transportation process may be efficient because it is comfortable and low-cost (low energy consumption) but not effective because the transportation time is inadequate. Contrariwise, it may be possible to provide quick transportation, thus obtaining an effective result but to spend a lot of energy in the transportation using inadequate, uncomfortable and high-cost means.

With thus understanding, when we deal of a water distribution network, we must try to define such terms in an adequate way so that we can, during their analysis, to propose improvements to increase the EFFICIENCY, that is to say, to reduce the costs of the water distribution process and to obtain EFFECTIVENESS in the result, translated by CONSUMER SATISFACTION.

This difference is subtle and poorly perceived in the decision phases of an engineering solution; usually, "cheap solutions" requiring smaller financial resources for the construction of a distribution network, as a result of the conception and of the choice of materials (pipes), may result in high process costs because of operational problems deriving from a poor execution or to excessive maintenance that are not evaluated in the design.

The result if ineffective when the consumer is not supplied in an adequate way because of excessive interruptions of the supply for maintenance actions.

Therefore, when it is wanted to obtain TOTAL QUALITY in the water distribution system, one must look both towards the hydraulic EFFICIENCY of the transportation process and towards EFFECTIVENESS of the results, that is to say, good quality water supply, at adequate pressures and without interruptions in the supply of the demands, at a RATE commensurate with the socio-economical conditions of the population.

## 4. Total quality

The hydraulic evaluation of a PRIMARY distribution network is restricted to operation simulation using computer programs for the calculations of flows in the Elements (ENOs) and heads at the NODEs for a given topological configuration of a network formed by several ENOs (pipes, valves, pumps, tanks) linked through the NODEs.

The hydraulic simulation makes it possible to check the pressure zones in the network for several demand configurations and, subsequently, through the comparison of the calculated values with the measured ones, to obtain:

i - Inadequate pressure zones;
ii - Short-circuits in the network;
iii - Excessive roughness;
iv - Operational conditions (transients).

### 4.1 INADEQUATE PRESSURE ZONES

The pressure zones are considered inadequate when:

a) the pressures are below the standardized limits and the demand is not met during all the period;

b) the pressures are above the standardized limits, damaging the consumer's hydraulic appliances and causing leaks.

In both cases the water distribution system is INEFFECTIVE and corrections of a technical nature are needed.

A) LOW PRESSURE zones are corrected through the installation of BOOSTERS (in line pumping) at appropriate locations in the network; its efficiency in the distribution process may be previously checked and analyzed through computational simulations in EXTENDED TIME.

The BOOSTER is actuated by the detection of pressures in the network, being started when the pressure falls, to keep it at values adequate to the satisfaction of the demand. The pressure control may be done by two alternate ways: by automatic control dissipating valves installed downstream from the BOOSTER or by speed variation. In this case, the speed is continuously adjusted to keep the pressure constant. When the pressure remains between the adequate limits the booster is automatically shut off and the demand is met through the BOOSTER's BY-PASS.

The installation problems of a BOOSTER are minimized by the selection of adequate pumps, requiring little maintenance. In secondary networks, multi-stage horizontally mounted "submersible pumps" have been used to allow the insertion of the pumping unit in a reduced space, with noise control so as not to affect the population. They are generally installed under the street pavement and automatically operated with pressure control through a dissipating valve.

In PRIMARY networks, the flows are greater and the BOOSTER is constructed as a PUMPING STATION, using urban space. In these cases, noise control is essential and the use of frequency variators is recommended to increase the efficiency of the pumping process with the reduction of the electrical energy consumption.

Sometimes, the construction of a BOOSTER and of an associated TANK to meet the maximum hourly demands (that coincide with the energy consumption "peak period") may be the best solution to satisfy global interests - those of the water utility and those of the power utility - and actually the consumer's interests through a low RATE.

B) HIGH PRESSURE zones occur in adequately designed networks during low consumption periods, generally at night. In these occasions, the pressure increase aggravates the leakage problems increasing water LOSSES.

Well-designed networks are divided into PRESSURE ZONES, in which adequate pressures are kept throughout the period by regulating tanks, adequately located and sized, normally operating downstream as surplus tanks (filling during the low consumption period). When they are installed upstream, they regulate the secondary network consumption absorbing a constant flow from the primary network.

In secondary networks located at unfavorable topographic conditions, pressure limitation is obtained with the installation of automatically operated PRESSURE CONTROL VALVES (PCV), commanded by the water that circulates through the network.

The use of PCVs is necessary and recommended as a resource for the control of pressure in networks and, consequently, of LOSSES; sometimes, the adequate installation of these valves in several positions will result in significant gains in the supply through the reduction of the volume of unaccounted-for water with the decrease of leaks that are usually aggravated during the night period.

## 4.2 SHORT CIRCUITS IN NETWORKS

The time distribution of chemicals in the water varies as a function of the time of permanence of the water in the network. Therefore, it is a consequence of the water velocity and path. If there is a "short circuit", that is to say, if the water circulates but is not consumed, or if the velocities are extremely small, the concentration of the chemicals varies and may reach values that are inadequate to guarantee the quality of the water for consumption, influencing taste and turbidity.

The models for the calculation of the water quality in a network make it possible to predict precisely the evolution of the concentrations, by incorporating, into the general equations, the conditions of transportation and diffusion of the concentration of chemicals in time. The model presented by Chaudhry and Islam (1) makes it possible to evaluate precisely the evolution of the concentrations at the NODES of a network, for a given period; form the analysis of these results, it is possible to check the network characteristics and to modify the topology if necessary, to obtain results that are adequate to the preservation of water quality.

The physical actions in the network are usually executed by shutting off certain reaches in order to modify the configuration of the distribution and of the path of the water, thus avoiding short circuit zones.

## 4.3 EXCESSIVE ROUGHNESS

As the pipe ages, its walls (or their lining) undergo changes (corrosion or solid deposition) and this is aggravated in short circuit zones. Solid deposition reduces the cross section of the pipes and increases the internal roughness, reducing the hydraulic efficiency of the network. Corrosion, besides weakening the mechanical resistance of the pipe, also increases its roughness.

The increase of roughness is detected by the reduction of the flows caused by the increase of head losses and may be checked by the measurement of pressures at the network NODES and of the flows in the reaches.

Computational simulation models furnish the means to evaluate the effects of the roughness increase in the distribution system and, therefore, the reduction of the efficiency. EFFICIENCY may be defined as the ratio between the heads necessary to satisfy the same demand, with increased roughness and with the normal roughness (new pipe).

Efficiency values decrease with the roughness increase and supply the indicators for the physical actions in the network aimed at restoring the design efficiency, by means of pipe cleaning by passage of scrapers (pigs) and internal relining.

In extreme cases, where the roughness and the age of the pipes are excessive and there are many leaks, the solution is radical, requiring pipe substitution.

There are many examples of pipes, with an efficiency reduction greater than 10% in less than ten years use, where pig scraping was necessary. It should be noted that the increase of the roughness points to the water production conditions and suggests corrections in the treatment.

The use of plastic materials (PVC, HDPE) that do not require internal lining and that, therefore, are not susceptible to corrosion nor to formation of solid deposits with the same intensity as in other pipes, has been widely used because of its advantages in terms of cost and execution.

However, it should be noted that, for these pipes, a thorough quality control is needed regarding the composition of the material (particularly with HDPE) that is subject to early ageing caused by thermal actions (inadequate storage) and dynamic actions (fatigue).

Indeed, for these materials, the resistance limit is significantly reduced by cyclical dynamic strains originating from pressure variations associated to the variations of demand. In networks with these materials it is recommended that attention be given to pressure control; the maintenance of an average pressure with limited variations is necessary to increase the useful life of the network. Otherwise, after a short time of use the leaks and the water losses get so bad that the network must be replaced!

## 4.4. OPERATIONAL CONDITIONS (TRANSIENTS)

The pipes of a network are sized to satisfy a demand profile in "quasi-permanent" state, that is to say, for slow and gradual variations of the flow. Pipe specifications conform to the static head limits, a margin being added to absorb the network maneuvers that cause higher transient heads.

However, usual practice does not consider the type of maneuver and uses only the extreme head values reached during a maneuver without paying attention to the evolution of transient heads resulting from the maneuver.

For a same maneuver, with equal extreme heads, it is possible to have a high frequency transient phenomenon (network without a protection system) or a low frequency phenomenon (adequately protected networks). But, if the maneuvers are rare, the frequency of the transient phenomenon is not relevant. However, if the maneuvers are usual (pump start-up and shut-down, valve opening and closure) and the transient phenomenon is a high frequency one, leading to several cyclical strains every time the maneuver is made, then the fatigue limit of the material is quickly reached.

Fragile materials break when this limit is reached and ductile materials undergo a lessening of its structural resistance and may break under smaller heads than those stipulated in the design.

Now, if this fact is not taken in consideration in the design phase and the network does not have adequate protection, serious problems may arise in the supply due to a high incidence of leaks requiring interruptions for their repair. Such facts are usual and there is little perception of the advantages of a protection system that modify the transient response to maneuvers.

It so happens that when one proposes the installation of a compressed air tank (AIR VESSEL) to reduce the frequency of transient phenomena, this meets resistance from the operation personnel that argue with the "difficulties" of handling an additional protection equipment, such as the air vessel that requires the installation of an air compressor (automatically operated). Many do not perceive the advantages of the installed protection because, at the design phase, they do not relate its effects to the reduction of leaks and to the increase of the useful life of the network.

Therefore, the HYDRAULIC EFFICIENCY IMPROVEMENT is strongly conditioned to the analysis of hydraulic transients and to the installation of protection elements that reduce the frequency of transient phenomena that result from a maneuver in the network.

The factors analyzed above (pressure zones, short circuits, roughness and transients) are fundamental because of their influence on the hydraulic efficiency and on the increase of the effectiveness of the operational results, resulting in the obtention of high indexes of TOTAL QUALITY of the water supply systems provided to the community.

## 5. Reliability and operational risks

Reliability of a water distribution system is defined as (3, pg. 472) "the probability of satisfying nodal demands and pressure heads for various possible pipe failures (breaks) in the water distribution system."

Computational models for network design involve sophisticated algorithms for the definition of an optimized solution, Martinez and Izquierdo (1, pg. 303-328); some of them include conditions of operational reliability and flexibility so as to extend the concepts of "optimization" based only on the MINIMUM COST for network construction.

The optimization parameters include the values of pipe diameters, the location and sizing of tanks and control valves and the operational scheme of the pumping stations to define the most adequate topological arrangement to satisfy the assumed nodal demand profiles.

Reliability criteria are difficult to formulate, just as the ones associated to costs and to the frequency of maintenance of the network elements.

Bouchart et al (3, chapter 14) present criteria, using non-linear programming, for the incorporation of reliability conditions subject to the probability of occurrence of failures in the network elements, in order to define the most adequate topology to satisfy a given demand configuration. The solution of the algorithm provides judgment elements by associating the costs to the probability of failures for a given topology, allowing the engineer to choose the most adequate solution, taking in consideration the available resources for investment and the strategic importance of the network in the satisfaction of demands.

Other algorithms are presented in the literature and, for each of them, hypotheses are formulated for the obtention of certain simplifications in the formulation of the mathematical model, as well as in the decision process, involving the optimization of an objective-function.

However, in all of them, it is clear that as the probability of failures is reduced, the costs increase and the choice of the solution to be adopted cannot be made solely with basis on the numerical results from the use of a given algorithm.

The possible solution(s) must always be analyzed with subjective considerations associated to the usual practices and to the local culture; frequently, it is not possible to quantify the probability of failure occurrence and the minimization of operational risks, but both may be minimized by:

- Adequate training of personnel (operation and maintenance);
- Use of control and protection elements for normal and emergency maneuvers;
- Rigorous specifications and inspection for the construction of the network;
- Adequate operation management, with constant exchange of information and good association of the work crews.

The last aspect, the managerial one, is simplified now by the resources provided by telematics. Indeed, the network data (topology register, mapping) may be computerized (for instance, on a GIS base) to provide ample access to the several work crews.

The design teams make simulations to figure out the hydraulic characteristics and the efficiency increase, as discussed in the previous item. The maintenance crews register the failures and the repair frequencies, providing data for the analysis of CAUSES and for the reformulation of design and operation. The values of the billed consumption data, compared to the volumes actually produced make it possible to investigate the LOSSES volume and to define the areas in which they occur with a greater intensity, allowing a statistical analysis associated to network parameters (pipe material, age, pressure levels, meter types).

Thus, without a management that promotes a SYNERGY among the several work teams, the success in rendering services to the community will be deficient and usually will not correspond to the results predicted by the simulation model. It is in this aspect that the modern information resources will benefit the operation and will allow the efficiency increase, since the diagnoses will be obtained with clarity, considering all the factors of influence in a holistic vision of the distribution network. With this, the risks of supply failures are minimized and the desired system RELIABILITY is obtained.

## 6. Quantical management - Loss control

The introduction of NEW INFORMATION TECHNOLOGIES (TELEMATICS) in public and private organizations has been justified by reasons associated to the techniques necessary to the increase of productivity, although ideological and corporativistic conceptions resist to the managerial transformations deriving from the incorporation of those new technologies.

The hierarchical management conceptions and the fear of losing positions in the organizational structure interrelate with the apparently paradoxical reality: "modernization initiatives characterized by the incorporation of new technologies and more powerful equipment show very poor results" (4).

Indeed, the management model based on the cause-effect relationship typical of the classic Newtonian mechanics allows the obtention of results useful to the management but the information coming from the interdependence between the various components of the organization is lost; the reason is that the management units are impervious and the exchange of information is restricted, either because of logistic (bureaucratical) difficulties or for subjective reasons deriving from individual personalities.

If attention is focused on quantical mechanics it will be perceived that there are no independent physical entities and that reality is a complex whole of CORRELATIONS and a permanent exchange of energy and information between the elements, which are "shaped" at each instant depending on the observer!

Thus, if instead of adopting the hierarchical Newtonian management model, one adopts the quantical management model, the organization starts to be understood as a "... multiplicity of agents that interrelate to produce global effects that also can not be explained only from those agents, isolately... In the organization, the key to success is the RELATIONSHIP between the agents" (5).

Quantical management has been employed in the Operational Programs of water distribution networks, establishing a relationship among the four essential agents that are necessary to improve efficiency and effectiveness, that is to say, to obtain TOTAL QUALITY in the services rendered:

The four agents (or work teams) are the following:

i - HYDRAULIC ANALYSTS

ii - NETWORK DATA REGISTER TEAM

iii- MAINTENANCE AND REPAIR CREWS

iv - BILLING AND CONTROL TEAM

## 6.1. HYDRAULIC ANALYSTS

They develop the SECTORAL NETWORKS ANALYSIS PROGRAM (using the computational calculation programs mentioned in this paper), perform steady state hydraulic simulations in extensive time and define the protection equipment that must be installed for hydraulic transient control.

Their job is to predict the operational conditions of the distribution networks and to design the enlargements necessary to satisfy increases of consumption; they also provide elements of analysis for LOSS CONTROL.

LOSS CONTROL (or unaccounted-for water control) requires sectoral BULK METERING of the demand during a given period, for comparison with the volumes measures at the connections located at a given CONSUMPTION BLOC that is a part of the consumption sector under analysis.

The analysis rest on up-dated network maps based on a GIS system, maps into which all the network data were inserted, as well as any alterations introduced by repairs and/or network expansion with the insertion of new elements.

## 6.2. NETWORK DATA REGISTER TEAM

One of the essential factors for the obtention of an adequate diagnosis of water consumption and of unaccounted-for water is the complete mapping of the network, including all its elements linked to the TOPOLOGY with the corresponding topographical elevations.

The maps must include:

A) For the PRIMARY NETWORK

i - Pipes (diameter, thickness, material, year of installation and updated measured roughness).

ii - Valves (type, diameter, drawings and components, year of installation).

iii - Tanks (dimensions, operational levels, piping lay-out, structures for flow control, measurement and overflow).

iv - Pumping stations (pumping units data, characteristic curves of pumps, year of installation, lay-out, electrical equipment).

v - Bulk water meters (type, location).

vi - Nodes with the demand for the secondary networks, that is to say, links to the secondary networks.

B) for the SECONDARY NETWORK

i - Sectoral consumption blocs.

ii - Network elements (as in A i, ii, iv).

iii - Location of consumer connections, identification of the type of consumer and of its meter (type, year of installation).

Usually such complete data are not available at all or they are out-dated, hampering the analysis.

The data register must be permanently up-dated in order to contain all the modifications introduced by the maintenance actions and included by decision of the hydraulic analysts.

Thus, the network data register team occupies a fundamental position in the managerial actions, since it updates information and puts it at the disposal of other teams that, in an interactive way and in accordance with standardized procedures, send in information about the modifications for register updating.

Usually, the monthly consumption data are sent to the network data register team that processes that information and produces statistical results from the comparison between the bulk metering data and the sum of the consumptions measured at each sectoral bloc.

Loss indexes are determined by bloc to allow the definition of priorities of the corrective actions aiming at the reduction of unaccounted-for water that will have to be carried out by the maintenance and repair crews.

## 6.3 MAINTENANCE AND REPAIR CREWS

Maintenace and repair work is done by specially trained crews, which install connections for new consumers, perform localized repairs and do leak detection work guided by information provided by the Network Data Register.

They are responsible for the maintenance of the bulk meters and the consumer meters and for the repair of visible leaks reported by the population. They should not perform design work (in charge of the hydraulic analysts) nor be allowed to introduce any modifications in the topology; but they should, in an active way, schedule preventive maintenance work and to perform corrective work in the network.

Any modification originating from repair work must be registered to feed the data bank, so as to supply information for the subsequent analysis in charge of the hydraulic analysts. Alterations of electro-mechanical equipment are also a responsibility of these crews, which work at previously defined areas that include the primary and secondary networks corresponding to a unique register basis transferred to all the other work crews.

## 6.4. BILLING AND CONTROL TEAM

This team is in charge of collecting consumption data at the various blocs through consumer meter reading. Activities of this type are sometimes contracted out but this practice has frequently presented inadequate results because the people in charge of it suffer influences tending on corruption and do not engage in the superior interests of the water corporation.

Modern resources make it possible, at the same time the meter reading is done, to print the bill as well as a local check of excessive consumption and/or leaks by the agent in charge of the reading.

In developing countries and in poor regions, there is also a SOCIAL FUNCTION that may be (and should be) the responsibility of the agent: to be a SANITARY AGENT, giving guidance regarding hygienic usages and habits as a means of preserving adequate HEALTH conditions of the people. Thus, these agents will have a multiple function of social guidance and should be duly trained for it. Alternatively, such guidance activities could be performed by university students, particularly by medicine students who could suggest and orient preventive actions regarding endemic diseases. It is a valuable experience for the professional formation and for the improvement of the sanitary conditions of the poorer classes.

## 6.5. ORGANIZATION CHART

The organization chart for quantical management proposed above, involving the four necessary work teams, must be interactive and may be developed in a "cyclical form" so as to show the need of relationship between the teams, as illustrated in the figure below.

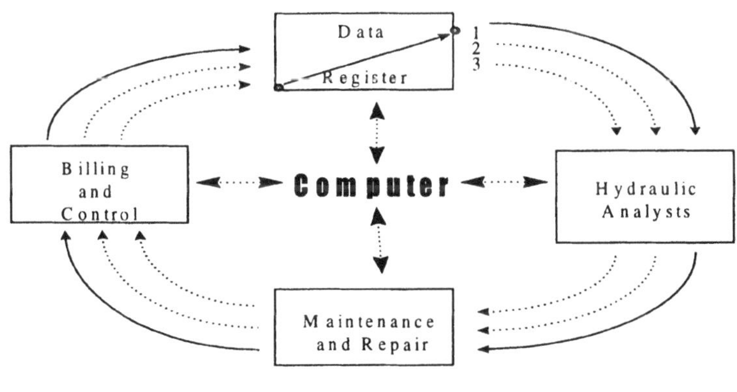


**Quantic Management**
**(loss control)**

Figure 1. Organizational Chart

the figure shows the cyclical action (1, 2, 3) in a temporal sequence, removing the hierarchies that are a characteristic of the Newtonian management model. The use of TELEMATICS (Telecommunications and informatics), with access to all work units and to the MANAGERIAL CONTROL, establishes engagements and denounces failures, by the attribution of the corresponding responsibilities.

The model proposed above is fundamental to increase the efficiency of the management process; it gives effective results and simplifies the transfer of the water corporations to the private sector, a transfer that may be done by water supply blocs or CONSUMPTION UNITS.

## 7. Conclusions

The IMPROVEMENT of the hydraulic efficiency of water distribution systems involves technical and managerial aspects in the search for TOTAL QUALITY of the services rendered to the consumers.

The future scenario of constant technological impacts and socio-economical globalization involves managerial aspects that must be considered in the first place, since the technical resources will be easily accessible at all levels due to the competition that occurs in the democratic countries. The human aspects and the work interrelationships must constitute the basic elements for investments in search of the best results in water supply.

## 8. Acknowelegements

My thanks to Engineer Miguel Zwi, for the critical analysis and for the opportunities for a constant exchange of ideas and to Engineer Edevar Luvizotto Jr., for the demonstrations of friendship and for the collaboration in the proofreading and presentation of this text.

## 9. References

(1) - CABRERA E., VELA A. (Edit.) - IMPROVING EFFICIENCY AND RELIABILITY IN WATER DISTRIBUTION SYSTEMS - KLUWER ACADEMIC PUBLISHERS - (1995).

(2) - ALEGRE, H., COELHO S. T. - HYDRAULIC PERFORMANCE AND REHABILITATION STRATEGIES - The use of levels-of-service and performance indices as decision support tools - pp. 267-282 in (1) - (1995)

(3) - MAYS, LARRY W. (Edit.) - RELIABILITY ANALYSIS OF WATER DISTRIBUTION SYSTEMS - ASCE (1989).

(4) - CASTRO, ARMANDO BARROS de - IMPACTOS DA NOVA TECNOLOGIA DE INFORMAÇÃO (Impacts of the new information technology) - SÃO PAULO EM PERSPECTIVA (pp. 105-108) - (October/December 1993).

(5) - NOBREGA, CLEMENTE - EM BUSCA DA EMPRESA QUÂNTICA (In search of the quantical corporation) - Ed. EDIOURO - Rio de Janeiro, 1996.

# MODELING WATER QUALITY IN DISTRIBUTION SYSTEMS

LEWIS A. ROSSMAN
*U.S. Environmental Protection Agency*
*Cincinnati, Ohio 45268*
*USA*

## 1. Introduction

The goal of a drinking water distribution system is to deliver sufficient quantities of water where and when it is needed at an acceptable level of quality. Although it is commonly thought that all water quality transformations end after water leaves the treatment works, in reality further changes in quality can occur as water travels through a distribution system. Indeed, the pipes and storage facilities of a distribution system constitute a complex network of uncontrolled chemical and biological reactors which can produce significant variations in water quality in both space and time. In the past, distribution systems were designed and operated mainly on the basis of hydraulic reliability and economics with little attention paid to water quality concerns, except when problems arose. This attitude is changing as more water suppliers realize the important influence that time spent in a distribution system can have on water quality.

Factors leading to water quality deterioration in distribution systems include:

- supply sources going on- and off-line,
- contamination via cross connections or from leaky pipe joints,
- dissolution of lead and copper from pipe walls,
- loss of disinfectant residual in storage facilities with long residence times,
- reactions of disinfectants with organic and inorganic compounds resulting in taste and odor problems,
- bacterial re-growth and harboring of opportunistic pathogens,
- increase in turbidity due to particulate re-suspension,
- formation of disinfection by-products, some of which are suspected carcinogens (e.g., trihalomethanes (THMs) and haloacetic acids).

It is difficult to use monitoring data alone to understand the fate and transformation of substances in drinking water as it moves through a distribution system. Even medium sized cities can have thousands of miles of pipes, making it impossible to achieve widespread monitoring coverage. The flow pathways and travel times of water



*E. Cabrera and J. García-Serra (eds.), Drought Management Planning in Water Supply Systems,* 103–127.

through these systems are highly variable due to the looped layout of the pipe network and the continuous changes in water usage over space and time. The common use of storage facilities out in the system makes things even more variable. At different times of the day a location might be receiving relatively new water from the treatment works when storage tanks are being re-filled or old water when storage tanks are being emptied. It is usually impractical to experiment on the entire distribution system by seeing how changes in pumping schedules, storage facility operation, or treatment methods affect the quality of water received by the consumer.

For these reasons, mathematical modeling of water quality behavior in distribution systems has become an attractive supplement to monitoring. These models offer a cost-effective way to study the spatial and temporal variation of a number of water quality constituents, including:

- the fraction of water originating from a particular source
- the age of water in the system
- the concentration of a non-reactive tracer compound either added or removed from the system (e.g., fluoride or sodium)
- the concentration and loss rate of a secondary disinfectant such as chlorine
- the concentration and growth rate of disinfection by-products such as trihalomethanes
- the numbers and mass of attached and free-flowing bacteria in the system.

The models can be used to assist managers to perform a variety of water quality-related studies. Examples include:

- calibrating and testing hydraulic models of the system through the use of chemical tracers
- locating and sizing storage facilities and modifying system operation to reduce water age
- modifying system design and operation to provide a desired blend of waters from different sources
- finding the best combination of pipe replacement, pipe relining, pipe cleaning, reduction in storage holding time, and location and injection rate at booster stations to maintain desired disinfectant levels throughout the system
- assessing and minimizing the risk of consumer exposure to disinfectant by-products
- assessing system vulnerability to incidents of external contamination.

Model development has progressed through several stages since the mid-1980's, beginning with the first steady-state, source blending models (Males et al., 1985), moving on to dynamic models for reactive species (Grayman et al., 1988), and most recently seeing the appearance of biofilm growth models (Servais et al., 1995). Many of today's most popular commercial network modeling packages contain water quality modeling capabilities (e.g., EPANET, H2ONET, PICCOLO, STONER WS, WATERCAD, WATERMAX, and WATNET).

## 2. Governing Equations

A water distribution system consists of pipes, pumps, valves, fittings, and storage facilities that are used to convey water from source points to consumers. The actual physical system is modeled as a network of links that are connected at nodes in some specified branched or looped configuration. Links represent pipes, pumps, or valves. Nodes serve as junction, source, consumption, and storage points. A network water quality model predicts how the concentration of a dissolved substance varies with time throughout the network under a known set of hydraulic conditions and source input patterns. Its governing equations rest on the principles of conservation of mass coupled with reaction kinetics. The following phenomena occurring in the distribution system are represented in a typical water quality model (Rossman et al., 1994; Rossman and Boulos, 1996):

### 2.1 ADVECTIVE TRANSPORT IN PIPES

A dissolved substance will travel down the length of a pipe with the same average velocity as the carrier fluid while at the same time reacting (either growing or decaying) at some given rate. Longitudinal dispersion is usually not an important transport mechanism under most operating conditions. This means there is no intermixing of mass between adjacent parcels of water traveling down a pipe. Advective transport within a pipe can be represented with the following equation:

$$\frac{\partial C_i}{\partial t} = -u_i \frac{\partial C_i}{\partial x} + r(C_i) \tag{1}$$

where $C_i$ = concentration ($M/L^3$) in pipe i as a function of distance x and time t, $u_i$ = flow velocity (L/T) in pipe i, and r() = rate of reaction ($M/L^3/T$) as a function of concentration.

### 2.2 MIXING AT PIPE JUNCTIONS

At junctions receiving inflow from two or more pipes, the mixing of fluid is taken to be complete and instantaneous. Thus the concentration of a substance in water leaving the junction is simply the flow-weighted sum of the concentrations from the inflowing pipes. For a specific node k one can write:

$$C_{i|x=0} = \frac{\sum_{j \varepsilon I_k} Q_j C_{j|x=L_j} + Q_{k,ext} C_{k,ext}}{\sum_{j \varepsilon I_k} Q_j + Q_{k,ext}} \tag{2}$$

where i = link with flow leaving node k, $I_k$ = set of links with flow into k, $L_j$ = length of link j, $Q_j$ = flow ($L^3/T$) in link j, $Q_{k,ext}$ = external source flow entering the network at node k, and $C_{k,ext}$ = concentration of the external flow entering at node k.

## 2.3 MIXING IN STORAGE FACILITIES

Most water quality models assume that the contents of storage facilities (tanks and reservoirs) are completely mixed. Thus the concentration throughout the facility is a blend of the current contents and any entering water. At the same time, the internal concentration could be changing due to reactions. The following equation expresses these phenomena:

$$\frac{\partial(V_s C_s)}{\partial t} = \sum_{i \varepsilon I_s} Q_i C_{i|x=L_i} - \sum_{j \varepsilon O_s} Q_j C_s + r(C_s) \quad (3)$$

where $V_s$ = volume ($L^3$) in storage at time t, $C_s$ = concentration within the storage facility, $I_s$ = set of links providing flow into the facility, and $O_s$ = set of links withdrawing flow from the facility.

## 2.4 BULK FLOW REACTIONS

While a substance moves down a pipe or resides in storage it can undergo reaction with constituents in the water column. The rate of reaction can generally be described as a power function of concentration:

$$r = kC^n \quad (4)$$

where k = a reaction constant and n = the reaction order. Some examples of different reaction rate expressions are:

- r = -kC for chlorine decay (first-order decay)
- r = k(C* - C) for THM formation (first-order growth, where C* = maximum THM formation possible)
- r = 1 for water age (zero-order growth)
- r = 0 for conservative materials (e.g., fluoride)

## 2.5 PIPE WALL REACTIONS

While flowing through pipes, dissolved substances can be transported to the pipe wall and react with material such as corrosion products or biofilm that are on or close to the wall. The overall rate of this reaction will also be influenced by the amount of wall area available for reaction and the rate of mass transfer between the bulk fluid and the wall. The former factor is determined by the surface area per unit volume, which for a pipe equals 2 divided by the radius. The latter factor can be represented by a mass transfer coefficient whose value depends on the molecular diffusivity of the reactive species and on the Reynolds number of the flow (Rossman et al, 1994). For first-order kinetics, the rate of a pipe wall reaction can be expressed as:

$$r = \frac{2k_w k_f C}{R(k_w + k_f)} \tag{5}$$

where $k_w$ = wall reaction rate constant (L/T), $k_f$ = mass transfer coefficient (L/T), and R = pipe radius (L). If a first-order reaction with rate constant $k_b$ is also occurring in the bulk flow, then an overall rate constant k ($T^{-1}$) which incorporates both the bulk and wall reactions can be written as:

$$k = k_b + \frac{2k_w k_f}{R(k_w + k_f)} \tag{6}$$

Note that even if $k_b$ and $k_w$ were the same throughout a system, the apparent rate k could still vary from one pipe to the next due to variations in pipe size and flow rate.

2.6 SYSTEM OF EQUATIONS

When applied to a network as a whole, Equations 1-3 represent a coupled set of differential/algebraic equations with time-varying coefficients that must be solved for $C_i$ in each pipe i and $C_s$ in each storage facility s. This solution is subject to the following set of externally imposed conditions:

- initial conditions that specify $C_i$ for all x in each pipe i and $C_s$ in each storage facility s at time 0,
- boundary conditions that specify values for $C_{k,ext}$ and $Q_{k,ext}$ for all time t at each node k which has external mass inputs
- hydraulic conditions which specify the volume $V_s$ in each storage facility s and the flow $Q_i$ in each link i at all times t.

An example of these equations applied to a simple network is shown in Figure 1.

## 3. Solution Methods - Steady-State Models

There are two general classes of water quality models -- steady-state and dynamic. Steady-state models compute the spatial distribution of water quality throughout a pipe network under the assumptions that hydraulic conditions do not change, at least over the length of time needed to move water from all entry points to all exit points in the network, and that storage does not affect water quality. They can be derived from the general mass conservation equations by setting all time derivatives to zero and requiring that all other coefficients be invariant with time. As an example, for the network shown in Figure 1, the steady-state quality in link 1 as a function of distance x is:

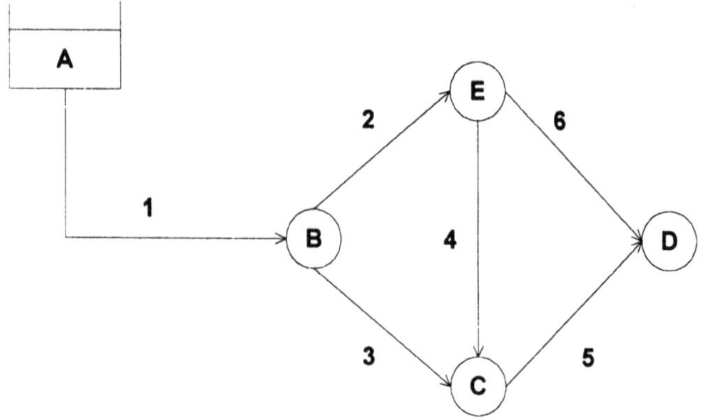


$$\frac{\partial C_1}{\partial t} = -u_1 \frac{\partial C_1}{\partial x} - kC_1 \qquad C_{1|x=0} = C_A$$

$$\frac{\partial C_2}{\partial t} = -u_2 \frac{\partial C_2}{\partial x} - kC_2 \qquad C_{2|x=0} = C_B = C_{1|x=L_1}$$

$$\frac{\partial C_3}{\partial t} = -u_3 \frac{\partial C_3}{\partial x} - kC_3 \qquad C_{3|x=0} = C_B = C_{1|x=L_1}$$

$$\frac{\partial C_4}{\partial t} = -u_4 \frac{\partial C_4}{\partial x} - kC_4 \qquad C_{4|x=0} = C_E = C_{2|x=L_2}$$

$$\frac{\partial C_5}{\partial t} = -u_5 \frac{\partial C_5}{\partial x} - kC_5 \qquad C_{5|x=0} = C_C = \frac{Q_3 C_{3|x=L_3} + Q_4 C_{4|x=L_4}}{Q_3 + Q_4}$$

$$\frac{\partial C_6}{\partial t} = -u_6 \frac{\partial C_6}{\partial x} - kC_6 \qquad C_{6|x=0} = C_E = C_{2|x=L_2}$$

$$C_D = \frac{Q_5 C_{5|x=L_5} + Q_6 C_{6|x=L_6}}{Q_5 + Q_6}$$

**Figure 1**. Example water quality equations for modeling a substance subject to first-order decay. Arrows indicate actual flow direction. C is concentration, t is time, x is distance, L is length of link, u is flow velocity, and k is reaction rate constant.

$$C_1 = C_A \exp(-ku_1 / x) \tag{7}$$

which when solved for $C_{1|x=L1} = C_B$ gives:

$$C_B = C_A \exp(-ku_1 / L_1) \tag{8}$$

Simplifying a network in this manner leaves only the concentrations at the nodes as unknowns.

The solution to such models is found by performing a topological sort on the network (Boulos and Altman, 1993). This means finding a re-ordering of the nodes such that for any given node, the pipes with flow into the node are connected to nodes which appear earlier in the ordering. As an example, for the network in Figure 1, concentrations at both nodes B and E must be computed before a value for node C can be calculated. The final sorting for this network would be A-B-E-C-D. An efficient computer algorithm is available for performing a topological sort of a network (Weiss, 1993). Its execution time is proportional to the number of links plus nodes. Although they are simple to set up and solve, the restrictive assumptions of steady-state models limit their applicability.

## 4. Solution Methods - Dynamic Models

Dynamic models of water quality in distribution systems take explicit account of how changes in flows through pipes and storage facilities occurring over an extended period of system operation affects water quality. These models thus provide a more realistic picture of system behavior. Solution methods for dynamic models can be classified spatially as either Eulerian or Lagrangian and temporally as either time-driven or event-driven. Eulerian approaches divide the pipe network into a series of fixed, interconnected control volumes and record changes at the boundaries or within these volumes as water flows through them. Lagrangian models track changes in a series of discrete parcels of water as they travel through the pipe network. Time-driven simulations update the state of the network at fixed time intervals. Event-driven simulation updates the state of the system only at times when a change actually occurs, such as when a new parcel of water reaches the end of a pipe and mixes with water from other connecting pipes.

Each of these approaches assumes that a hydraulic model has determined the flow direction and velocity in each pipe at specific intervals over an extended period of time. These intervals are referred to as hydraulic time steps and are typically one hour for most applications. Within a hydraulic time step the velocity within each pipe remains constant. Constituent transport and reaction proceeds at smaller intervals of time known as the water quality time step. Adjustments are made at the start of a new hydraulic time step to account for possible changes in flow velocity and direction. Brief descriptions of four different solution methods for dynamic models, two which are Eulerian and two which are Lagrangian, follows.

## 4.1 EULERIAN FINITE DIFFERENCE METHOD (FDM)

FDM is an Eulerian approach that approximates the derivatives in Equation 1 with their finite difference equivalents along a fixed grid of points in time and space (Islam et al., 1997). There are many choices available for converting (1) into a difference equation. The Lax-Wendroff scheme is a popular method for solving hyperbolic differential equations arising in fluid flow problems (Smith, 1978). Under this scheme, the finite difference form of (1) becomes:

$$C_{i,s}^{t+\Delta t} = .5\alpha(1+\alpha)C_{i,s-1}^{t} + (1-\alpha^{2})C_{i,s}^{t} - .5\alpha(1-\alpha)C_{i,s+1}^{t} + R(C_{i,s}^{t})$$

where $\Delta x$ = distance between each spatial grid point; $\Delta t$ = water quality time step; $\alpha = u\Delta t/\Delta x$, and $C_{i,s}^{t}$ = concentration at grid point s of link i at time t. This scheme is stable for $0 < \alpha \leq 1$ and has second order accuracy (Smith 1978).

In a similar fashion, a forward difference approximation in time is used to re-express the mixing condition (3) occurring at tanks. The end result is a series of algebraic equations for the entire network that can be solved in explicit fashion by marching forward in time and down the length of each pipe. At the start of each hydraulic time step, a new grid spacing is chosen in each link so that $\alpha$ is kept as close as possible to 1 (i.e., the number of distance intervals equals the largest integer less than or equal to $L/(u\Delta t)$, where L is the link length). Concentrations at the new grid points are found by linearly interpolating from the old ones. The accuracy of FDM is dependent on the size of the water quality time step. Because it is unlikely that $\alpha$ will exactly equal 1, the method is subject to numerical dispersion as sharp concentration fronts get spread out among adjacent grid points.

## 4.2 EULERIAN DISCRETE VOLUME METHOD (DVM)

DVM (Rossman et al., 1993) divides each pipe into a series of equally-sized, completely-mixed volume segments. At the end of each successive water quality time step, the concentration within each volume segment is first reacted and then transferred to the adjacent downstream segment. When the adjacent segment is a junction node, the mass and flow entering the node is added to any mass and flow already received from other pipes. After these reaction/transport steps are completed for all pipes, the resulting mixture concentration at each junction node is computed and released into the first segments of pipes with flow leaving the node. (See Figure 2.)

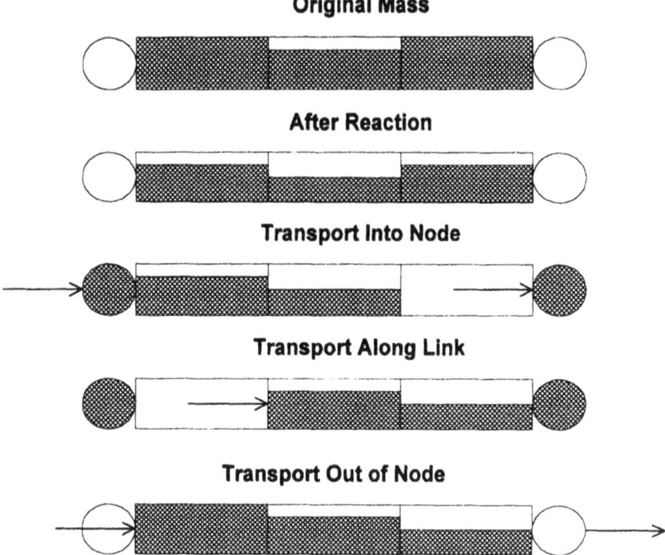


**Figure 2**. Computational steps of the Discrete Volume Method

This sequence of steps is repeated until the time when a new hydraulic condition occurs. The network is then re-segmented to reflect changes in pipe travel times, mass is re-apportioned from the old segmentation to the new one, and the computations are continued. For a specified water quality time step, the number of volume segments in a pipe is the largest integer less than or equal to its travel time divided by the water quality time step. This produces the same grid spacing as used by FDM.

The accuracy of DVM will depend on the size of the water quality time step used. The method avoids numerical dispersion within each hydraulic time step because the contents of adjacent segments are never blended together. When a link is re-segmented at the start of a new hydraulic time step there will be some degree of blending in the case when fewer segments are used (i.e., flow velocity increases). The method is subject to phase shift errors (i.e., errors in tracking the timing of abrupt concentration changes) because the volume of flow during a time step will most likely be less than the volume of a link segment (i.e., when the ratio of the travel time in a link to the time step is not a whole number).

### 4.3 LAGRANGIAN TIME-DRIVEN METHOD (TDM)

This method tracks the concentration and size of a series of non-overlapping segments of water that fill each link of the network (Liou and Kroon, 1987). As time progresses, the size of the most upstream segment in a link increases as water enters the link while an equal loss in size of the most downstream segment occurs as water leaves the link. The size of the segments in between these remains unchanged. (See Figure 3).

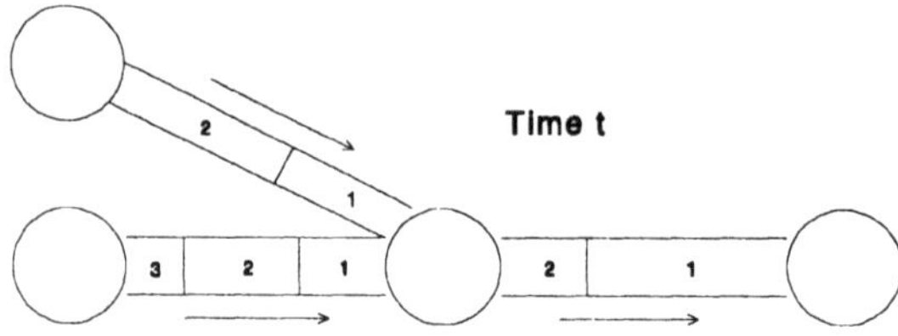


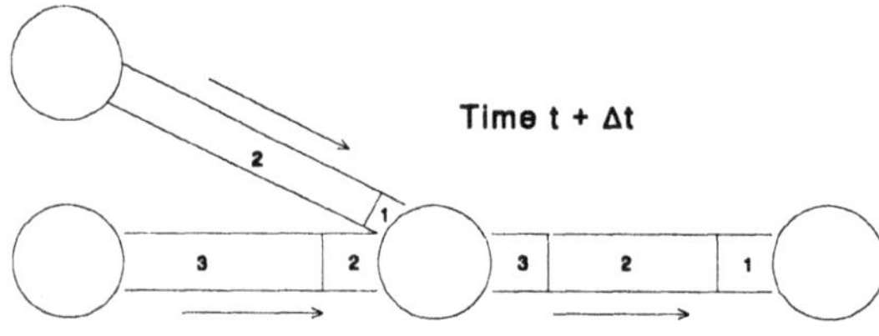


**Figure 3**. Behavior of segments in the Lagrangian solution methods

For each water quality time step, the contents of each segment are subjected to reaction, a cumulative account is kept of the total mass and flow volume entering each node, and the positions of the segments are updated. New nodal concentrations are then calculated and new segments are spawned at the start of links with flow leaving each node. Excessive segment generation is controlled by only creating new segments when the existing segment downstream of a node differs in concentration by a specified tolerance. The process is then repeated for the next water quality time step. At the start of the next hydraulic time step the order of segments in any links that experience a flow reversal is switched. Otherwise no other adjustment is necessary.

The Lagrangian nature of TDM avoids any numerical dispersion within the interior length of links. However some artificial mixing between segments can be introduced at downstream nodes when more than the leading segment in the link is consumed during a time step. The accuracy of this method will depend on the choice of a time step and the concentration tolerance used to limit the generation of new segments.

## 4.4 LAGRANGIAN EVENT-DRIVEN METHOD (EDM)

EDM is similar in nature to TDM except rather than update the entire network at fixed time steps, individual link/node conditions are updated only at times when the leading segment in a link completely disappears through its downstream node (Boulos et al., 1995). It requires that an ordered list be maintained of the projected lifetime of the leading segment in each link (i.e., the time, based on the current flow velocity and parcel size, until the segment disappears through its downstream node). The next "event" occurs for the segment at the head of this list, the one with the shortest projected lifetime. At the time of this next event the following actions take place: i) the "event" segment is destroyed and the simulation clock time is updated, ii) a new concentration is recorded at the node consuming the "event" segment as the next segment in line replaces it and mixes with the water in the leading segments of other connecting links, iii) if the change in concentration at the "event" node is above a specified tolerance, new segments are generated at the start of all links with flow leaving the node, with their concentration set equal to that of the node, and iv) projected lifetimes for all leading segments are adjusted and the event list is re-ordered accordingly.

This process continues until the end of the current hydraulic time step. At that time all segment positions and concentrations are updated. At the start of the next hydraulic event the order of segments in any links that experience a flow reversal is switched. Then a new ordered event list is generated and the sequence of event processing continues.

EDM is free of numerical dispersion and phase shift errors. Its accuracy is not dependent on any time step limitation, but only on the concentration tolerance used to limit segment generation. Some additional error can be introduced when flow reversals occur for reactive constituents, depending on how one decides to treat the reversal of the concentration profile that exists within each segment. See Boulos et al. (1995) for further discussion of this issue.

## 4.5 COMPARISON OF METHODS

Rossman and Boulos (1996) made a numerical comparison of the four solution methods just described. The EPANET simulation model (Rossman, 1994) was used as the testing vehicle. EPANET is a public domain software package developed by the U.S. Environmental Protection Agency that simulates extended period hydraulic and water quality behavior within water distribution systems. The current version of EPANET uses the Eulerian DVM method in its water quality module. Additional modules were written to implement the FDM, TDM, and EDM procedures in a special experimental version of the software. Tests of the four methods were made for the following cases:

- accuracy comparisons for analyzing two simple networks where it was possible to compute exact, analytical solutions - one network modeled sharp concentration fronts propagating from 3 different sources under constant flow conditions; the other modeled chlorine decay in a system first fed from a reservoir and then from a storage tank thus inducing flow reversal in several pipes,
- relative comparisons of model results against field data taken from two actual distribution systems, one for modeling fluoride tracer and the other for chlorine decay,
- execution time and memory usage comparisons for analyzing chlorine decay and water age in five actual systems of varying sizes for a 24-hour period of operation.

The results of these tests, based on using a 3-minute water quality time step for each method (where applicable), can be summarized as follows:

- When tracking several sharp moving concentration fronts on the first simple test network, FDM produced some small smearing of one front while DVM produced a 3 minute phase error (i.e., early arrival) on two of the fronts. Solutions from both TDM and EDM matched the analytical results exactly. For the second test network, all four methods produced virtually identical results which matched the analytical solution.
- Results from one of the networks where a fluoride tracer study was made were essentially the same for the four methods. Good agreement with observed fluoride levels over time was obtained. For the second network, which was modeled for chlorine decay, DVM showed some small disparities with the other methods. An example of the results obtained for one of the sampling locations is plotted in Figure 4.
- Figure 5 compares execution times against network size for the four methods applied to chlorine decay. The two Lagrangian methods provide a significant advantage over the Eulerian methods. However as shown in Figure 6, for water age simulations this advantage disappears for the Lagrangian EDM method.

Based on these results it appears that the Lagrangian time-driven method (TDM) is the most efficient and versatile of the methods available for solving dynamic water quality network models. As such, plans are underway to incorporate this method into the next version of the EPANET software.

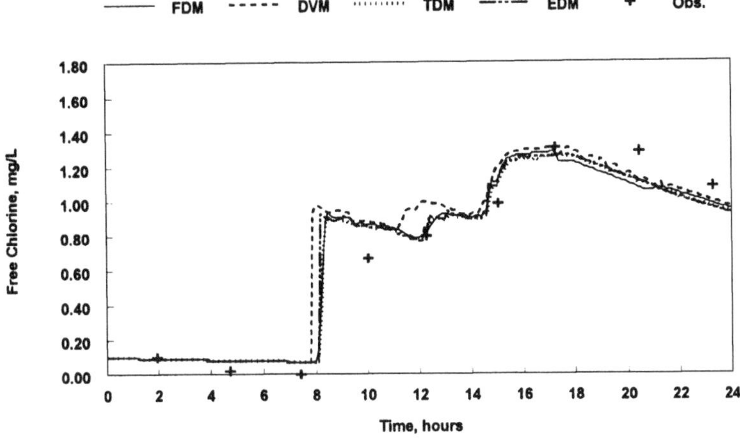


**Figure 4**. Comparison of solution methods with field data

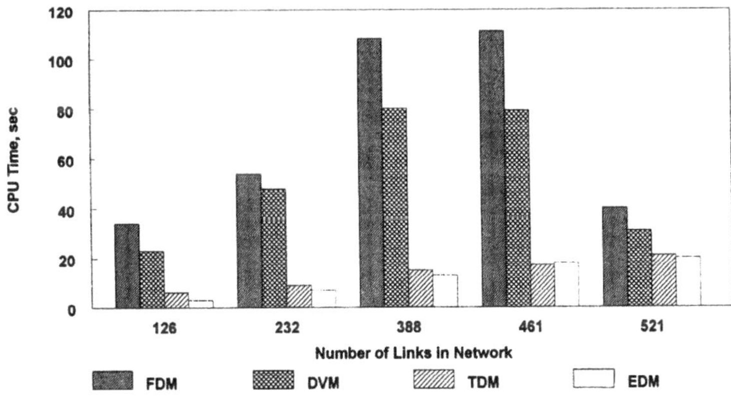


**Figure 5**. Computation time comparisons for chlorine decay analysis

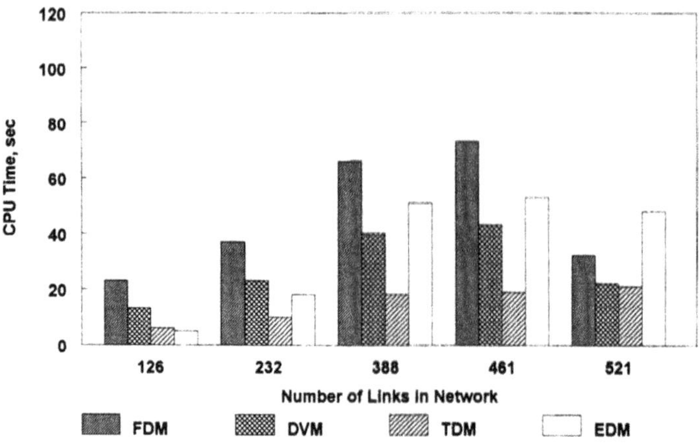


**Figure 6**. Computation time comparisons for water age analysis

## 5. Data Requirements

Data requirements for water quality models fall into the categories of hydraulic, water quality, reaction rate, and field data. A brief discussion of the requirements in each of these areas follows.

### 5.1 HYDRAULIC DATA

A water quality model uses the flow solution of a hydraulic model as part of its input data. Steady-state models require only a single, steady-state flow value for each pipe. Dynamic models utilize a time history of flow in each pipe and of volume changes in each storage facility. These quantities are determined by making an extended period hydraulic analysis of the system being studied. Most modeling software packages have the capability of integrating the hydraulic and water quality analyses together into a single operation. This relieves the analyst from having to manually supply flow data to the water quality solver. Having a good hydraulic understanding of a network is essential for computing accurate water quality results. A poorly calibrated hydraulic model will invariably lead to a poorly performing water quality model.

## 5.2 WATER QUALITY DATA

Dynamic models require a set of initial water quality conditions to start the simulation with. There are two basic approaches for establishing these conditions. One is to use the results from a field monitoring survey. This approach is often used when calibrating the model to field observations. Locations in the model corresponding to sampled sites can have their initial quality set to the measured value. Initial conditions for other locations can be estimated by interpolating between the measured values. When using this method it is important to get good estimates of quality conditions within storage facilities. Model results can be sensitive to these values, which can be slow to change during the simulation due to the usually slow replacement rate of water in storage. This approach cannot be used when modeling water age as there is no way to directly measure this parameter.

The other approach is to start the model simulation with arbitrary initial values and run it for a sufficiently long period of time under a repeating hydraulic loading pattern until the system's water quality behavior settles into a periodic pattern. Note that the length of this pattern might be different than the length of the hydraulic pattern. Results from the last period would then be taken to represent the system's response to the imposed hydraulic loading. Good estimates of initial conditions in the storage facilities can reduce the time needed for the system to reach a dynamic equilibrium.

In addition to initial conditions, the water quality model needs to know the quality of all external inflows into the system. This data can be obtained from existing source monitoring records when simulating existing operations or could be set to desired values when investigating operational changes.

## 5.3 REACTION RATE DATA

The specific form of reaction rate data needed to run a water quality simulation will depend on the constituent being modeled. It is essential that this data be developed on a site-specific basis since research has shown that reaction rates can differ by orders of magnitude for different water sources, treatment methods, and pipeline conditions.

First-order rate constants for chlorine decay in the bulk flow can be estimated by performing a bottle test in the laboratory. Water samples are stored in several amber bottles and kept at constant temperature. At several periods of time a bottle is selected and analyzed for free chlorine. At the end of the test the natural logarithms of the measured chlorine values are plotted against time. The rate constant is the slope of the straight line through these points. There is currently no similar direct test to estimate wall reaction rate constants. Instead, one must rely upon calibration against measured field data.

A similar bottle test can be used to estimate first-order growth rates for trihalomethanes (THMs). The test should be run long enough so that the THM concentration plateaus out to a constant level. This value becomes the estimate of the maximum THM formation potential. A plot is then made of the natural logarithm of the difference in the formation potential and measured THM level versus time. The slope of the line through these points is the growth rate constant.

## 6. Example Applications

Several example applications will be discussed to illustrate how water quality modeling is being used in water distribution systems. All of the results discussed were produced using the EPANET modeling software.

### 6.1 HYDRAULIC CALIBRATION

The first example demonstrates how a water quality model can be used to confirm the calibration of a hydraulic model. This requires that a non-reactive tracer chemical be monitored over time at several locations after it has been added (or removed) from the system. For systems with multiple supply sources, this can be done during normal operation providing that the tracer compound is present at different concentrations in the different sources. Such a case occurs in the North Marin Water District in northern California. A schematic of the system is displayed in Figure 7. In addition to its normal supply of bank-filtered water from the Russian River, the district also treats water from Stafford Lake during periods of high demand. The treatment plant at the lake uses sodium hydroxide for pH adjustment. This results in a finished water sodium concentration of about 23 mg/L compared to 9 mg/L in the Russian River water. The blending of sodium throughout the system was modeled over a 42-hour period of operation which occurred in July 1993 (Vasconcelos et al., 1996). At the same time, periodic sampling was made at several locations, three of which are shown in Figure 7. During this period of operation, the Stafford Lake treatment plant was operating for only 8-9 hours during the day. This resulted in widely fluctuating sodium concentrations throughout the system.

The existing hydraulic model of the district was updated to reflect water demands recorded during the study period. Initial sodium levels at the nodes of the model were assigned by interpolating from the initial samples taken at the sampling stations. Source input concentrations of sodium were kept at 23 and 9 mg/L at the lake and river supplies, respectively. Because sodium is a non-reactive constituent, no other information was needed to run a water quality analysis. Figure 8 compares measured sodium levels with model-computed ones for several locations in the network. The close correspondence between these values indicates that the model is hydraulically well-calibrated, since flow is the only factor affecting the mixing of sodium from the two sources.

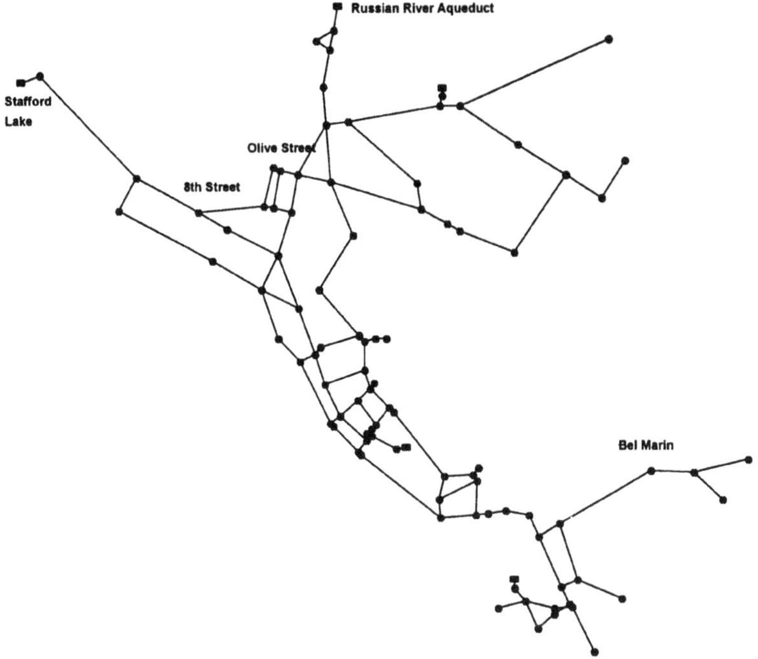


**Figure** 7. Zone I of the North Marin Water District

## 6.2 CHLORINE DECAY

The second example illustrates how a chlorine decay model might be applied to a network. The system studied is the Oberlin district of the United Water Resources service area in Harrisburg, Pennsylvania. It served as a study site for a recently completed research project on modeling chlorine decay in distribution systems (Vasconcelos et al., 1997). A schematic of the network is shown in Figure 9. It is an isolated zone receiving all of its water from the Oberlin booster pump station. It is entirely residential and contains many pipes that are unlined galvanized iron or steel, 6 inches (150 mm) or less in diameter, and many are 30 to 50 years old. It was anticipated that these characteristics would contribute to a significant pipe wall chlorine demand, as evidenced by a drop in chlorine levels from about 1.0 mg/L entering the zone to less than 0.2 mg/L at the end of the zone which was only some 1.5 miles (2.4 km) away.

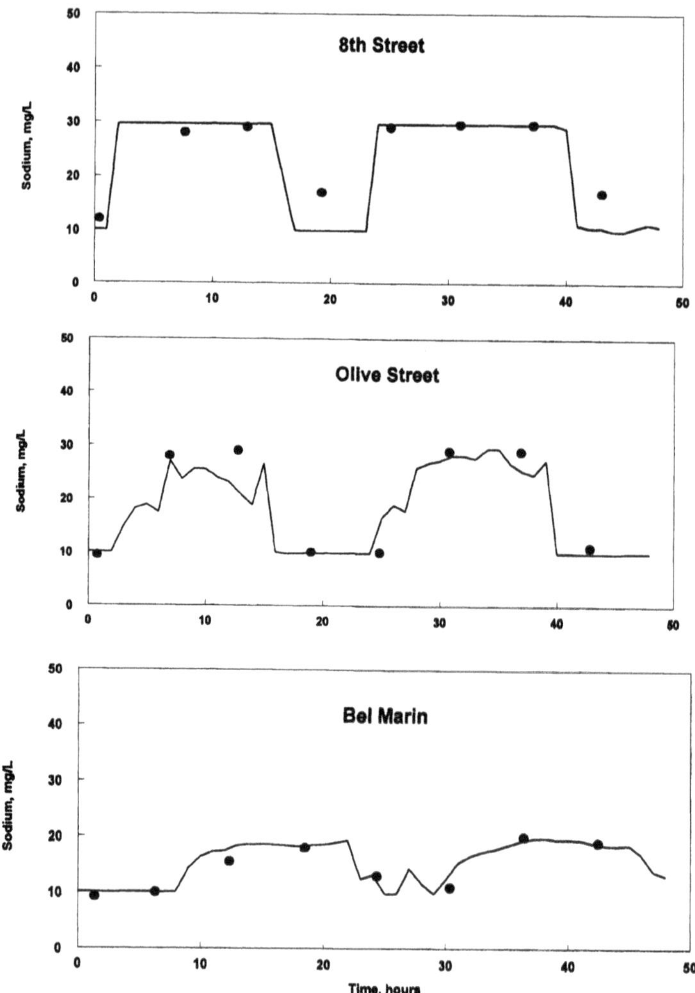


Figure 8. Comparison of computed and measured sodium levels in North Marin Water District Zone I

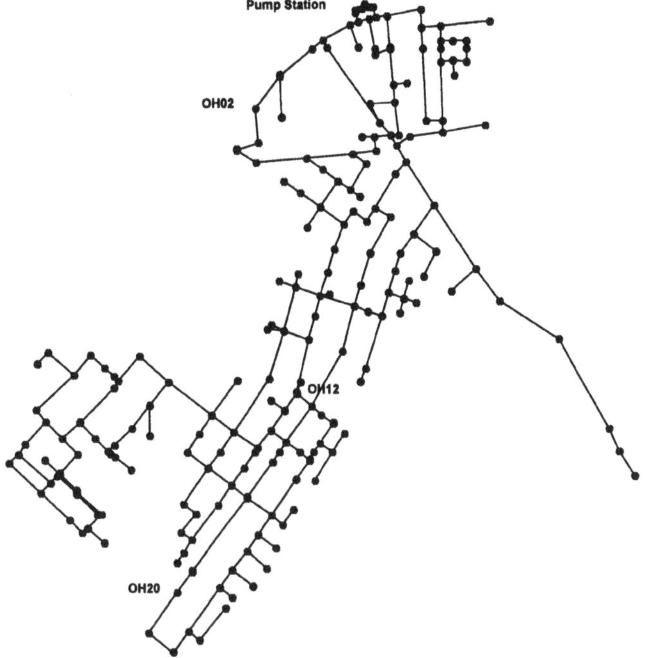


**Figure 9**. Schematic of the Oberlin pipe network

Data were collected from a sampling study performed on October 11-13, 1993. Free chlorine was measured at approximately hourly intervals at numerous locations over a 35-hour period. In addition, a chlorine decay bottle test was made on the water entering the zone. An existing hydraulic model of the zone was calibrated to reproduce hydraulic conditions observed during the sampling study. Initial nodal chlorine levels for the model were estimated from the first samples taken during the study. Chlorine levels entering the zone over the study period were obtained from the measurements made at the pump station.

The initial model run assumed that chlorine was being consumed only by a first-order reaction within the bulk flow. The rate constant for this reaction, $k_b$, was set equal to 0.232 $day^{-1}$ as determined from bottle decay test data. Figure 10 shows the comparison between computed and measured chlorine values for three sites in the zone -- one near the top (OH02), one near the middle (OH12), and one near the bottom (OH20). Clearly, more chlorine decay occurred within the pipe environment than could be explained by reactions within the bulk water alone.

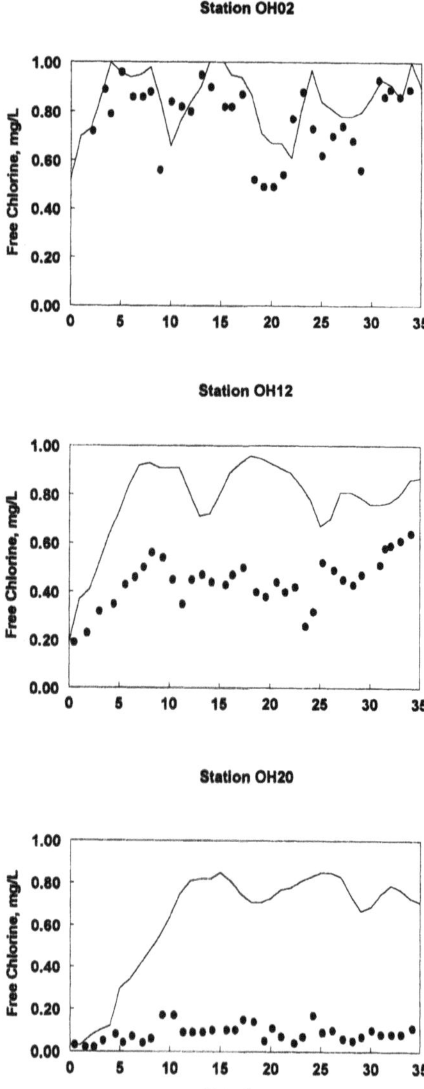


**Figure 10.** Comparison of computed and observed chlorine residuals in Oberlin - bulk demand reactions only.

The next phase of the calibration determined values for a first-order pipe wall reaction rate constant, $k_w$, that would improve the model fit. Recall that the actual rate of wall reaction is affected not only by $k_w$, but also by the rate of mass transfer which depends on the pipe's diameter and Reynolds number. Because there is no direct way to measure $k_w$, a systematic numerical search was performed on $k_w$ to find the value that minimized the summed errors between observed and computed chlorine levels. A value of 0.27 m/day resulted in an average absolute error of 0.11 mg/L. However using a single $k_w$ value for all pipes in the network tended to under-predict concentrations at the top end of the network and over-predict them at the far end.

The final phase of the calibration tried to improve the model fit by allowing $k_w$ to vary spatially across the network. The hydraulic roughness coefficients (Hazen-Williams C-factors) used for the pipes in the model had been assigned on the basis of pipe material and age, with older more deteriorated pipes having lower coefficients. Because these pipes should also have higher reactivity with chlorine, it seemed reasonable to assume that $k_w$ should be inversely proportional to a pipe's C-factor. Using another iterative search, a value for this proportionality factor of 24.4 was found to produce the smallest estimation error. Although this overall error was again 0.11 mg/L, it did produce a more even distribution of errors among the various sampling locations. The final fit of the model to the data for stations OH01, OH12, and OH20 is shown in Figure 11.

Similar calibration studies were performed at four other sites as part of this research project. Results showed that both bulk and wall chlorine decay coefficients can vary widely from one site to another and that in some systems only a portion of the pipes appear to exhibit any wall demand. The error of these models in replicating point measurements was between 0.05 to 0.15 mg/L, which translated to relative errors between 17 and 31 percent. The models were more accurate in matching time-averaged chlorine concentrations at sampling locations, with correlations between computed and measured means ranging from 85 to 98 percent.

## 6.3 TRACKING A CONTAMINATION EVENT

In December of 1993 a waterborne outbreak of salmonellosis occurred in Gideon, Missouri. Almost 600 of the town's 1104 residents were affected with diarrhea and seven elderly persons died. The incident was investigated and reported on by Clark et al. (1996). As part of their investigation, a network modeling study was performed to shed light on a possible cause of the outbreak. The suspected source of the *Salmonella* contaminating the system was bird infestation of a storage tank in disrepair. The first illnesses coincided shortly after all 50 fire hydrants in the system were flushed sequentially over a 12-hour period in response to taste and odor complaints. As a result of this flushing, the contents of the suspect tank were rapidly drained down and emptied into the distribution system. The modeling study sought to determine if the spread of this tank's water into the system was consistent with the locations where the first illnesses were reported.

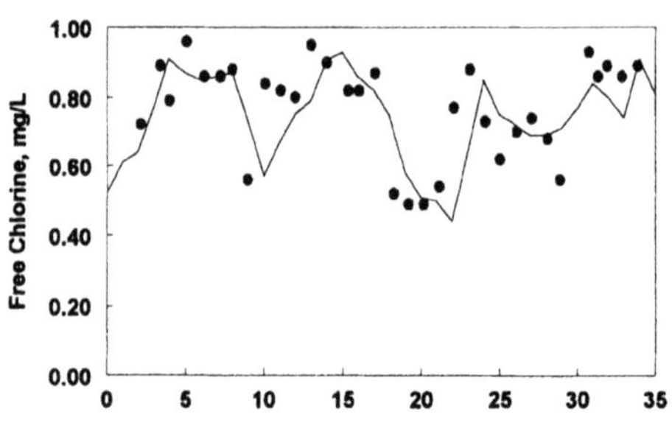


OH12

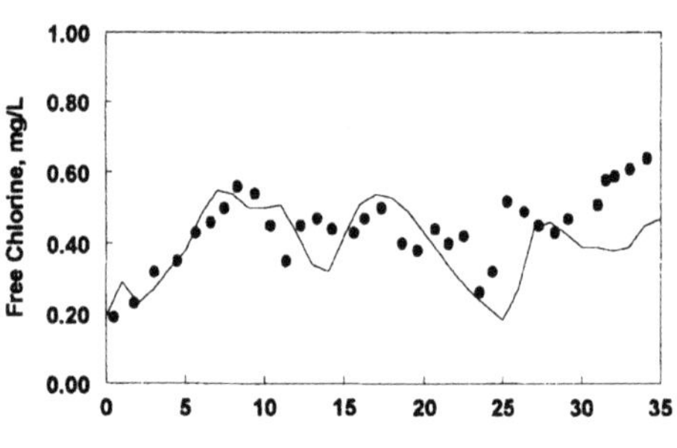


OH20

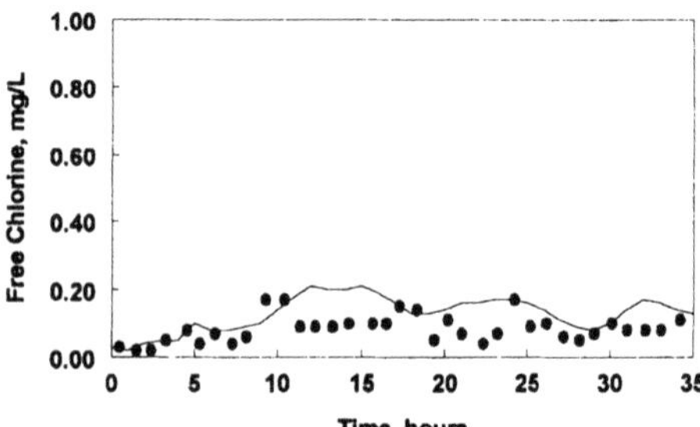


**Figure 11**. Comparison of computed and observed chlorine residuals in Oberlin - bulk + wall reactions.

A hydraulic model was developed for the Gideon system and was run to simulate the conditions occurring during the flushing program. The source tracing option of the EPANET software was used to track what percent of the water reaching each node of the network originated from the tank. Figure 12 shows an example of the results obtained 6 hours into the flushing event. The black circles show locations where more than 50 percent of the water originated from the tank. By examining similar maps generated at other time periods it became apparent that most of the locations which experienced the first signs of the outbreak were within the zone of influence of the suspect tank.

Similar types of retrospective studies using network water quality models are being carried out to estimate long-term population exposures to suspected carcinogenic compounds, such as trichloroethylene, which have entered water supply systems from contaminated groundwater aquifers (Aral et al., 1996).

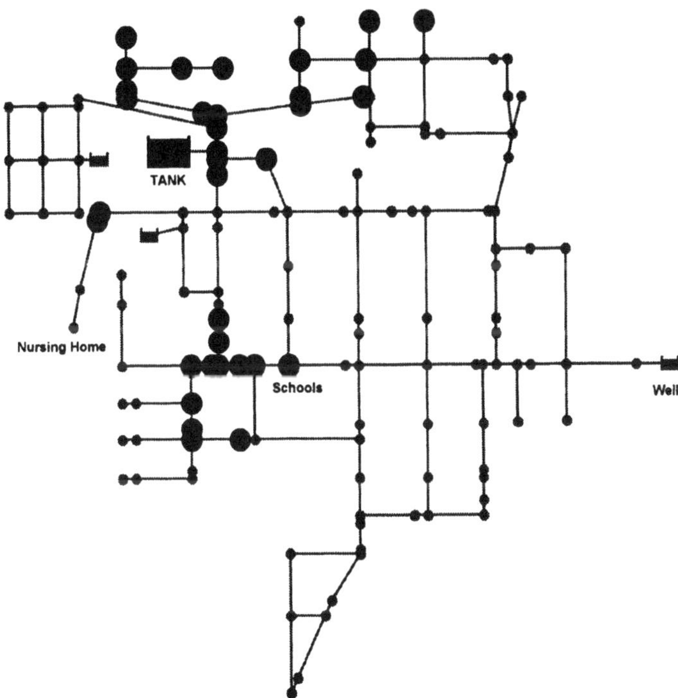


**Figure 12**. Model results showing sites in Gideon receiving 50% or more of their water from the suspected contaminated tank six hours into the hydrant flushing program of 10 November 1993.

## 7. Conclusions

The transport, mixing, and transformation of constituents in finished water traveling through a distribution system can be modeled using well-established principles of mass conservation and reaction kinetics. Applying these principles to each pipe, junction, and storage facility in the distribution network leads to a coupled set of differential/algebraic equations with time-varying coefficients. A simplified version of these equations can be applied under steady-state flow conditions, but the reasonableness of this restriction is questionable. Several approaches exist for solving the dynamic version of the model. Numerical comparisons between the approaches suggests that the Lagrangian Time Driven method is the most efficient and versatile.

In applying water quality models to actual distribution systems it is critical that a well-calibrated hydraulic model be used along with site-specific reaction rate data. The most common uses of water quality models to date have been for performing hydraulic model calibration using tracers, analyzing source blending issues, estimating and reducing water age within systems, optimizing chlorine disinfection application, and performing both short- and long-term contaminant exposure studies. The accuracy of tracer, blending, and water age analyses will depend principally on the accuracy of the hydraulic models used to drive such studies. Limited field calibrations of chlorine decay models suggest that they can match individual grab sample measurements with average errors of 17 to 30 percent and can provide even more accurate estimates of time-averaged concentrations at fixed monitoring locations. The accuracy and relevance of models used for exposure studies will be influenced primarily by the ability to accurately reconstruct historical loading and operational conditions within the system being analyzed. The state of water quality modeling for distribution systems will continue to be further refined and advanced as more system managers recognize the potential these models have for improving and maintaining the quality of water delivered to the consumer.

## 8. References

Aral, M.M., Maslia, M. L., Ulirsch, G.V., and Reyes, J.J. (1996). "Estimating exposure to volatile organic compounds from municipal water-supply systems: Use of a better computational model", *Archives of Environmental Health*, 51, 300-309.

Boulos, P.F. and Altman, T. (1993). "Explicit calculation of water quality parameters in pipe distribution systems", *Civil Eng. Syst.*, 10, 187-206.

Boulos, P.F., Altman, T., Jarrige, P.A., and Collevati, F. (1995). "Discrete simulation approach for network water quality models", *J. Water Resour. Plng. and Mgmt., ASCE*, 121, 49-60.

Clark, R.M., Geldreich, E.E., Fox, K.R., Rice, E.W., Johnson, C.H., Goodrich, J.A., and Barnick, J.A. (1996). "Tracking a Salmonella serovar typhimurium outbreak in Gideon, Missouri: role of contaminant propagation modelling", *J. Water SRT - Aqua*, 45, 171-183.

Grayman, W.M., Clark, R.M., and Males, R.M. (1988). "Modeling distribution system water quality: dynamic approach", *J. Water Resour. Plng. and Mgmt., ASCE*, 114, 295-312.

Islam, M.R., Chaudhry, M.H., and Clark, R.M. (1997). "Inverse modeling of chlorine concentration in pipe networks under dynamic condition", *J. Envir. Engrg., ASCE*, 123, 1033-1040.

Liou, C.P. and Kroon, J.R. (1987). "Modeling the propagation of waterborne substances in distribution networks", *J. AWWA*, 79(11), 54-58.

Males, R.M., Clark, R.M., Wehrman, P.J., and Gates, W.E. (1985). "Algorithm for mixing problems in water systems", *J. Hydr. Div., ASCE*, 111, 206-219.

Rossman, L.A., Boulos, P.F., and Altman, T. (1993). "Discrete volume-element method for network water-quality models", *J. Water Resour. Plng. and Mgmt., ASCE*, 119, 505-517.

Rossman, L.A. (1994). *EPANET - Users Manual*, EPA-600/R-94/057, U.S. Envir. Protection Agency, Risk Reduction Engrg. Lab., Cincinnati, Ohio.

Rossman, L.A., Clark, R.M., and Grayman, W.M. (1994). "Modeling chlorine residuals in drinking-water distribution systems", *J. Envir. Engrg., ASCE*, 120, 803-820.

Rossman, L.A. and Boulos, P.F. (1996). "Numerical methods for modeling water quality in distribution systems: A comparison", *J. Water Resour. Plng. and Mgmt., ASCE*, 122, 137-146.

Servais, P., Laurent, P., Billen, G., and Gatel, D. (1995). "Development of a model of BDOC and bacterial biomass fluctuations in distribution systems", *Rev. Sci. Eau*, 8, 427-462.

Smith, G.D. (1978). *Numerical solution of partial differential equations: Finite difference methods*, 2nd. Ed., Oxford University Press, Oxford, England.

Vasconcelos, J.J., et al. (1996). *Characterization and modeling of chlorine decay in distribution systems*, AWWA, Denver, Colorado.

Vasconcelos, J.J., Rossman, L.A., Grayman, W.M., Boulos, P.F., and Clark, R.M. (1997). "Kinetics of chlorine decay", J. AWWA, 89(7), 54-65.

Weiss, M.A. (1993). *Data structures and algorithm analysis in C*, Benjamin/Cummings Publishing.

*Acknowledgment:* This material is based on work partially funded by the U.S. Environmental Protection Agency. It has not been subject to agency review and does not necessarily reflect the agency's views.

# RELIABILITY AND RISK IN A WATER SUPPLY SYSTEM EMPHASISING DROUGHT PERIODS

I C GOULTER
Swinburne University of Technology
Hawthorn
Melbourne, Victoria 3122
Australia

## 1. Introduction

While the issue of defining risk and the associated problem of providing acceptable levels of reliability in urban water distribution networks has received a considerable degree of attention over the last few decades, there is still no universally accepted and computationally feasible measure for assessment of reliability of water distribution network nor an agreement or understanding of what constitutes an acceptable level of reliability (Goulter, 1995). In order to provide comprehensive interpretations of the reliability performance of the network, most approaches to consideration of risk and reliability of water supply systems attempt to examine, implicitly or explicitly, the complete range of events which contribute to, or detract from, the reliability of the system. Extreme events such as droughts which are major determinants in the assessment of risk and reliability in water distribution networks are clearly contained in the spectrum of events which must be considered in these approaches. This paper examines

(i) the risk to water distribution networks that droughts constitute,
(ii) the associated impacts of those risks on the reliability of the system, and
(iii) new concepts and strategies for the management, design and operation of water supply systems which can be employed to improve their performance, in terms of reliability and risk, in the face of drought situations.

In undertaking this examination it is important to note that water distribution systems can be broken up into three major subsystems, namely, the supply system, the transmission system, and the distribution system or network (including the treatment plants) as shown in Figure 1. Assessment of risk and reliability in a water distribution system should therefore take into consideration risk and reliability issues associated with each of these sub-systems as broadly categorised in Table 1.



*E. Cabrera and J. García-Serra (eds.), Drought Management Planning in Water Supply Systems,* 128–147.

TABLE 1. Typical risk aspects associated with each sub-system in a water supply system

| System | Nature of the Risk |
|---|---|
| Source system | The stochastic nature of the hydrologic processes that provide the water supplies at the source |
| Transmission system | Mechanical failure of components of the system, Deterioration of the system with age. |
| Treatment and distribution system | Cyclic variations and overall increases in the demands which have to be met by the system, Failure of components in the distribution network and/or treatment plant, Deterioration in the hydraulic capacity of the distribution network. |

This paper focuses primarily on the risks associated with the last of three subsystems, namely, the treatment facilities and distribution network and more specifically on ways in which the reliability of the overall system in the face of these risks can be improved by management of both the demands themselves and the consumers' expectations of what constitutes an acceptable level of service in meeting those demands.

## 2. Definitions of Reliability in Water Distribution Networks

Before proceeding with an examination of risk and reliability in water distribution networks specifically in relation to droughts it is necessary to review how risk and reliability is defined in more general terms. Goulter (1995) uses a definition of reliability of a water distribution network derived from the work of Cullinane *et al.* (1992), namely, "the ability of a distribution system to meet the demands that are placed on it where demands are specified in terms of:

1. the flows to be supplied; and
2. the range of pressures at which these flow rates must be provided."

Cullinane *et al.* (1992) extended this definition to "... the ability of the system to provide service with an acceptable level of interruption in spite of abnormal conditions".

This interpretation of reliability acknowledges that failure of the network can arise from either the flow rate associated with a demand not being met, the flow rates associated with the demand being met but at delivery pressures lower than the minimum specified or accepted level, or a combination of neither the flows nor the required delivery pressures associated with the demands being met. This definition also acknowledges the role that abnormal or extreme events, such as droughts, play in reliability in that the ability to meet the volumetric requirements associated with the design flow rates over extended periods of time is clearly a function of the total water supply available to meet the demand.

Wagner et al. (1986) also provide twenty different aspects of reliability which were able to be considered in their simulation, as opposed to analytical, based approach for assessment of reliability.

**Event Related**

Type of event (failure or repair)
Inter failure times and repair durations
Total number of 'events' in the simulation period
System status during each event (normal, reduced service, failure)

**Node Related**

Total demand during the simulation period
Shortfall (total unmet demand)
Average head
Number of reduced service events
Duration of reduced service events
Number of failure events
Duration of failure events

**Link Related**

Number of pipe failures
Total duration of failure time for each pipe
Percentage of failure time for each pump
Percentage of failure time for each pipe
Total duration of failure time for each pump

**System Related**

Total system consumption
Total number of component breaks (pipe, pump, etc)
Maximum number of breaks per event

Such definitions also imply that the period of time, i.e., duration, over which a network is unable to meet its demands is another important aspect of reliability. Total, or partial failure, of the network therefore occurs when the network is unable to deliver all or part of the design demand because of

i) MECHANICAL FAILURE:
reduced hydraulic capacity of the system due to failure of a component in the system or overall deterioration of the hydraulic capacity of the network with age, or

ii) DEMAND VARIATION:
the demand itself increases beyond the delivery capacity of the network due to a) increase in population served by the network, b) changes in per capita water consumption arising from an increased standard of living, or c) the design value itself being exceeded, e.g., the system may have been designed for the 20 year maximum daily flow and the 50 year maximum daily flow occurs. (The fact that a design value of the 20 year minimum daily flow is used is in itself an implicit recognition of the risk to the system arising from the occurrence of higher flow values which occur with known probabilities.)

The impacts of both the above situations can be modelled probabilistically, i.e., through the probability of component failure or probability of the demand being higher than the design demand.

These interpretations of reliability are premised on sufficient water being available to meet the design demands. However, during a period of drought, particularly severe drought, it is not appropriate to assume that the total supply to the network is sufficient to meet the demand. Reduced levels of supply to the network arising from drought events, and the associated probabilities of such events, have to be explicitly considered in such situations. The definition of network failure provided earlier, i.e., a failure was assumed to occur through an inability of the network to meet the flow rates associated with the demands and/or an inability to supply those demands at minimum acceptable pressure, is able to handle these situations. Extension of the definition to explicit consideration of droughts, wherein the supply of water to the network is not sufficient to meet the demands even if the network is working perfectly to design specifications and the demand is less than the demand values, is conceptually easy. However, its implementation adds a further layer of some considerable analytical complexity to the consideration of risk and interpretation of reliability. The probability of the bulk supply to the network being less than the total demand on the network needs to be integrated into the probabilistic analysis associated with consideration of the component failure or demand variation scenarios described briefly above. It is useful to note at this stage that demands on the water supply can also increase during periods of drought even for urban systems. This factor adds further complexity to the process required to model demand variation under normal conditions.

The difficulties of interpreting reliability, or perhaps more precisely determining acceptable levels and types of reliability noted previously, are aggravated by situations demonstrated by the following cases taken from Goulter (1995):

Case 1: The flow demand is delivered at acceptable (greater than or equal to minimum) pressure 95% of the time, while 5% of the time the required flow is only able to be supplied at pressure heads which are below the minimum.

Case 2: 95% of the required flow demand is met at minimum pressures 100% of the time.

Case 3: 100% of the required flow demand is met 100% of the time but at pressures which are only 95% of the minimum acceptable level.

Case 4: 100% of the flow and pressure demand is satisfied 100% of the time at 95% of the demand nodes and there is no supply at all at 5% of the nodes.

There is considerable difficulty in identifying which of these four cases constitutes the most reliable situation. All could be interpreted as having the same reliability in purely percentage terms, namely 95%. However, while Case 4 is clearly unacceptable on a practical, "non purely numerical," basis, it is more difficult to differentiate between, or rank, the other cases. These cases clearly show that simple measures such as total deficits in volume of supply, or periods over which supply is delivered at pressures below the minimum acceptable, are clearly inadequate measures of the reliability performance of networks. Nevertheless, measures which reduce deficits in supply, or reduce the periods over which delivery pressure is below the minimum, will contribute to improved levels of reliability.

An important issue in the analysis and management of reliability in water distribution systems which is not stated explicitly above is the opportunity and, at times, need to manage the demand side as well as the supply side. Such options have been considered in both the water and electricity supply industries and have lead to the reliability cost and reliability worth approaches summarised graphically in Figure 2 (Billinton and Lakhanpal, 1996). An important feature of these approaches is that they recognise that the costs associated with reliability are distributed across, or borne by, different constituencies, namely the consumer and the supply utility. Furthermore, the share or proportion of the costs between these two constituencies varies as the reliability varies. As shown in Figure 2, the higher the reliability of the system (network) the greater the costs to the utility and the lower the cost to the customer and visa - versa. This scenario probably does not hold as rigorously in privatised water supply systems where costs incurred by the supply utility can be passed directly to the customer. However, recognition of where the costs lie, combined with the ability to manage demand as represented by the customer, and supply as represented by the utility, are important issues in maximising reliability under drought conditions from a total societal context.

There are a number of other important features in these types of approaches. It can be seen in Figure 2 that, under this conceptual approach, both customer perspectives and system costs are addressed. The optimal system reliability is therefore defined in terms of the lowest total societal cost which occurs at the point at which a balance between reliability cost and reliability worth is achieved, namely, R* in Figure 2. Both features are important. Customer perspectives on what constitutes reliability (or an acceptable level of reliability) may change, or be changed by education or publicity, during severe droughts. Similarly the system cost curve may also shift due to reduced water supply or lack of water of suitable quality. These situations, both of which can be caused by drought conditions, can give rise to a shift in the optimal societal cost and associated optimal reliability. The other important feature of this approach, and one which has particular relevance to reliability specification for water distribution systems, is that it highlights the conceptual weakness of traditional approaches to reliability where a reliability level is selected *a prior* (e.g., C in Figure 2) and the system design and operational strategy which achieves this level of reliability at minimum cost then determined.

The range, effectiveness and efficiency of measures which minimise total deficits in volumetric supply and minimise periods over which supply is delivered at pressures below the minimum acceptable while also recognising the opportunity to vary the acceptable societal levels of reliability, e.g., to allow for flow rates and delivery pressure levels which fall temporary below normal minimum acceptable levels though management of both supply and demand during drought periods, are examined in later sections.

## 3. Definitions of Risk in Water Distribution Networks

Risk is associated with decision making in stochastic environments. The stochastic environment in water distribution networks operate arises from a number of sources, broadly categorised as follows:

i) factors 'external' to the system, e.g., uncertainty in the hydrologic regime which provides the water supply to system and which also impacts upon the demands on the system, and
ii) factors 'internal' to the system, e.g., uncertainty in the performance of the components of the system.

Risk differs from reliability in that it is a statement of the probabilities of occurrence of events and the impacts of those events, whereas reliability describes how a system responds or reacts to events. An important feature of risk is that the consequences of events may be described in quantitative or qualitative terms.

Management of a system under risk is directed at reducing the probabilities of events with undesirable consequences or reducing the impacts of the undesirable consequences associated with events. In the context of planning and operation of water distribution networks in periods of drought the predominant component to the risk obviously arises from the reduced volume of bulk water supply available to the system.

Of particular importance or relevance to consideration of risk in urban water distribution systems is the fact that public perception of risk, and therefore the associated political responses to risk, do not provide a sound basis for managing risk (Hambly and Hambly, 1994). The problems surrounding public perception of risk require engineers to improve their own ability to evaluate and manage risk. It also leads to the need for engineers to "educate" the public about the true nature of risk as encapsulated in the probabilities associated with events and the objective assessment of the consequences of those events (Hambly and Hambly, 1994). Recall that the customer cost curve in Figure 2 reflects the customers perception of risk. The need, or perhaps more correctly the ability, to educate the public about the true nature of risk provides the opportunity to manipulate this customer cost curve and thereby achieve improved levels of reliability from the perspective of the customer. However such a philosophy requires a systematic approach for specifying and determining the costs associated with different levels of inadequate water supply. The techniques used by the electricity supply industry are further advanced than those of the water supply industry in this regard and are examined in more detail in the following sections.

The importance of education in managing water shortages in water supply systems was highlighted by Wilchfort and Lund (1997). However, their work focussed on educating users on the consequences of water shortage as a means of managing the resources through conservation rather than on clarifying the probabilistic aspects of the risk associated with the occurrences of those shortages.

A further consideration in the treatment of risk is the need, when making decisions with respect to systems exposed to risk, to consider the complete risk curve rather than surrogate or summative metrics of risk (Bouchart, 1996). In the context of designing and managing water distribution systems for drought conditions this requirement means that simple summative measures or surrogates such as expected deficit, expected duration of below minimum pressures, or a single probability value of the system not being able to meet the demand are inadequate. However appropriate consideration of risk by decision makers goes beyond the need to address the complete spectrum of consequences arising from a particular management strategy. Recent work by Bouchart and Goulter (1998) on the management of risk in water resources management has acknowledged the problems associated with distortion of the criteria used to determine the desirability of decision options caused by 'biased' interests of stakeholders, be they the consumers or the decision makers.

Interestingly, Bouchart and Goulter (1998) also discuss the tendency for decision makers to distort the information provided by models to reinforce their individual preferences (biases) and thereby produce what is known as cognitive dissonance (Festinger, 1957). In these cases emphasis is placed on the positive outcomes of a decision with the negative consequences being devalued. Russo et al. (1996) noted that situations which are susceptible to the problem of information distortion are those which are "loosely structured both in the nature of the information and the clarity of the criteria for a successful solution". Risk management in water supply systems clearly falls into the category for a number of reasons.

i) water supply systems operate in a stochastic hydrologic environment with all the associated uncertainties,
ii) customer perceptions of risk are poorly defined and there is no well defined or widely accepted measure of reliability or standard for reliability and
iii) costs of system failure (or conversely the benefits of high level performance) are incurred in both financial and social terms and therefore difficult to measure with any degree of accuracy or confidence.

An additional important feature of the Bouchart and Goulter (1998) study is their observation that risk perceptions as well as a more complete representation of risk needs to be incorporated in the decision making process. This assertion impacts on both the curves shown in Figure 2. Risk perceptions of the customer define the customer cost curve while risk perceptions of the decision maker are included in the development of the supply system cost curve. Any change in either perception, e.g., changes in risk perceptions and associated customer costs arising from pro-active education or re-active desensitisation of customers to lack of service (water) during periods of extended drought, will change the location of the optimal reliability for the particular water supply and climatic, social and economic circumstances. Similarly more knowledgeable operation of the supply side arising from more sophisticated (or 'less biased'?) operation of the system by the water authority will move the supply cost curve with a corresponding change in the nature and value of the optimal societal reliability.

## 4. Specification of Reliability in Water Distribution Networks

As mentioned earlier, there is no generally accepted standard for what constitutes acceptable levels of reliability in water distribution networks nor method(s) for calculating that reliability, particularly if the full range of probabilistic contributors to reliability are considered. The electrical power supply industry has faced similar problems but appears to be further advanced than the water supply industry in establishing standard approaches to the determination of what constitutes acceptable reliability. Consider that the two sectors share many common features, namely,

- there are social and economic benefits associated with delivery of their product/service,
- the product/service is delivered through networks to geographically dispersed customers with a range of demand characteristics,
- the delivery networks have redundant features, and
- the demands vary in time (daily, seasonally, etc) and generally increase with time
- the networks have redundant features as demonstrated by the presence of loops which provide continuity of supply should a component in the network failure.

In spite of these similarities, there appears to have been very little transfer of the concepts employed in the electricity supply industry to the water supply industry. It is useful therefore to review some of the developments in the electricity supply industry as a framework for examining how the question of acceptable levels of reliability might be more appropriately and effectively addressed in the water supply industry.

It is important at this point to differentiate between 'interruption' or 'outage' costs and shortage costs. Outage costs occur as a result of component failure and consist of the economic and non-economic consequences of short-run service reductions where the consumers do not have sufficient warning to plan for or implement steps to minimise the impacts of the service reductions (Sanghvi, 1990). Outage costs may include direct and indirect consequences. Shortage costs arise from more chronic situations such as droughts and are therefore able to be planned for, even if only over relatively short planning horizons. Shortage costs include the cost the customers incur in undertaking mitigation actions, ie., the adaptive costs, plus the 'outage' costs which still arise after the adaptive measures have been implemented. Shortage costs tend to be lower than the outage costs because of the ability to manage or plan for them. (If this was not so consumers would not implement adaptive measures as they would not be economical.) This difference between the magnitudes of outage and shortage costs is important to the planning and operational management of water supply systems in drought conditions as it recognises that the costs of a particular reduced level of service can be less in a 'shortage'-type drought scenario than a 'outage'-type component failure scenario.

Another important distinction which the electricity supply industry has recognised explicitly in its approaches to the assessment of reliability is the difference between the value of service (VOS) and the more traditional cost of service (COS) model (Burns and Gross, 1990). VOS defines the worth of a product or service. COS, on the other hand, represents what the cost of acquiring a product or service. VOS is greater than COS as most consumers "buy" a product or service if the benefit or worth derived from its use exceeds its cost. This situation is particularly true for water.

It is useful to review Figure 2 again from the perspective of VOS and COS. COS is clearly represented by the supply cost curve while VOS has features which are related to the customer cost curve. The optimal level of reliability in this figure occurs at the point corresponding to the minimal total societal cost where a balance between the cost of reliability and the value or worth of reliability is achieved, i.e., there is a balance between the cost of service and the value of service. The problem that remains, however, is that while the COS for water supply reliability (represented by the supply cost curve in Figure 1) as determined by the cost of supplying the required amounts of water of adequate quality at acceptable pressures, is relatively easy to define the corresponding VOS value, with its significant social component is considerably more difficult to define.

The procedures for of calculating VOS, particularly the social components, have troubled the electricity supply industry. Sanghvi (1990) provides a useful review of these problems and their potential solutions for the electricity supply industry. The particular problems, or more precisely questions, addressed by Sanghvi (1990) which have relevance to the application of the VOS concept to the water supply sector are:

i) *"Service should not be planned for interruptions"*

As noted previously, most, if not all, water supply systems recognise at least the possibility of service reduction relative to the demand in the choice of the demand flow. The probability of the actual demand being greater than the design flow, and therefore the probability of service reduction, may be low but it does exist and is acknowledged implicitly in the design process.

ii) *"We have an 'obligation' to serve our customers. Therefore we cannot and commissioners will not allow us to plan economic reliability standards that will lower reliability"*

Sanghvi (1990) argues that one of the major reasons for employing engineering reliability standards is to produce what is essentially a complaint or failure free service. This approach, or philosophy, assumes all customers would, given the choice, agree to pay the cost of the reliability chosen or imposed by the supply utility no matter how high that level of reliability is set. It also implicitly assumes the VOS is greater than the COS for all levels of reliability. Sanghvi (1990) asserts that, intuitively and on the basis of a growing body of empirical literature, this is not the case for the electrical supply industry. It can be argued that the situation is similar in the water supply industry, where the diversity and flexibility of end users, particularly domestic customers in responding to outages, particularly brief outages, may be even greater than in the electrical supply industry. While industrial users are often able to implement measures which reduce their long run demand they tend to be far more vulnerable to outages caused by acute short term events such as component failure or even short-term shortages (long-term outages) caused by the most critical phases of extreme droughts. Domestic consumers on the other hand are generally able to handle outages with some ease and have some capacity or flexibility to implement conservation steps in the face of droughts or as a long term demand management strategy. However, interestingly, depending on the level of conservation measures previously implemented in their households, domestic consumers may in some cases actually be less able to respond to shortage events such as extended droughts due to what has been termed "demand hardening"(Wilchfort and Lund, 1997). (Demand hardening is discussed in more detail later in this paper in the review of conservation measures.)

iii) *"This type of problem is not suited to a cost benefit analysis"*

Unlike the electricity supply industry where a significant portion of the demand is often associated with industrial or commercial activities, and the economic aspects can therefore be identified in a relatively systematic manner, there is some merit in this argument for urban water supply systems. Nevertheless, Sanghvi (1990) asserts that new methods such as the 'contingent valuation method' could be used to estimate outage costs for domestic situations. A similar argument could also be used to assess the costs, or more precisely the worth, of reduction in water supply service.

iv) *"Consumers cannot estimate their outage costs (or how can one produce credible estimates of outage costs)"*

and

*"Consumers don't really know their preferences for different service features, so how much confidence can be placed in VOS estimates"*

These are serious barriers to the use of the VOS approach in the water supply industry. However, if the argument that VOS is more appropriate than COS for determining the appropriate level of reliability for urban water supply systems is accepted, then the problem becomes one of finding methods to overcome these barriers rather rejecting what is potentially a more desirable and useful methodology.

v) *"Outage costs should be estimated following an outage rather than as a hypothetical response to a hypothetical posed situation"*

Again this is a concern about methodologies for determining inputs to the method rather than about the concept of VOS. Notwithstanding the need for improved methodologies for estimating the outage costs due to water demands not being met, it should be recognised that consumers' views of risk, and implicitly their views of probability, are heavily weighted by their recent experience. Consider a water supply system which is expected to fail on average twice in every 100 years. The interpretation of the reliability of this system will vary greatly depending on the timing and sequencing of the failures. For example, consider the scenario where the system has failed twice in the first five years of operation but does not fail again in the next 95 years. This scenario would be assessed very differently, and probably less sympathetically by the public, i.e., it would be viewed as having resulted in higher costs, than the scenario wherein the same system failed for the first time in its 80th year of operation and then failed again twice in the period between the 90th and 100th years of operation even though this second system has actually performed from a purely analytical perspective at a lower level than the first system. In other words, while the user is able to interpret costs and other implications of a failure immediately after experiencing the event, these estimates may also have bias. Furthermore, while the exact sequence of events leading up to, and following, the failure impacts greatly on that interpretation, that sequence of events may never be repeated again in conjunction with a failure of similar magnitude. Thus a hypothetical posed situation may in fact have some advantages over actual experience in estimating domestic outage costs.

vi) *"Users might give incorrect responses to purposely bias the estimates"*

*and*

*"VOS data can be used to any end, i.e., to help justify some decisions or to conclude the reverse"*

These problems were explicitly acknowledged in the earlier discussions on interpretation of risk and exist for any technique in which risk is an issue. In fact it is a problem in any decision making environment and hence VOS should be not singled out for rejection because of these factors.

vii) *"Outage costs are much higher if it is the fault of the utility"*

This is a very important aspect of planning and management of water supply systems and relates directly to the culture of expectation for water supply systems. Sanghvi (1990) reports that in the electricity supply industry, service interruptions (outages) are viewed as being beyond the control of the utility and "reliability" in fact is measured by the time taken for service to be restored. Similar situations arguably exist in water supply systems and have particular relevance to the way in which authorities attempt to improve reliability, or perhaps more correctly, attempt to improve public perception and valuation of reliability, particularly reliability associated with outages/shortages caused by component failure or short-term demand variation. It has less relevance to the management of reliability during shortages caused by events such as extended periods of drought.

Sanghvi (1990) also indicates consumers are far more forgiving of interruptions of service or failures that are caused by "Acts of God" than of situations where the utility is seen as the prime cause. Education is a key factor in this process if any degradation in service occurring as a result of management response to, or in preparation for, drought conditions are correctly interpreted by the public. Of course such a strategy requires the water supply system authority to have implemented steps which plan for and minimise the impacts of drought and that the public clearly understand both that such steps have been taken and the short term impacts (normally costs) and long term benefits of those steps.

Sanghvi (1990) similarly asserts that there is also some implication in this statement that the system should be designed to provide a uniform level of reliability to all customers. A more forward looking and appropriate approach would unbundle the various types of users and provide different levels of service (reliability) to the different users. Such a situation already exists implicitly in the ways water is provided, at various levels of service, to essential services such as hospitals, to industry, and to domestic customers. This option is particularly applicable to long-term planning for drought where separate networks for water of different qualities might be an option. It should be noted, however, that Wilchfort and Lund (1997) have recently reported that the concept of networks for different quality water did not improve options for management of shortage for their case study unless high quality water was not available and further conservation measure were also not available. Such preconditions may begin to become more prevalent as population and demand for water increases while existing supplies remain fixed or effectively diminish due to decrease in the quality of the water supplies.

viii) "*Will existing engineering reliability standards be too high when compared to economic reliability standards?*"

[Engineering reliability is defined as the reliability specified implicitly or explicitly through the standards to which the system must be designed with the system then being engineered (designed and/or operated) to meet those standards at minimum cost. Economic reliability standards on the other hand are those where the reliability selected for the system is that which minimises the total societal cost. (See Figure 2)]. The experience of the electrical supply industry suggests that the differences between economic and engineering reliability standards will be relatively small. A similar situation is likely to hold for the water supply industry. However, once again the response to this concern requires a careful examination of all the customers and possibly a differentiation between the reliability requirements of the different types of users.

Of more importance, however, is the question of reliability standards for catastrophic impacts associated with short term events, e.g., serious component failure or demand variation, or long term supply shortages occurring as a result of extreme drought. If the characteristics of the water supply industry are similar to those of the electrical supply industry, engineering and economic reliability standards are likely to converge in such catastrophic situations. This convergence tends to arise through the 'economic' reliability standards increasingly recognising and incorporating the higher level of social costs in some implicit or explicit fashion as the 'catastrophe' becomes more serious. In fact the economic reliability standards may in fact become more stringent than the corresponding engineering reliability standards in extreme catastrophic situations.

A very important issue in the establishment of reliability standards and planning for catastrophic events identified by Sanghvi (1990) is that aggregation of outage costs for individual users is usually an inadequate estimate of the total costs of system wide interruption. Sanghvi (1990) also asserts that society at large may be willing to pay "disproportionately more to avert low - likelihood but catastrophic events than to avert an equivalent impact from smaller events that occur with much higher probability". The collection of input data for determination of reliability in these circumstances is, however, particularly prone to the problems discussed previously in relation to the estimation of social costs for the VOS method.

## 5. Strategies and Methodologies for Management of Reliability and Risk

Management of reliability and risk in water supply can be divided into two major complementary strategies, demand management and supply enhancement as noted in Figure 1. Wilchfort and Lund (1997) have listed the measures available for demand management as modification of consumption patterns and reduction in total demand through education, low volume water fixtures, water rationing, tiered water pricing and landscape control. The measures available for supply enhancement were listed by Wilchfort and Lund (1997) as developing new supplies from reclamation and desalinisation, water transfers, improved operation of the system and increased use of ground water.

These measures can be further categorised into long-term and short-term measures. Wilchfort and Lund (1997) assert that long-term measures such as conservation strategies (education and fixtures), transfer contracts, additional water treatment and water reuse, have a long life span, relatively fixed cost and usually have to be implemented in advance of the shortage occurring. Short term measures were defined as temporary responses to specified levels of outage. A very important point of the work of Wilchfort and Lund (1997) which is supported by Weber (1993) is that short term conservation strategies for management of emergency shortages lose their effectiveness as long term conservation strategies which deal with shortage in a more planned and holistic sense are implemented. In other words, a hardening in demand occurs with implementation of long term conservation measures and therefore conservation measures to handle outages represented by emergency shortages become less effective.

The type of short and long term measures available for demand management and supply enhancement and their overall effectiveness as described by Wilchfort and Lund (1997) are:

**Conservation and its impacts on risk**

Water conservation for urban water supply system include changes in landscaping practices, water fixture retrofits, lawn watering practices. Cameron and Wright (1990), however, have shown that education that reinforces the public benefit (recall the total societal cost concept in Figure 2) of conservation and informs the public of the consequences of severe water shortage is critical in the effective implementation of conservation measures. Interestingly two different studies on the effectiveness of conservation measures in reducing water consumption came to two different conclusions. Nieswiadomy (1997) concluded that conservation practices were not effective in reducing water consumption in the United States while Briassoulis (1994) found that the application of conservation measures in Athens, Greece was effective in reducing consumption.

Conservation measures can be divided into short term responses to specific shortage events e.g., reduced watering, and long term measures such as retrofitting of water fixtures and landscaping regulations. Both approaches can reduce the risk of failure of a water supply system and correspondingly increase the reliability of the system through a reduction in the amount of water to be supplied. Implementation of conservation measures assumes of course that the savings, or perhaps more correctly the increase in total societal good, accruing through the reduction in water requirements are greater than the cost of the conservation measures. The actual 'improvement' in reliability, and reduction in cost of providing that reliability, occurring through water conservation measures can arise in two ways in a reliability worth / reliability cost framework. The supply cost curve in Figure 2 can move to the right where cost might now be considered in terms of \$ / satisfied customer rather than \$ / unit volume of water supplied. Alternatively, the customer cost curve can move to the left reflecting the situation where a reduced level of supply, as a surrogate for a reduction in reliability, represents, in the mind of the customer, a similar level of satisfaction or acceptable cost as a higher level of supply.

Supply enhancement alternatives such as water re-use and water transfers, which were among the other options considered by Wilchfort and Lund (1997) in their work on modelling shortage management, have somewhat different impacts on the cost curves shown in Figure 2. Both options increase the level of supply, but generally at costs greater than the cost of existing supplies. In these cases the design level of reliability is maintained but at increased cost of supply, i.e., the supply cost in Figure 2 moves vertically and under an implicit assumption mentioned earlier that customers or society are willing to 'pay' an increased amount to maintain the existing level of service. A further observation of Wilchfort and Lund (1997) was that small increases in demand do not change the nature of the long and short term decisions used to address the increase, but rather affect the extent to which an option is employed and, correspondingly the cost of that option.

## 6 Summary

This paper has proposed a new conceptual framework for consideration and management of risk and reliability in urban water supply systems. This new conceptual framework draws on developments in the electricity supply industry for management of reliability and uses economic principles rather than engineering standards to define the optimal reliability for a system. The application of these economic principles involves use of "total societal cost", as defined by the sum of the cost to the utility of supplying the water at specified levels of reliability of service and the cost to the customers of that level of service, as the means of identifying the optimal level of reliability for a system. The new framework also incorporates aspects of the value of service (VOS) and the cost of service (COS) concepts which differentiate between the cost of providing a level of reliability (service) and the value to the consumers of that level of reliability (service).

The new framework, which is able to address a range of reliability issues arising from the hydrologic uncertainty associated with droughts, deterioration and mechanical failure of the distribution system, and temporal variation in the demands on the system has the potential to result in more economical and socially acceptable solutions to determination of appropriate (optimal) levels of reliability in urban water supply systems.

## References

Billinton, B., and Lakhanpal, D. (1996) Impacts of demand side management on reliability cost/reliability worth analysis, *IEE Proceedings, Generation Transmission and Distribution,* 143(3), 225 - 231.

Bouchart, F. J-C. (1996) *Incorporating Risk Attitudes in an Irrigation Reservoir Management Model,* Ph.D Thesis, Central Queensland University, Rockhampton, Australia, 247pp.

Bouchart, F.J-C. and Goulter, I. (1998) New developments in the consideration of risk in water resources management, *Proceedings of Second International Conference on Environmental Management,* 10 - 13 February, 1998, Wollongong, Australia.

Briassoulis, H. (1994) Effectiveness of water-conservation measure in Greater Athens area, *Journal of the American Water Works Association, 12016* 764 - 778.

Burns, S., and Gross G. (1990) Value of service reliability, *IEEE Transactions on Power Systems,* 5(3), 825-830.

Cameron, T. A., and Wright, M.B. (1990) Determinant of household water conservation retrolit activity - a discrete choice model using survey data, *Water Resources Research,* 26(2), 179-188.

Cullinane, M., Lansey, K., and Mays, L. (1992) Optimisation - availability based design of water distribution networks, *Journal of Hydraulic Engineering, ASCE,* 118(3), 420-441.

Festinger, L., (1957) *A Theory of Cognitive Dissonance,* Stanford University Press, Palo Alto California, 291pp.

Goulter, I. C. (1995) Analytical and simulation models for reliability analysis in water distribution systems, In E. Cabrera and A. Vela (eds), *Improving Efficiency and Reliability in Water Distribution Systems,* Kluwer Academic Publishers, Dordrecht, 235 -266.

Hambly, E.C. and Hambly, E. A. (1994) Risk evaluation and realism, *Proceedings of the Institution of Civil Engineers, Civil Engineering,* 102, 64-71.

Nieswiadomy, M.L., (1992) Estimating urban residential water demand-effects of price structure, conservation, and education, *Water Resources Research,* 28(3), 604-615.

Russo, J.E., Medvec, V.H., and Meloy, M. G. (1996) The distortion of information during decisions, *Organisational Behaviour and Human Decision Processes,* 66 (1), 102-110.

Sanghvi, A.P. (1990) Measurement and application of customer interruption costs/value of service for cost-benefit reliability evaluation: some commonly raised issues, *IEEE Transactions on Power Systems,* 5(4), 1333-1342.

Weber, J. A., (1993) Integrating conservation targets into water demand projections, *Journal of the American Water Works Association,* 85, 63-70.

Wagner, J., Shamir, U., and Marks, D. (1986) Water distribution reliability: simulation methods, *Journal of Water Resources Planning and Management, ASCE,* 114(3), 276-293.

Wilchfort, O. and Lund, J. R. (1997) Shortage management modelling for urban water supply systems. *Journal of Water Resources Planning and Management, ASCE,* 123(4), 250 – 258.

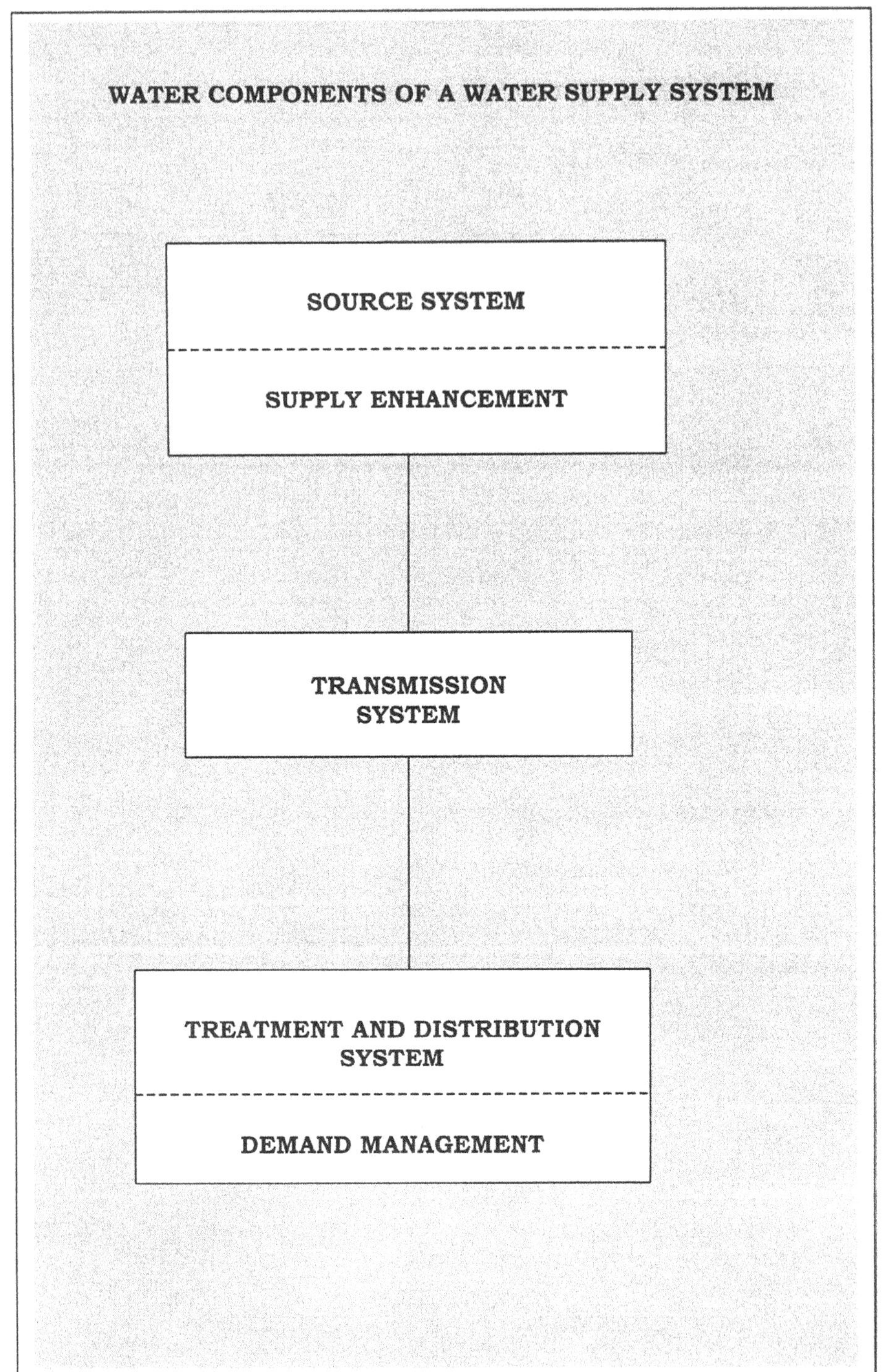


**Figure 1** Schematic of the major sub-systems of a water supply system

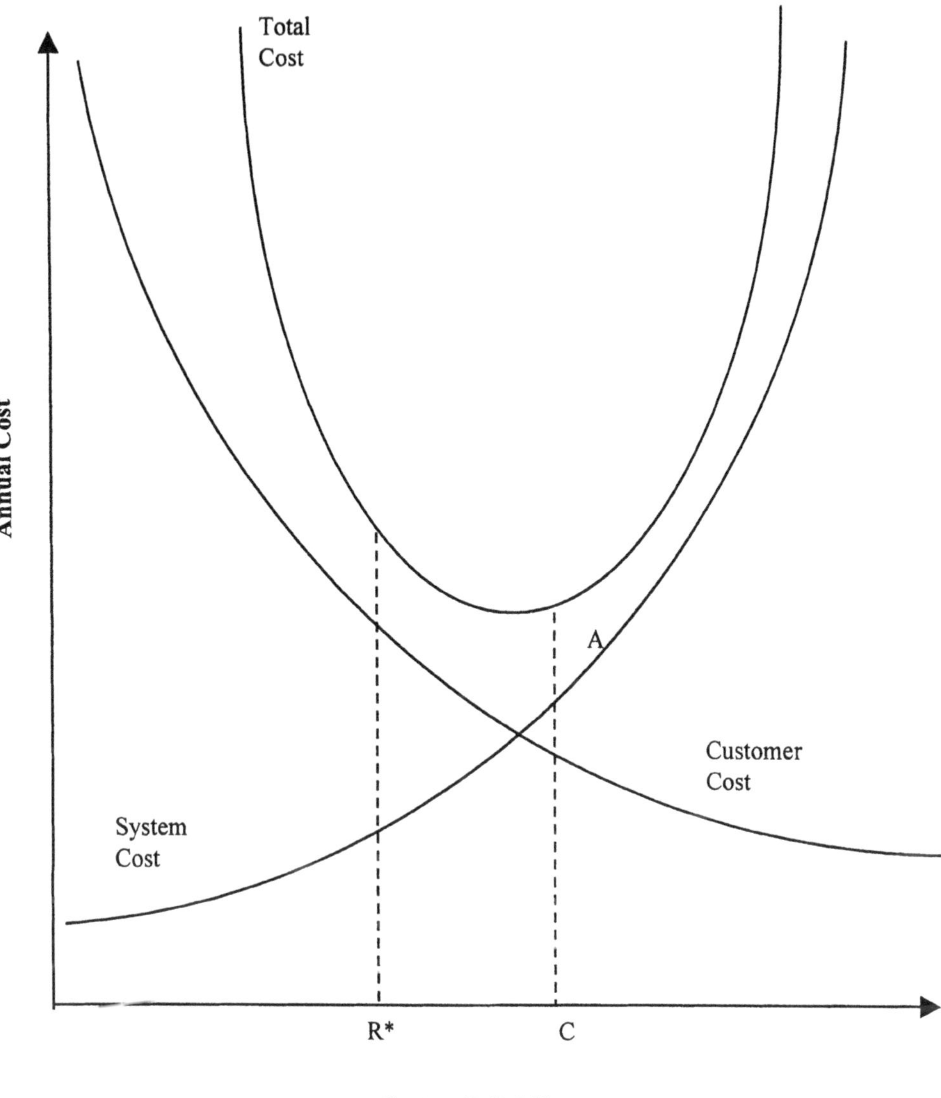


**Figure 2** Optimal reliability level from reliability cost/reliability worth approach
Source: Billinton and Lakhanpal (1996)

# PERFORMANCE INDICATORS FOR WATER SUPPLY SYSTEMS

***Current trends and on-going projects***

HELENA ALEGRE
*Leader of the IWSA Task Force on "Performance Indicators", Operations and Maintenance Committee; Senior Research Officer at the National Civil Engineering Laboratory, Av. do Brasil, 101, 1799 Lisboa Codex, Portugal; tel.: + 351 1 848 21 31; fax.: + 351 1 847 86 14; e-mail: halegre@LNEC.pt*

## 1. Introduction

*Performance indicators* are measures of the *efficiency* and *effectiveness* of the water utilities with regard to specific aspects of the utility's activity and of the system's behaviour. *Efficiency* is a measure of the extent to which the resources of a water utility are utilised optimally to produce the service, while *effectiveness* is a measure of the extend to which the targeted objectives (specifically and realistically defined) are achieved. Each performance indicator expresses the level of actual performance achieved in a certain area and during a given period of time, allowing for a clear cut comparison with targeted objectives and simplifying an otherwise complex analysis. They can be applied at different levels. Within the water utility, the use of performance indicators is growing in importance as the basis for assessing the efficiency (ratio between input consumed and output achieved) and effectiveness of the utility as a whole and of its management units. Such measures can be used to trace the trend in the performance of a given unit over a period of time as well as to make performance comparisons between similar units. Within the country, performance indicators are applicable at a regional or at a national level. Within the supra-national regions, performance indicators allow for comparing different countries of the same or of different regions of the world.

The objective of this paper is two-fold: to provide a synthetic overview of the international trends regarding the development and use of performance indicators for water supply systems and to present the on-going work developed in the scope of the International Water Services Association (IWSA). It summarises the author's views and experience about the matter, and is partially based on Alegre (1997), Alegre & Baptista (1997) and Alegre *et al.* (1997).

In the early 90's, the International Water Supply Association (IWSA) selected the topic "Performance Indicators" for one of its world congresses. No abstracts were submitted on this topic and it had to be cancelled. The subject did not seem to raise much interest. However, just three or four years later, the response to an inquiry held



*E. Cabrera and J. García-Serra (eds.), Drought Management Planning in Water Supply Systems,* 148–178.

in the scope of IWSA to about 150 senior members of water utilities from all over the world clearly showed that performance indicators and unaccounted-for water are by far the two topics of greatest interest in the scope of the water transmission and distribution systems. Such rapid evolution deserves some thought.

Independently from their nature (private, public or combined) and geographical extent, all water utilities comply with a managing logic whose main philosophy may be stated as follows: *greater satisfaction of a greater number of consumers and concerned entities, with the best use of the available resources* (Faria and Alegre, 1996). In terms of the utilities, this is equivalent to: *greater efficiency and effectiveness of the management.* As Peter Allison comments on a recent editorial in the journal *World Water*, under the title "Keeping the customer happy" (Allison, 1997), "*Efficiency seems to be the buzzword of the moment in the water industry. Utilities and municipalities are casting anxious eyes over excessive personnel costs, deficient metering and huge water waste while foreign concessionaires - lured by the prospects of high returns on their investments - lurk menacingly, poised with pen in hand, to sign contracts with the public sector*".

Conversely, due to the monopolistic nature of the water supply services, there is a need to create artificial competition mechanisms and promote *effectiveness.* The implementation of standardised procedures to collect information, assess well devised performance indicators and report the results is being more and more broadly recognised as a key step for improving the quality of service provided to the consumers.

However, the need to improve the efficiency and effectiveness is not new and does not explain by itself the current interest in the assessment of performance indicators. A very brief analysis of some key events in the world can help to understand this trend.

## 2. User-dependent motivations to use PI

Performance indicators may have various objectives, depending on who the end-users are. For the *water utilities,* the assessment of performance indicators brings new management perspectives:

- considering that decision-making processes are based on the available information, the use of reliable performance indicators allows for better quality and more timely response from the water utility managers; in parallel, performance indicators allow for an easier monitoring of the effects of management decisions;
- performance indicators provide key information to the water utility, allowing for a reinforcement of a pro-active approach to management, as opposed to the more traditional reactive approach, usually relying on apparent system malfunctions;
- as performance indicators clearly bring to light the existing strengths and weaknesses of water utility departments, they are an incentive for the adoption of corrective measures, such as the re-allocation of human resources in order to improve productivity and modernise traditional procedures and routines;

- when water utilities are interested in implementing a Total Quality Management approach, performance indicators may have a relevant role to play, as a way of emphasising all-round quality and efficiency throughout the organisation;
- performance indicators allow the water utilities to implement benchmarking routines, either internally, for comparing the performance at different geographical areas, or externally, for comparing themselves with others; this type of practice creates a healthy competition that naturally promotes the improvement of performance;
- performance indicators facilitate auditing, as they constitute a good technical *language* for the auditing team to understand the company's workings and conversely express their recommendations on the financial, administrative and operational areas.

For the national or regional *policy-making entities*:

- when a common set of performance indicators is adopted within a region or a country, policy-makers can easily get a global and comparative view of the performance of the water utilities, which allows for easy identification of relevant weaknesses and, therefore, promotion of corrective measures; examples may be the allocation of funds, both in normal and in crisis situations, or the improvement of regulations;
- performance indicators can be used to support the formulation of regional of national policies for the water sector, within the integrated management of water resources; this is particularly important in the scopes of planning water resource allocations, investments, and the need for new regulating tools.

For the *regulators*:

- considering this sector is a monopoly by nature and has a very important impact on public health and the well being of populations, the eventual opening of the water industry to the private initiative requires a special care with regard to its regulation; in this case, performance indicators may have a key role as monitoring tools of the activity of the private entities, detecting deviations as compared to the contracted goals.

For the *financial agents*:

- performance indicators are important for assessing investment priorities, project selection and subsequent investment follow-up.

For the *users*:

- as performance indicators translate complex processes into simple-to-understand information, they are *the* adequate means of transmitting a measure of the quality of service provided to the users.

## 3. Conceptual framework model for calculating performance indicators

A conceptual framework model for calculating performance indicators can be conceived as a layered pyramid structure, to be defined from bottom to top in successive layers of increasing aggregation. The bottom layer corresponds to the raw data that will feed the performance indicators system. The second layer contains the basic performance indicators, in as much detail as the water utility's departments need to support their decisions. The top layer contains the most synthetic indicators that the system in use expects to deploy. There can be a varying number of layers between the second and the top layer, corresponding to the aggregation of the basic performance indicators according to end-users' needs.

The number and level of indicators managed by each hierarchical level in the organisation should decrease as this level increases. At one extreme, policy-makers, regulatory agencies and general managers would probably be satisfied with the information on the overall level in a utility and with a periodical trend analysis of the major indicators. At the operational level, the information should be more disaggregated to serve the different units in the organisation, in the context of their specific functions.

The application of performance indicators is already a successfully adopted practice by some water utilities, or at least by some of their departments. In these cases the selection of indicators depends only on their specific needs, and on technical and financial constraints. However, to broaden the application to more than one utility, a common framework is required. It is not feasible to compare performance which is assessed in different ways, and therefore may have different meanings.

## 4. Key requirements of performance indicators

A global system of performance indicators must comply with the following requirements:

- to represent all the relevant aspects of the water utility performance, allowing for a global representation of the system by a reduced number of indicators;
- to be suitable for representing those aspects in a true and unbiased way;
- to be clearly defined, with a concise meaning and a unique interpretation for each indicator;
- to include only non-overlapping performance indicators;
- to require only measuring equipment that targeted utilities can afford; the requirement of sophisticated and expensive equipment should be avoided;
- to be verifiable, which is specially important when the performance indicators are to be used by regulating entities that may need to check the results reported;
- to be easy to understand, even by non-specialists – namely, by consumers;
- to refer to a certain period of time (one year is the basic assessment period of time recommended, although in some cases other periods are appropriate);
- to refer to a well limited geographical area;
- to be applicable to utilities with different characteristics and stages of development;
- to be as few as possible, avoiding the inclusion of non-essential aspects.

## 5. Recent historical evolution

### 5.1 PUBLIC / PRIVATE WATER SERVICES - WORLD TRENDS

It is not the aim of this paper to discuss the comparative advantages and disadvantages of the public and private water services. The objective is to discuss the reasons why the use of performance indicators is becoming more and more important nowadays.

Most water services in the world are public. The main reason for this is the nature of the service: human life requires the consumption of water, and the health of the population is greatly affected by the quality and availability of this water. It is therefore a public service. However, many countries came to the conclusion that the management of public companies is not efficient, as these companies are often under-funded, and there is a need to change the situation. Enhancing the private sector participation was the solution pointed out in many of these cases. An extreme situation of full privatisation occurred in England and Wales. Although there are no parallels to this extreme case, examples of this trend can be found in many countries all over the world. The increasing number of systems run by French companies in Europe, America, the Far East and Africa is a clear demonstration of this trend.

The requirement to report systematically the performance achieved by means of a framework of performance indicators is not a general practice in most existing cases. The most relevant exception is what happens in England and Wales, where the utilities must assess and publish their levels-of-service. The contracts focus mainly on financial aspects and on the physical conditions of the assets, and the private companies are subject to the same legislation as the public companies in terms of service requirements. This option goes in line with the tradition in France, where private companies have for many years been responsible for the management of important water supply systems, in coexistence with public companies. Only very recently these companies started to demonstrate some interest in the use of performance indicators. However, it has to be taken into account that the French privatisation process was very slow, and there was time enough to stabilise the procedures. Nowadays, these processes tend to be much more rapid, there are normally several companies competing when there is a call for tenders for the management of water and sewerage systems, and there is a consensus of the benefits arising from clear rules and objectives. This is another very important reason that justifies the increasing interest for performance indicators.

### 5.2 OBJECTIVE-ORIENTED MANAGEMENT AND BENCHMARKING

Another factor that has greatly influenced this trend is closely related with the evolution of management procedures in most industrial sectors. Management techniques have changed all over the world, and nowadays there seems to be a general agreement that the implementation of objective-oriented management procedures is an indispensable step for the success of most companies. This approach requires: the establishment of clear objectives to be achieved within given deadlines; the comparison between targets and results; and the correction of the causes for the deviations, so that the company's efficiency may improve. Performance indicators are a rather powerful tool in this context, as they allow for clear and quantified comparative measures. Some companies have realised that if they compare themselves to the best ones, and correctly identify the reasons for any discrepancies achieved, they can improve their performance significantly. This is how benchmarking appeared and was successful used in many industrial sectors (for example, photocopying companies, car manufacturers). Benchmarking is starting to become popular in the water industry as well, and it is obvious that the comparison between different companies requires the use of standardised performance indicators.

## 6. An overview of international experiences

### 6.1 LESSONS ARISING FROM THE UK PRIVATISATION PROCESS

It was mentioned earlier that the privatisation process held in England and Wales is a rather special one. It has advantages and disadvantages; it has supporters, but also detractors. One of the criticisms often pointed out to the system is the fact that the British companies spend enormous amounts of money in publicity campaigns to improve their external image and keep the customer happy, instead of re-enforcing the investment for improving their physical assets. However, it has to be recognised that the history of water services in England and Wales since the 1970's has had a major impact in the water industry in the world. For this reason, its analysis is fundamental for any country interested in enhancing the private sector partnership.

In 1970 the British government decided to merge several hundreds of medium or small size water and wastewater companies into 10 Water Authorities. The scale effect thus achieved allowed the British water industry to undergo a tremendous technical evolution. The new bodies could afford the human expertise and equipment that the previous companies could not reach due to their small size. During this phase many modern management procedures were implemented, and the efficiency and effectiveness achieved were assessed by the use of performance indicators.

However, many of the existing systems were rather old, heavy investments were required, and the Water Authorities could not solve the difficulties without external financial support. It was within this context that the government decided to privatise all water and wastewater services in England and Wales in 1990. Regardless of its advantages and disadvantages, this was an unparalleled initiative world-wide, at least to the author's knowledge. It was a major challenge, particularly due to the monopolistic nature of the industry. There had been some unsuccessful experiences with private companies in the English water sector, and the British government, under criticism from many sides, tried to protect itself by creating a legal framework that would ensure an efficient control of the situation (the 1990 Water Act is the key legal instrument of that time). Before the private players started to perform, all the "rules of the game" were defined. New regulatory bodies were created, and the new water service companies had to report regularly the levels-of-service achieved to OFWAT (Office of the Water Services). Therefore, a clear and coherent way to assess those levels-of-service had to be defined. The previous work already developed by some water authorities inspired the new assessment system. Three main alternatives were available:

- to allow the private companies to act just as the previous public authorities, subject only to the existing legislation;
- to create mechanisms to control the activity developed inside the company;
- to create mechanisms to control the output of the company.

The first option, adopted by other countries in privatisation processes, was not acceptable due to the reasons previously presented. The second alternative has some major disadvantages:

- the private companies do not accept easily a continuous interference on their activity, and therefore would not support this option;
- it is a mistake to support any control mechanisms on data that cannot be easily verifiable; in order to control all the internal procedures in a reliable way, the regulators would need to have huge teams, and the system would hardly be effective.

The third alternative, selected by the British government, has some important advantages:

- the companies can freely keep their "business secrets", provided that the service actually delivered is good enough;
- the consumers, together with the media, help to control the whole process, as actual managing partners;
- the number of variables to monitor and control is much smaller and these variables are verifiable by any audit, whenever appropriate.

The decision to limit the reporting requirements to the output indicators, in terms of the actual service delivered, is one of the key reasons why the British experience is so relevant to other countries. This means that the companies were free to adopt the management procedures they considered most appropriate, provided that the tariffs were fair and the service provided to the consumers was good enough. This apparently simple decision represents a major milestone in the history of privatisation of water supply systems. Nothing similar had been implemented before. This decision is based on the principle that the regulator should focus its attention on the effectiveness of the utility. It is a target of any private agent to improve its efficiency, in order to increase its profits. Therefore, the regulator does not need to care too much about this side of performance, particularly because the physical assets belong to the private managing body. When this is not the case, complementary measures are necessary, such as the need for periodic independent audits of the physical conditions of the assets.

The presentation of the British experience within this context does not mean it is necessarily an example to follow. On the contrary, it is a singular experience, implemented due to a number of circumstances that are unlike to happen in any other country. Therefore, it cannot be directly exported elsewhere. However, there are some lessons that any country should learn from it, particularly if it is going through the process of enhancing the private sector participation in the water industry. In the author's view, the major input of the British experience to the outside world is:

- the implementation of a clear regulatory framework applicable to all the water service companies, that ensures that the key needs of the population are safeguarded;

- the systematic publication of the results achieved with regard to the quality of service provided to the population by each company, contributing to create a new culture of "open-doors" and transparency;
- the implementation of competition mechanisms in a sector of natural monopoly;
- the implementation of procedures for data quality control and the publication of the degree of confidence associated with each result reported;
- the decision of controlling only the output performance, leaving the companies with a reasonable degree of freedom to implement their own management strategies;
- the capacity for focusing the control mechanisms in a relatively small number of performance indicators;
- the recognition of the increasing role and rights of the consumer.

The information published yearly by OFWAT is divided into two main reports: "Report on levels-of-service for the water industry in England and Wales" and "Report on the cost of water delivered and sewage collected". The former receives also inputs from two other regulatory bodies, the Drinking Water Inspectorate, that controls drinking water quality, and the Environment Agency, that enforces environmental quality standards and reports on compliance, particularly with regard to the sewerage companies. This reports focus on the following performance indicators, designated by levels-of-service:

- properties at risk of low pressure;
- properties subject to unplanned supply interruptions of 12 hours or more;
- population subject to flooding incidents;
- properties at risk of flooding (sewerage services);
- billing contacts not responded to within 5 working days;
- written complaints not responded to within 10 working days;
- bills not based on meter reading[1].

## 6.2 THE INITIATIVES OF THE AMERICAN WATER WORKS RESEARCH FOUNDATION

The American Water Works Association Research Foundation promoted two applied research studies that represent relevant contributions for the development of performance indicators:

- Distribution System Performance Evaluation (concluded in 1995);
- Performance Benchmarking for Water Utilities (concluded in 1996).

The objectives of the project on distribution system performance evaluation were:

- to identify and define distribution system performance criteria and measures;
- to develop procedures to evaluate system performance using performance measures;

[1] In England and Wales most domestic water consumption is not metered.

- to develop guidelines for utility managers to evaluate the overall condition of their distribution systems, establish target levels of performance, and identify system improvements needed to achieve these targets levels.

As a result, three distribution system performance criteria are recommended on the basis of customers' needs. *Adequacy* refers to the delivery of an acceptable quantity and quality of water to the customer. *Dependability* measures the ability of the distribution system to consistently deliver an acceptable quantity and quality of water. Finally, *efficiency* reflects how well resources such as water and energy are utilised. The performance measures proposed corresponding to each performance criterion are as follows:

*Adequacy:*
- pressure
- flow
- water quality
- customer complaints
- responsiveness to customer complaints
- customer satisfaction

*Dependability*
- service interruptions
- inoperable valves and hydrants
- main breaks
- water quality violations of extended duration

*Efficiency*
- unaccounted-for water
- pumping efficiency

With regard to the project on performance benchmarking for water utilities, the objectives were:

- to identify those water utilities processes suitable for benchmarking;
- to illustrate the use of metric benchmarks by quantifying a limited number of possible benchmark measures;
- to investigate the availability of key data for *metric benchmarking*;
- to identify data shortfalls and to prepare a program for the collection of additional data;
- to investigate the extent of, and challenges to, benchmarking in the US water industry;
- to illustrate *process benchmarking* by preparing a case study, performed with the active participation of the Philadelphia Water Department.

The authors define *metric benchmarking* as a quantitative comparative assessment that enables utilities to track internal performance over time and to compare this performance against that of similar utilities. Areas of relatively good performance compared to that of the other utilities can be identified, as can those where there is particular room for improvement in performance. In addition, through the comparison process, target levels of performance can be established.

Still adopting the authors' definition, *process benchmarking* involves first identifying specific work procedures to be improved through a step-by-step "process mapping" and then locating external examples of excellence in these process elements for standard setting and possible emulation. This is also known as "Xerox-style" benchmarking after work undertaken in the Xerox Corporation from 1979 onward.

Traditionally, metric analysis is mostly limited to simple ratios that do not explicitly account for more complex "explanatory factors". The report concludes that while ratios may illuminate trends, they are poor measures for inter-utility comparisons. The use of more sophisticated metric analysis based on existing databases resulted in the development of a series of econometric models that allow to overcome these limitations. However, sound, consistent and up-to-date data are critical to any benchmarking activity. At present, data collected by one utility may not match the specifications of the data collected by another. Overcoming this problem requires thorough definition of the data to be collected and, perhaps, incentives to prepare data in a uniform and precise manner.

Process benchmarking, unlike metric benchmarking, provides a tool by which utilities can change the way they work by introducing improvements in efficiency and service. The methodology, when properly applied and supported throughout an organisation, provides utility staff with a full understanding of the current work process and its strengths and weaknesses. Visits to a partner outside one's own utility, specially outside the water industry, are an essential part of the process benchmarking, since these experiences generate a new perspective toward best practices and facilitate the development of innovative solutions to process problems.

This latter AWWARF project is currently having a follow-up that includes the practical application of benchmarking procedures to a good number of North American water utilities.

## 6.3 INITIATIVES FROM FINANCIAL AGENTS

### *6.3.1 Water and Sanitation Division of the World Bank*

It was referred previously that performance indicators are a powerful tool for financial agencies, as a means for assessing investment priorities, project selection and subsequent investment follow-up. In fact, agencies such as the World Bank and the Asian Bank seem to be fully aware of this fact and started the development and use of performance indicators long ago.

Yepes and Dianderas (1996) is a World Bank publication focusing on water and wastewater services indicators. The structure adopted is as follows:

| | **OPERATIONAL INDICATORS** |
|---|---|
| Water consumption | unit consumption (based on metered consumption) |
| | water consumption and metering (relationship between % of metering and per capita consumption) |
| | distribution of water consumption (as function of the number of connections) |
| | water consumption by main user category |
| | ratio of peak day to average day |
| | water price and income elasticities |
| | short run water price elasticity for domestic users |
| | water price elasticity for industrial users |

| | **OPERATIONAL INDICATORS** *(cont.)* |
|---|---|
| Water distribution system | length of water piped systems |
| | storage volume |

| | |
|---|---|
| | pipe breaks<br>pipe breaks as a function of pipe material |
| Unaccounted-for water | total water losses (defined as UFW)<br>composition of UFW<br>UFW effective reduction programs<br>sustainability of UFW reduction programs |
| Wastewater collection systems | length of sewer systems<br>infiltration flows in sewer systems<br>wastewater treatment<br>typical composition of untreated municipal wastewater<br>typical constituent removal efficiencies for primary and secondary treatment<br>removal of micro-organisms |
| Personnel | number of staff<br>staff composition<br>training effort |
| Miscellaneous indicators | vehicles/1000 water connections<br>meter reading<br>meter maintenance & replacement practices |
| **FINANCIAL INDICATORS** | |
| Efficiency | working ratio<br>operating ratio<br>accounts receivable/collection period<br>percentage contribution to investment |
| Leverage indicators | debt service coverage ratio<br>debt equity ratio |
| Liquidity indicator | current ratio |
| Profitability indicators | return net fixed assets<br>return on equity |
| Operational ratios | personnel (subdivided into personnel costs and staff productivity index)<br>composition of operational costs<br>unit operational cost |
| **TARIFF STRUCTURE** | |
| Tariff structure | |
| Domestic tariff | |
| Average charges and average incremental cost | |
| Rate discrimination by consumer group | |
| Water billing, consumption and users | |

*6.3.2 The Water Utility Partnership for Capacity Building in Africa*

In many African countries, the water and sewerage infrastructures suffered a significant degradation during the post-independence period. Water is often provided at low cost or free of charge, with little or no revenue for the government water services. The water is frequently highly subsidised by the government, although not sufficiently to allow for an adequate maintenance and rehabilitation of the systems, and the quality of the service has gradually decreased.

It is in this context that some African governments and financing agencies such as the World Bank seem to agree that the management procedures must change significantly. It is therefore natural that one of the six priority projects of the Water Utility Partnership for Capacity Building Program 'W.U.P', an organisation sponsored by the World Bank, is on "Performance Indicators for African Utilities". The objective of this project (Djerrari, 1997) is to develop performance appraisal tools for use by African water and sanitation utilities' managers to evaluate their performance level and to launch, as early as possible, necessary corrective measures to avoid failure in achieving their objectives. This would lead to the increase of efficiency and performance within all water and sanitation utilities.

This project has three main objectives :

- to develop a practical methodology for the elaboration of performance indicators to be used by all the utilities and/or sanitation organisations in Africa;
- to promote the use of the proposed system as a management tool to improve operational efficiency of African water utilities and/or sanitation organisations, and to encourage them to carry out regular performance evaluation of their operations;
- to compile and update a performance data base on African water and/or sanitation organisations and make it available in form of a « digest ».

This project will try to define some specific indicators capable to evaluate a utility. It is within WUP and the World Bank plans that the development of this projects proceeds in the same track as the work being developed by the International Water Services Association.

Several workshops have already been organised by the program. The most recent one was held in Uganda (Iliyas, 1997, Alegre, 1997 and Hirner, 1997) and brought together utility representatives from about 20 African countries.

So far, an agreement has been reached on the main definitions of the data and a questionnaire has been developed and was sent to all African utilities. Some data start to be available.

### *6.3.3 The Asian Development Bank*

A recent publication of the Asian Development Bank (McIntosh & Iñiguez, 1997) compares in a well structured way the performance of 50 utilities of the Asian and Pacific Region. It is divided in three parts: Part I: Sector profile; Part II: Regional profiles; and Part III: Water utility and city profiles.

A multidimensional analysis procedure is developed: each indicator is analysed individually for the set of utilities, in terms of trend analysis, and each company/city is presented and analysed through the whole range of indicators. Information includes both explanatory information and indicators. The indicators used to summarise the results achieved are:

| | | | |
|---|---|---|---|
| Private sector participation | (*short description*) | Metering | (%) |
| | | Operating ratio | (-) |
| Production/population | ($m^3/d/c$) | Staff/1000 connections ratio | (-) |
| Coverage | (%) | Management salary | (*US$*) |
| Water availability | (*hours*) | New connection | (*US$*) |
| Consumption | (*l/c/d*) | Accounts receivable | (*months*) |
| Unaccounted-for water | (%) | Grant financing | (%) |
| Non-revenue water | (%) | Commercial financing | (%) |
| Average tariff | ($US\$/m^3$) | Local bond financing | (%) |
| Water bill | (*US$/month*) | Capital expenditure / connection | (*US$*) |
| Power/water bill ratio | (-) | Annual report | (*none, type script or glossy covered report*) |
| Public taps | (*yes/no*) | | |

## 6.4 THE DUTCH CONTACT CLUB FOR WATER COMPANIES AND THE 6-CITIES GROUP OF THE NORDIC COUNTRIES

A rather interesting example can be found in the Netherlands (van der Willigan, 1997). In 1992, the 13 larger Dutch water companies created a "contact club for water companies" and a common standard to be used for benchmarking has been introduced. These companies agreed to define indicators to assess their own performance and compare the results achieved once a year. There are no goals set or yearly targets. According van der Willigen (1997), the following main categories of performance indicators are being considered:

- *general*: includes connections, personnel, length of main pipes and connections;
- *production*: include internal production, external production, delivery external, delivery to distribution, unaccounted-for water, net sales, sales in volume, sales per connection;
- *costs*: include production, distribution, sales, general, total, income and result and is broken down into costs total, per $m^3$ sales and per connection;
- *personnel*: include connection per person, production, distribution, sales, general, salary per person, absenteeism, short term absenteeism and long term absenteeism.

A similar initiative is under way in the Nordic countries (Adamsson, 1997), involving in a first stage a group of six cities: Stockholm, Gothenburg, Malmö, Copenhagen, Oslo and Helsinki. The aim of the project is to develop and apply performance indicators, which should make comparisons between the cities possible. Parallel to the work of the 6-Cities Group VAV, the Swedish Water & Wastewater Association started a preliminary study with the aim to survey the key figure area and to propose a number of key management figures focusing on quality and costs. The first report regarding a group of 25 water services is due to be published before the end of 1997.

## 6.5 PORTUGUESE STUDIES

### *6.5.1 Context*

In Portugal, a significant positive evolution can be observed over the last twenty years with regard to the level of supply coverage. However, the quality of service provided to consumers is still not good enough in many areas of the country, and its improvement is step by step being recognised as a fundamental issue, particularly with regard to continuity of supply, water quality and reliability.

Significant capital investments have recently been and are expected to be carried out in Portugal during the forthcoming years in order to improve the current levels-of-service, with the emphasis on a correct establishment of priorities and development of a new culture towards management efficiency and consumer satisfaction. A wide variety of measures of national scope is clearly required, including the definition of a coherent framework for sustainable development in the forthcoming decades.

A number of initiatives aiming at contributing to that goal can currently be recognised, such as the publication of new legislation, the implementation of National and European funding schemes, and development of applied research focusing on the tools to support the technical management of water utilities (Alegre, 1994, Baptista, 1983, 1994, Alegre & Coelho, 1990, 1992, and CSOPT, 1991).

### *6.5.2 Technical and social-economical indicators in the scope of water supply, wastewater and solid waste systems*

In 1993, the National Civil Engineering Laboratory held a study for the Directorate-General for the Environment on technical and social-economical indicators in the scope of water supply, wastewater and solid waste systems (Matos *et al.*, 1993). Its objective was to develop a tool to support the analysis of proposals and the monitoring of the results of waterworks funded by European funding schemes. A hierarchical structure of performance indicators is proposed, including four main groups of indicators. For each group a varying number of indicators are defined and a sub-set of key indicators are selected as the ones to be used in the upper layer of the performance indicators structure:

- *demographic*: regard the population to supply, its geographical distribution and recent grades of evolution, and include aspects such as population density, size of villages and towns, and population growth (natural and due to migration movements);
- *technological*: deal with the physical and operational conditions of the systems, and include coverage of population supplied, availability of water resources, condition of physical assets (percentage of protected water impounding, unaccounted-for water, storage capacity, length of transmission and distribution pipes, population served per system and per water utility) and degree of ageing of the systems;
- *efficiency*: measure the consumers' satisfaction, including aspects of water quality (number of samples analysed and results achieved as compared with the legal requirements), continuity of supply and water pressure and consumer complaints (technical and accounting);

- *economical and financial*: include capital investment per capita and as a percentage of the GNP, operations and maintenance costs, financial inputs of the utility from the consumers' bill.

*6.5.3 Framework for the assessment of levels-of-service*

Two independent initiatives regarding the assessment of levels-of-service were held in 1994 and 1995 and were finalised during the first semester of 1995. One of the projects was promoted by the Portuguese Association of Water Resources (APRH, 1994). The other project was contracted by the Portuguese Government to the National Civil Engineering Laboratory (Alegre & Almeida, 1994). It is part of a more general project entitled *"Tools to Support a Policy of Sustainable Development in Basic Sanitation[2] Systems"*. The two teams worked in close contact and co-ordination (Faria and Alegre, 1996) and proposed an hierarchical structure of levels-of-service in four main categories:

- *organisation*: reflects the efficiency and effectiveness of the interface management between the utility and the society and includes coverage of the population by the service, attendance to the public (complaints, information) and health, safety and professional qualification of the workers;
- *engineering*: reflects the efficiency and effectiveness of technological resources use and includes reliability of service delivery and quality of drinking-water supplied to consumers;
- *environmental*: reflects the efficiency and effectiveness of water resources use and includes environmental impact and inter-active effects in water resources, both quantitative and qualitative;
- *capital*: reflects the efficiency and effectiveness of financial resources use and includes the acceptance of charges by the public.

This set is aggregated to give rise to a global, top layer indicator, and is sub-divided into the indicators that form the third layer, treated as a pivotal one, with its eight basic levels-of-service as follows:

---

[2] Basic sanitation is an expression adopted in Portugal and Brazil to include the domains that more closely affect public health: drinking water supply, urban wastewater and urban solid waste.

| | |
|---|---|
| NU - | Attendance to the public (complaints, information) |
| NF - | Health, safety and professional qualification of the workers |
| NC - | Coverage of the population by the service |
| NR - | Reliability of service delivery |
| NQ - | Quality of drinking-water supplied to consumers |
| NH - | Inter-active effects in water resources, both quantitative and qualitative |
| NA - | Environmental impact |
| NT - | Acceptance of charges by the public |

These eight levels-of-service were split into auxiliary levels as a first step in the wide universe of the normal operation and management activities.

In order to illustrate the method with an application, let us take a partial view by considering the level-of-service *NR - Reliability of service delivery.*

This indicator is divided into three auxiliary levels-of-service as follows:

- NR 1 - Hydraulic performance
- NR 2 - Continuity of delivery
- NR 3 - Repairs in the system

Upwardly, this main level is aggregated with levels *NC - Coverage of the population* and *NQ - Quality of drinking-water* to form capital level *nP - Performance* and subsequently, will contribute to global level G (see Figure 1).

Down the line, the three above mentioned auxiliary levels are broken down into ten calculation levels, as follows:

NR 1.1 - Complaints on pressure deficiencies
2 - Minimum pressure violations
3 - Maximum pressure violations
4 - Pressure fluctuation violations
NR 2.1 - Complaints on interruptions of service delivery
2 - No. of consumers affected by interruptions
NR 3.1 - Complaints on repairs delay
2 - Response time to interruptions
3 - Construction time of the repairs
4 - Cost of the repairs

The assessment of the situation should result into a single value for each of the ten levels-of-service listed. Zooming in on item NR 1, for instance, the variables selected to best quantify the hydraulic performance under assessment are the following:

NR 1.1 - Number of complaints of a percentage of total number of complaints in the NR field

NR 1.2 - Percentage of properties supplied with a mean minimum pressure below the minimum allowable value (dependent on the network area)

NR 1.3 - Percentage of properties supplied with a mean maximum pressure above the maximum allowable value (60 m, according to the Portuguese regulations).

NR 1.4 - Percentage of properties that suffer from mean daily pressure variation above the allowable limit (30 m, according the Portuguese regulations).

The global value of the level-of-service NR 1 is assessed as a weighted average of those values after their conversion into a given classification (in our case, from 0 to 100 points), as follows:

$$\mathbf{R} = w_1\ \mathbf{R1} + w_2\ \mathbf{R2} + w_3\ \mathbf{R3} + w_4\ \mathbf{R4}$$

$w_1$, $w_2$, $w_3$, and $w_4$ being weighting factors;
R, the value of level-of-service NR1;
R1, R2, R3 & R4, the values of levels-of-service NR1.1, NR1.2, NR1.3 & NR1.4.

*6.5.4 An engineering approach for performance assessment*

Coelho (1997A&B) addresses the engineering side of the problem by using a flexible framework based on an array of performance indices. The Performance Evaluation System developed is a technical analysis tool, designed to shift the focus of management and operations of water systems to a wider, more rigorous, performance-oriented view. It is based on a system of penalty curves, flexible enough to accommodate individual sensitivities and interpretations. The methodology allows for a rapid gain in sensitivity to the network's behaviour, providing a standardised means of diagnosis, and is a valuable aid for planning, design, operation and rehabilitation of water distribution systems.

Similarly to the other type of performance indicators, these also contemplate

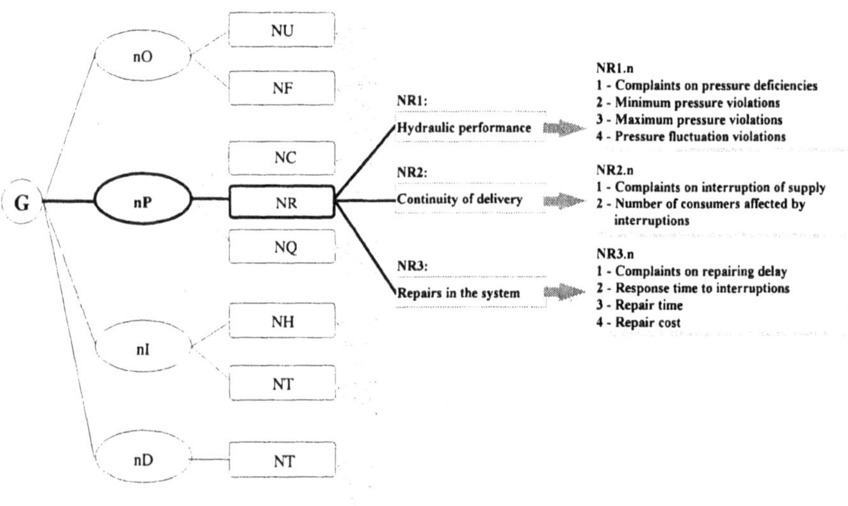


***Figure 1*** *- Partial view of the level-of-service framework*

quantitative indicators based on the analysis, from specific viewpoints, of the network's characteristics and behaviour. However, the former type of performance analysis is mainly based on sound records, such as damage rate or maintenance costs, or on data obtained from systematic or *ad hoc* field surveys. This latter type requires the use of hydraulic and water quality dynamic simulation tools, in order to infer the system's behaviour if a given remedial solution were implemented.

The performance indicators developed by the author are classified under three generic terms, although the methodology is applicable to other viewpoints:

- *hydraulic performance*: focus on pressure head and pressure head variation;
- *water quality performance*: focus on concentrations and travel times;
- *reliability performance*: focus on network topological reliability.

The limit values for everyone of these performance indicators must be defined case by case by the undertaker, according the local characteristics, legal constrains and its management strategy.

The performance evaluation can be accessed on an empirical and non-systematic basis or on a more systematic way. In this latter situation, the framework can be defined by three types of entities, established for each aspect to be analysed:

- A relevant *state variable*, that is, the quantity which translates the network behaviour or properties at the network element level, from the point of view taken into consideration. The network element is the node or the pipe, as conventionally used in network analysis, from which all modelling assumptions are inherited. Examples of such state variables are: nodal head, to check compliance with the pressure requirements; damage rate, to be compared to reference values defined for each type of material and class of diameter; or nodal concentration of a particular constituent as compared to the water quality guidelines.
- A *penalty function*, which scores the values of the state variable against a scale of index values (6 - optimum through 0 - no service). The penalty curves are designed by the user, and translate as much as possible a common-sense grading of performance in the addressed domain.
- Finally, a *generalising function*, used for extending the element-level calculation across the network, producing zonal or network-wide diagnoses. The indices are intended to have both local and network-wide meaning. In the case, for example, of a pressure index, the proposed generalising function is the weighted average across the network. Other types of operator may be used, such as those which focus on maximum or minimum values, etc.

The technique is applied both to extended period operational scenarios and to a range of load factors, displaying appropriate *dispersion bands* (25% stepped percentiles).

The above considerations about the nature of the indices imply a careful selection of the variables that may be eligible for such a treatment. In hydraulic terms, for instance, the variables used to assess system's energy performance are total pumping energy, total energy dissipated, and the difference between the actual potential energy and the minimum potential energy required to supply every node with the minimum allowable pressure.

## 6.6 OTHER STUDIES AND INITIATIVES

Many water utilities are already using their own performance indicators. However, the most interesting experiences to be analysed from the conceptual viewpoint are the ones regarding groups of utilities. Beyond the benchmarking processes previously referred, such as in the Nordic countries and in North America (involving the US and Canada), other are currently starting to be developed and implemented in other areas of the world. In Africa, for instance, a process conducted by Umgeni Water, from South Africa, started recently. Also the World Bank established some months ago a project on benchmarking, under the leadership of Augusta Dianderas.

A study that cannot go without reference is Malaysian Water Association (1996), an excellent document that suggests that the performance indicators are organised under three generic terms, as follows:

- *physical*: give an indication of size and coverage of physical aspect of water supply and extent of the underground assets, and is useful for future planning;
- *service*: focus on consumer services and productivity in an environment where consumers' expectations are increasing; they measure the utility's output against staff resources, assessment of level-of-service such as staff ratios, interruptions to mains supply, quality of water, water losses, and consumer's complaints on an annual basis;
- *financial*: assess the financial, human and physical aspect of efficiency, covering costs and the economic utilisation of assets, manpower, plant and equipment.

Another interesting experience is on-going in Mexico, held by Mario Buenfil, IMTA, Mexican Institute of Water Technology. The objective is the definition of a short number of indicators and of applicability subsets, depending on the size and level of development of the utility.

Also from the scientific point of view the topic of performance indicators has been attracting the attention of researchers in different places of the world. Apart from the studies previously mentioned, it is rewarding to know that several PhD students are developing their research theses on the topic. This is the case of Laetitia Guerin, supervised by professors of Ecole Nationale des Mines de Paris and of Ecole Nationale du Génie Rrural des Eaux et Fôrets, Montpelier (France), of Margareta Lundin, from the Teknisk Miljöplanering, Chalmers Tekniska Högskola, Göteborg (Sweden), and, more recently, of Adriana Cardoso, from LNEC, Lisbon (Portugal).

## 6.7 INITIATIVES OF THE INTERNATIONAL WATER SERVICES ASSOCIATION (IWSA)

### *6.7.1 IWSA motivations*

As the potential of performance indicators has already been recognised and widely used in the scope of various sectors of industry, many IWSA members have expressed the view that their Association should define guidelines on the information to be collected and how it should be presented.

This is currently a major challenge that IWSA is willing to tackle and win. It requires the definition of a common reference of performance indicators, as well as an adequate model of aggregation that fits the basic common needs of the key types of users of performance indicators.

*6.7.2 Lisbon workshop*

One of the first steps was the organisation of some brainstorming sessions in the form of a workshop on "Performance indicators for water transmission and distribution systems". This workshop was organised by the Operations and Maintenance Committee of the IWSA Distribution Division and by the National Civil Engineering Laboratory (LNEC), as the local organiser in Lisbon, Portugal (May 1997). Several assessment models were proposed by different authors, the experience from several countries was presented, and participants had a good opportunity to share their experiences, difficulties, needs and expectations. The following papers were presented and discussed:

- "A general framework of performance indicators in the scope of water supply", by H. Alegre (Secretary of the IWSA Distribution Division), J. Melo Baptista (Chairman of the IWSA Distribution Division), and A. Lobato Faria, from Portugal; this paper presents and discusses a general framework of performance indicators in the scope of water supply, and aimed to be a starting point for the analysis and systematic discussion of the problem within IWSA. It synthesises many of the ideas expressed in the other papers, combined with the authors' own views. The water utility context and challenge are described as a base to justify the need for performance indicators. The concept of performance indicators, the potential uses of performance indicators, the levels of applicability of performance indicators and the key requirements of performance indicators are presented. The paper concludes with a discussion about the structure of performance indicators and the methodology for the development of a framework model.
- "Technical, operational and economic performance indicators of water utilities", by G. Hansen, former leader of the Task Force "Performance indicators" of the IWSA Statistics and Economics Committee, from Luxembourg; the author refers that the objectives a water utility has to fulfil are not only purely economic but, as far as possible also qualitative (drinking water quality), quantitative (quantity of water and pressure limits), reliability (guarantee the service even in the worst conditions) and ecological (protection of natural environment). The conditions *sine qua non* for any performance indicator are to be linked to a phenomenon upon which we have influence, to be easily measurable and to link basic indicators to the performance indicators.
- "Technical, operational and economic performance indicators of water utilities", by W. Hirner, Chairman of the IWSA Operations and Maintenance Committee, from Germany; the author presents his view on the reasons for performance indicators and presents a proposal for its classification, divided in quality of service and operation, supply structure, plant structure, personnel and productivity and costs, to serve as a proposal for further discussion.
- "Information systems: a key factor to manage performance indicators", by F. Cubillo, Chairman of the IWSA Information Systems Committee, from Spain; this paper analyses the feasibility of the use of information technologies in the

different areas related to the management of performance indicators, the advantages to be obtained and the requirements for their implementation. Because the specific characteristics of performance indicators in distribution networks mean that it is impossible to manage them without the aid of information systems, the new technologies applied to the management of water supply systems facilitate the assessment and use of these indicators as a value added to technological innovation. Geographical Information Systems and mathematical models for analysing the distribution network are the basic pillars for carrying this out.

- "Understanding and satisfying customer expectations", by J. W. Oatridge, Chairman of the IWSA Public Relations Committee, from the United Kingdom; the paper highlights the need to measure customer satisfaction levels related mainly with the drinking water quality, the results of which are tracking water utilities improvements.
- "The UK experience of PI from a regulatory and operational viewpoint", by C. Harries, from the United Kingdom; the author explains that due to the monopolistic nature of the business of the English and Welsh water companies, a comprehensive and sophisticated set of indicators have been developed recently. Those indicators form a critical role in company operation and extend to all corners of the company, influencing strategy, investment decisions, comparative efficiency and day to day operation. The indicators are aimed at measuring the company's service to the customer and the promotion of competition. For this reason, this paper focuses, not only on the technical measures, but also the measures relating to specific customer interfaces and the penalties which go with these.
- "The Dutch experience and viewpoints", by F. van der Willigen, from the Netherlands; this paper presents the above mentioned "contact club for water companies" and the author refers that it is believed that the Dutch water utilities can improve their performance by 20-30%.
- "The Swiss experience with PI and special viewpoints on water networks", by B. C. Skarda, from Switzerland; Although in the Swiss water industry no general performance indicators are defined and applied, this paper presents the Swiss experience under seven special viewpoints: unaccounted-for water, water losses in distribution networks, pipe breakage, preventive maintenance, network renewal, experienced pipe materials for communities and water quality aspects. A special view is done over the distribution network, due to the economical importance of this water supply component.
- "The Swedish experience and viewpoints", by J. Adamsson, from Sweden; the paper reports on the previously mentioned co-operative project under way in 6 Nordic cities: Copenhagen, Oslo, Helsinki, Stockholm, Gothenburg and Malmö.
- "Water distribution system performance assessment", by A. Deb and L. Cesario, from the USA; the paper refers to the project funded by the AWWA Research Foundation on performance of distribution systems.
- "Monitoring indicators for project design and implementation", by G. Yeppes and A. Dianderas, from the World Bank; the authors refer that comparison of

unaccounted-for water among utilities needs to be approached with care as there is no universally accepted definition, and there can be significant cost and benefit differentials. Therefore, the evolution of unaccounted-for water over time is often more important than the absolute level at one point in time. Monitoring indicators are important tools to understand the composition of unaccounted-for water and to help in the design and follow up abatement programs.

- "Managing benefits of mains rehabilitation through structured surveys", by D. P. Parsons, from the United Kingdom; this paper tracks the use of performance indicators from structured surveys in planning the rehabilitation of a large asset base of distribution mains. Through assessment of historical data and a structured programme of water quality sampling, use of automated sample filtration equipment and customer questionnaires the pattern of problems is determined in the pre-survey which enables the most suitable rehabilitation techniques to be selected. In the post-survey the same sampling sites are revisited and a statistical assessment of the results is made.
- "Performance analysis in water distribution", by S. T. Coelho, from Portugal; this papers summarises the methodology proposed by the author and previously mentioned.

The main workshop conclusions can be summarised as follows:

- it is worth proceeding with the task of trying to develop a common framework for performance indicators;
- IWSA can have a important role promoting this task;
- a hierarchical structure of performance indicators is more appropriate to fit a great variety of circumstances;
- it should be possible to develop general structures for different degrees of development of water supply utilities;
- comparisons between utilities should be based on the whole range of performance indicators, including operational and quality of supply levels.

This workshop, while presenting and discussing some key aspects that have to be taken into account in the definition of a general framework of performance indicators in the scope of water supply, aimed only to be a starting point for the analysis and systematic discussion of the problem within IWSA. New actions are needed to achieve a useful tool for the utilities.

### *6.7.3 Follow-up of the Lisbon workshop*

To meet this target, it was decided to create a new Task Force on Performance Indicators under the umbrella of the Operations and Maintenance Committee of the Distribution Division of IWSA, with the following goals:

- preparation of an IWSA draft proposal to circulate among members and to be improved according to the comments received (1997/98);
- organisation of a workshop to allow for a broader discussion of this proposal (1998);
- edition of the workshop post-prints, beginning with Volume I - "Introduction to performance indicators" of a new IWSA series on performance indicators (targeted for 1997);

- edition of Volume II - "Proposed performance indicators: structure, concepts and definitions, and guidelines for application" (targeted for 1998);

It is expected that further volumes continue to be edited beyond the end of the Task Force, starting with Volume III - "Application of performance indicators" (targeted for 1999-2000); the contents should be statistics resulting from the practical application of the performance indicators proposed in Volume II.

This work is to be developed in close contact with three other IWSA task forces:

- Task Force "Performance indicators", Statistics and Economy Committee (Management Division), formerly lead by H. Hanssens, from Luxembourg, and currently lead by Renato Parena, from Italy;
- Task Force "Benchmarking, Management", Organisation, Training and Education Committee (Management Division), created in September 1997 and lead by R. S. Hodge, from the United Kingdom;
- Task Force "Leakage management terminology and prefered performance indicators for international use", Operations and Maintenance Committee (Distribution Division), lead by Allan Lambert, from the United Kingdom;
- Task Force "Principles for operation of transmission and distribution systems", Operations and Maintenance Committee (Distribution Division), leaded by Michael Buckler, from Germany.

In the end, the IWSA expects that a robust and well devised system of performance indicators emerges, able to attract water utilities to use it as a normal tool in their activities.

It is important to stress that the IWSA proposal aims to cover the basic needs of different types of users, with special emphasis for the utilities themselves. The results should be applicable to utilities with different levels of development, and different climatic, demographic and cultural characteristics.

## 7. Preliminary framework of performance indicators

### 7.1 STRUCTURE

The water supply systems are formed by a set of physical infrastructures planned, built, operated and maintained by the water utility. The use of input resources (environmental resources, human resources, technological resources and financial resources) is required in order to supply a satisfactory service to consumers.

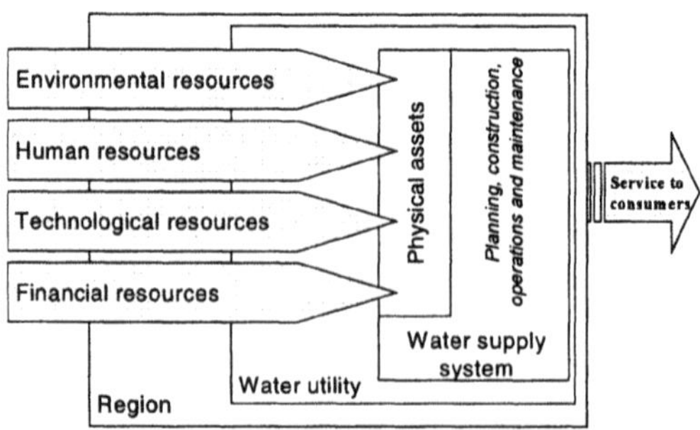

The performance indicators should focus on the *efficiency* of use of resources and of physical assets and on the *effectiveness* achieved in terms of quality of the service provided to consumers. However, the performance indicators cannot be adequately interpreted without taking into account a number of external explanatory characteristics that managers must deal with. These include external characteristics (geographic, demographic, economic, ...) and internal ones regarding the water utility and the system on focus. Therefore, the proposed structure includes a group of explanatory information, and a group of performance indicators, as follows:

| | | |
|---|---|---|
| Explanatory information | i | Utility profile |
| | ii | System profile |
| | iii | Region profile |
| Performance indicators | vi | Water resources indicators |
| | v | Personnel indicators |
| | iv | Technologic indicators |
| | iiv | Physical indicators |
| | [illegible] | Operational indicators |
| | xi | Levels-of-service |
| | x | Financial indicators |

The listing of indicators presented in the following sections aims to be sufficiently detailed to be effective for the water utility departments. It is still a preliminary proposal to be discussed and improved. When the listing is stabilised (early 1998), than a second step consists of selecting the most important ones for each type of utility and/or user, in order to establish a basis for comparison of utilities in a less detailed way. The exact meaning of the terms used for every indicator definition is part of other Task Forces of the Operations and Maintenance Committee of IWSA (see section 0).

## 7.2 EXPLANATORY INFORMATION

The explanatory information considered in the current preliminary version is as follows.

| | UTILITY PROFILE | |
|---|---|---|
| | Water utility identification | (-) |
| | Mission statement | (-) |
| | Type of activity (*) | (-) |
| | Type of organisation (*) | (-) |
| | Private sector participation | (-) |
| | Total water supplied | ($10^6$ $m^3$/year) |
| | Staff (*) | (No.) |
| | Annual operational costs (*) | (US$) |
| | Annual revenue (*) | (US$) |
| | Annual investment (*) | (US$) |
| | **SYSTEM PROFILE** | |
| | System identification | (-) |
| | Mission statement | (-) |
| Production/distribution | Population | (inhabit.) |
| | Population supplied | (inhabit.) |
| | Supply area | ($km^2$) |
| | Protection area | ($km^2$) |
| | Supplied water | ($10^6 m^3$/year) |
| | Maximum daily supply | ($m^3$/day) |
| | Storage volume | ($m^3$) |
| | Length of transmission and distribution mains | (km) |
| | Pumps (*) | (No.) |
| | Pumping power | (Mw) |
| | Service pipes | (No.) |
| | Meters | (No.) |
| | Hydrants | (No.) |
| Service connections | Total number of service connections (*) | (No.) |
| | Persons/service pipe | (No./ service pipe) |
| | Public taps | (No.) |
| | Persons/public tap | (No./ public tap) |

(*) - This indicator has subdivisions.

| | REGION PROFILE | |
|---|---|---|
| Demography | Population density | (inhabit./$km^2$) |
| | Population growth rate (*) | (%) |
| | Population seasonal peak | (-) |
| | Population average income | (US$ per capita/year) |
| | Yearly working time | (hours) |
| | Economic activities(*) | (%) |
| Environment | Rainfall (*) | (mm) |
| | Air temperature (*) | ($^oC$) |
| | Topography | (m) |
| | Raw water quality(*) | (%) |

## 7.3 PERFORMANCE INDICATORS

| | WATER RESOURCES INDICATORS | |
|---|---|---|
| Water balance | Water abstraction (*) | ($m^3$ per capita/year) |
| | Water production | ($m^3$ per capita/year) |
| | Imported water | ($m^3$ per capita/year) |
| | Supplied water | ($m^3$ per capita/year) |
| | Exported water | ($m^3$ per capita/year) |
| | Authorised consumption (*) | ($m^3$ per capita/year) |
| | Water losses | ($m^3$ per capita/year) |
| Per capita consumption and peak factors | Production/population | (l per capita/day) |
| | Total per capita consumption (*) | (l per capita/day) |
| | Water consumption peak factors (*) | (-) |
| | **PERSONNEL INDICATORS** | |
| Number of employees | Employees per volume of water | (No./$10^6 m^3$) |
| | Employees per connection | (No./$10^3$ service pipes) |
| | Employees per pipe length | (No./$10^2$ km) |
| Employees per activity | Administration employees | (No./$10^2$ km) |
| | Commercial and customer service employees | (No./$10^2$ km) |
| | Planning, construction, operations and maintenance employees (*) | (No./$10^2$ km) |
| Qualification of employees | University degree employees | (%) |
| | Skilled workers | (%) |
| | Other workers | (%) |
| Training | Total training (*) | (hours/employee/year) |
| Salaries | Wages of a university degree worker | ($US/year) |
| | Wages of a skilled worker | ($US/year) |
| | Wages of a non-skilled worker | ($US/year) |
| | **TECHNOLOGIC INDICATORS** | |
| | Information systems(*) | (yes/no) |
| | Automation and control(*) | (yes/no) |
| | Mapping(*) | (yes/no) |

(*) - This indicator has subdivisions.

| PHYSICAL INDICATORS | | |
|---|---|---|
| Water intake | Catchment protection | (%) |
| | Abstraction capacity | ($m^3$ per capita/day) |
| | Intake capacity | ($m^3$ per capita/day) |
| Treatment | Treatment capacity | ($m^3$ per capita/day) |
| | Level of treatment (*) | (%) |
| Storage | Reservoir capacity | (days) |
| | Storage tanks capacity | (days) |
| | Private tanks density | (%) |
| Pumping | Pumping capacity (*) | ($W/m^3$) |
| | Fixed speed pump performance (*) | (various) |
| | Fixed speed pump performance (*) | (various) |
| Transmission and distribution | Transmission and distribution capacity | ($m^3$/day) |
| | Network delivery rate | ($m^3$/km/year) |
| | Service connections / network length | (No./km) |
| | Pipe length per capita | (m per capita) |
| | Pipe materials (*) | (%) |
| | Pipe diameters (*) | (%) |
| | Pipe age (*) | (%) |
| | Average pipe age | (years) |
| | Valves density | (No./km) |
| | Hydrant density | (No./km) |
| | Service pipe density | (No./km) |
| | Service pipe materials (*) | (%) |
| | Private pumping systems density | (%) |
| | Public taps density | (No./km) |
| | Meters (*) | (No./service pipe) |
| **OPERATIONAL INDICATORS** | | |
| Routine operations | Energy consumption | (kWh/$m^3$/year) |
| | Staff mobility | (vehicles/km) |
| Inspection and maintenance | Transmission inspection | (%/year) |
| | Pumping inspection | (%/year) |
| | Storage tank cleaning | (%/year) |
| | Network inspection | (%/year) |
| | Leakage detection | (%/year) |
| | Hydrant inspection | (%/year) |
| | Instrumentation calibration (*) | (%/year) |
| | Electrical equipment inspection (*) | (%/year) |
| Water quality monit. | Samples tested (*) | (No. /$10^6$ $m^3$/year) |
| Preventive maintenance | Pipe rehabilitation (*) | (%/year) |
| | Service pipe replacement | (%/year) |
| | Meter replacement | (%/year) |
| | Pumps rehabilitation (*) | (%/year) |
| Water losses | Total water losses | (*) (l/km/year) |
| Failures | Failures and repairs | (No. /km/year) |
| | Pipe failures | (No. /km/year) |
| | Service pipe failures | (No. /service pipe/year) |
| | Power failures | (hours/pumping station/year) |

(*) - This indicator has subdivisions.

| | **OPERATIONAL INDICATORS** *(cont.)* | |
|---|---|---|
| Metering | Domestic meter reading frequency | (No./meter/year) |
| | Domestic reading efficiency | (%) |
| | Meter replacing frequency | (%) |
| | **LEVEL-OF-SERVICE INDICATORS** | |
| Service | Supply coverage (*) | (%) |
| | Pressure of supply adequacy | (%) |
| | Continuity of supply | (hours/day) |
| | Water restrictions | (%) |
| | Water interruptions | (%) |
| | Quality of supplied water | (%) |
| Customer complaints | Service complaints (*) | (complaints/connection/year) |
| | Billing complaints | (complaints/connection/year) |
| | Other complaints | (complaints/connection/year) |
| | Response to written complaints | (days/complaint) |
| | **FINANCIAL INDICATORS** | |
| Efficiency indicators | Working ratio | (-) |
| | Operating ratio | (-) |
| | Accounts receivable/collection period | (months equivalent) |
| | Investment ratio | (%/year) |
| | Percentage contribution to investment | (%) |
| | Billing efficiency | (%) |
| Leverage indicators | Debt service coverage ratio | (%) |
| | Debt equity ratio | (-) |
| Liquidity indicators | Current ratio | (-) |
| Profitability indicators | Return on net fixed assets | (%) |
| | Return on equity | (%) |
| Operational indicators | Unit operational costs | (US$/m$^3$) |
| | *Composition of costs per activity:* | |
| | Running costs (*) | (US$/m$^3$) |
| | Capital costs (*) | (US$/m3) |
| | *Composition of costs per type of cost:* | |
| | Total internal personnel costs | (US$/m$^3$) |
| | Total energy costs | (US$/m$^3$) |
| | Total external operational and maintenance costs | (US$/m$^3$) |
| | Other costs | (US$/m$^3$) |
| | Average water charges | (US$/m$^3$) |
| | Total income (*) | (US$/m$^3$) |

(*) - This indicator has subdivisions.

## 8. Concluding remarks

The systematic assessment and use of performance indicators is a widely disseminated practice in many industrial sectors. Any modern productive industry strives to maximise internal efficiency in order to improve its profits, and monitors its customers' views and attitudes, with regard to the services provided, in order to adapt them to the customers' preferences. Performance indicators are an indispensable tool for those purposes and have for a long time been a central part of management procedures.

The water industry, due to its nature of public service and monopoly, is at an earlier stage of development compared to other industrial sectors, and only recently has begun to realise the potential benefits of this tool.

The implementation of a framework of performance indicators in a given company or group of companies is more important the greater the need for change of management procedures or installed routines; it allows managing decisions to be supported on the analysis of actual results and trends. Parsimony is required in the selection of the indicators to evaluate. It is most advisable that a small number of key indicators is selected in a first stage, and that the degree of complexity is increased slowly and gradually. The "levels-of-service" group should always be graded "top priority" and included in every phase of development, because it translates the actual output of the utility.

Governments may have a relevant role in the promotion of this new tool by associating the allocation of financial incentives to the need for reporting reliable output indicators, or levels-of-service. Experience shows that any company that has to comply with such requirement soon begins implementing its own internal, complementary system of indicators to help the establishment of diagnosis of systems malfunctions and prioritise investments.

In this paper, a general overview of the international situation on the use of performance indicators for water supply systems is presented, with a particular emphasis for the on-going work in the scope of the International Water Services Association.

## 9. REFERENCES


ADAMSSON, J. (1997) - *The Swedish experience and viewpoints*, IWSA Workshop on Performance Indicators for Transmission and Distribution Systems, Lisbon, Portugal.

ALEGRE, H. (1997) - Performance indicators - current trends, workshop "Public-private partnership for African Water Utilities", Water Utility Partnership, World Bank, Kampala, Uganda.

ALEGRE, H.; BAPTISTA, J. M. (1997) - *Indicadores de desempenho de sistemas de distribuição de água: uma questão de moda ou de necessidade?*, "Revista Indústria da Água", No. 25 (Oct./Nov./Dez. 1997), EPAL, Lisboa.

ALEGRE, H., HIRNER, W., BAPTISTA , J. M., CUBILLO F. (1997) - *Proposal for basic and performance indicators for water supply*, 2nd draft , September 1997, IWSA Operations and Maintenance Committee, Task Force 8 - "Performance Indicators", Lisbon, Portugal.

ALEGRE H., BAPTISTA, J. M., FARIA A. L. (1997) - *A general framework of performance indicators in the scope of water supply*, IWSA Workshop on Performance Indicators for Transmission and Distribution Systems, Lisbon, Portugal.

ALEGRE, H., ALMEIDA, M. C. (1995) - *Framework for the assessment of levels-of-service*, Vol. 12 of the series "Management of water supply, waste water and solid waste systems", ed. Jaime Melo Baptista and Maria Rafaela Matos, LNEC, Lisbon (167 pp.).

ALLISON, P. (1997) - *Keeping the customer happy*, editorial of the Journal "World Water and Environmental Engineering", Vol. 20, No. 7, July 1997.

COELHO, S. T. (1997A) - *Performance analysis in water distribution*, IWSA Workshop on Performance Indicators for Transmission and Distribution Systems, Lisbon, Portugal.

COELHO, T. (1997B) - *Performance indicators in water supply - a systems approach*, Series Water Engineering and management systems, Research Studies Press, John Wiley and Sons, UK.

CUBILLO, F. (1997) - *Information systems: a key factor to manage performance indicators*, IWSA Workshop on Performance Indicators for Transmission and Distribution Systems, Lisbon, Portugal.

DJERRARI, F. (1997) - *Survey analysis on the main indicators for utilities*, IWSA Workshop on Performance Indicators for Transmission and Distribution Systems, Lisbon, Portugal.

DEB, A. AND CESARIO L. (1997) - *Water distribution systems performance assessment*, funded by the AWWA Research Foundation, IWSA Workshop on Performance Indicators for Transmission and Distribution Systems, Lisbon, Portugal.

FARIA, A. L., ALEGRE, H. (1996) - *Paving the way to excellence in water supply systems: a national framework for levels-of-service assessment based on consumer satisfaction*, The Maarten Schalekamp Award - 1995, AQUA, Vol. 45, n. 1, February 1996, IWSA, London, UK (pp. 1-12) and .

HANSEN, G. (1997) - *Technical, operational and economic performance indicators of water utilities*, IWSA Workshop on Performance Indicators for Transmission and Distribution Systems, Lisbon, Portugal.

HARRIES C. (1997) - *The UK experience of performance indicators from a regulatory and operational viewpoint*, IWSA Workshop on Performance Indicators for Transmission and Distribution Systems, Lisbon, Portugal.

HIRNER, W. (1997a) - *Technical, operational and economic performance indicators of water utilities*, IWSA Workshop on Performance Indicators for Transmission and Distribution Systems, Lisbon, Portugal.

HIRNER, W. (1997b) - *Unaccounted-for water*, workshop "Public-private partnership for African Water Utilities", Water Utility Partnership, World Bank, Kampala, Uganda.

ILIYAS, M. (1997) - *WUP Project No. 2: Performance indicators project*, workshop "Public-private partnership for African Water Utilities", Water Utility Partnership, World Bank, Kampala, Uganda.

IWSA (1997) - *International statistics for water supply*, published for the IWSA Congress 1997, Madrid, Spain.

JANSSENS, J., PINTELON, L., COTTON, A., GELDERS, L. (1994) - *Development of a framework for the assessment of operation and maintenace (O&M) performance of urban water supply and sanitation*, IWSA Specialised Conference on "The quality of service (or Managing a Water Company)", Amsterdam, the Netherlands.

JERUSALEM WATER UNDERTAKING (1995) - *Performance prospects*, Jerusalem Water Undertaking - Ramallah District, Palestinian Territories.

MALAYSIAN WATER ASSOCIATION (1996), *Performance indicators for water supply, a proposal from the Malaysian Water Association for the consideration of member countries of ASPAC.*

MATOS, M. R.; BICUDO, J. R.; ALEGRE, H. (1993) - "*Technical and social-economical indicators in the scope of water supply, wastewater and solid waste systems*", in *Preparatory study for the definition of selection criteria for environmental projects to be funded by the European Communities*, Vol. 2, project contracted by the European Commission, report 107/93, LNEC, Lisbon, Portugal (43 pp.).

MCINTOSH, A. C.; YÑOGUEZ, C. (1997) *Second Water Utiliies Data Book -Asian and Pacific Region*, Asian Deveplmen Bank's Regional Technical Assistance No. 5694, Manila, Philippines.

OATRIDGE, J. W. (1997) - *Understanding and satisfying customer expectations*, IWSA Workshop on Performance Indicators for Transmission and Distribution Systems, Lisbon, Portugal.

OFWAT (1996) - *1995-96 Report on levels-of-service for te water industry in England and Wales*, Office of Water Services, United Kingdom.

OFWAT (1996) - *1995-96 Report on the cost of water delivered and sewage collected*, Office of Water Services, United Kingdom.

PARSONS, D. P. (1997) - *Managing benefits of mains rehabilitation through structured* surveys, IWSA Workshop on Performance Indicators for Transmission and Distribution Systems, Lisbon, Portugal.

SKARDA, B. C. (1997) - *The Swiss experience with performance indicators and special viewpoints on water networks*, IWSA Workshop on Performance Indicators for Transmission and Distribution Systems, Lisbon, Portugal.

THE MALAYSIAN WATER ASSOCIATION (1996) - *Performance indicators for water supply, a proposal from the Malaysian Water Association for the consideration of member countries of ASPAC.*

VAN DER WILLIGEN, F. (1997) - *Dutch experience and viewpoints on performance indicators*, IWSA Workshop on Performance Indicators for Transmission and Distribution Systems, Lisbon, Portugal.

YEPPES, G., DIANDERAS, A. (1996) - *Water & wastewater utilities indicators*, Water and Sanitation Division, The World Bank.

YEPPES, G., DIANDERAS, A. (1997) - *The role of indicators to guide unaccounted-for water reduction programs*, IWSA Workshop on Performance Indicators for Transmission and Distribution Systems, Lisbon, Portugal.

# PERFORMANCE INDICATORS FOR WATER SUPPLY SYSTEMS. A CASE STUDY

MIGUEL ANDRÉS FOLGADO
*Aguas de Valencia S.A.*
*Gran Vía Marqués del Turia, 19*

## 1. Introduction

A water supply's main mission is to guarantee the delivery of water in sufficient quantity and quality 24 hours a day, 365 days a year.

To achieve such objective, great investments (both material and human) are usually necessary. Investments that need of a previous engineering work (during the design stages), an optimum construction and development and finally, an adequate maintenance program of the installations.

Considering the circumstances we could easily conclude that the price of such a service should be quite high, and yet it remains the cheapest of all public services given by the municipalities, or so it is in Spain. The question is: Why?. We shall not answer that question. Instead, we urge each reader to provide an answer and even, if he deems it necessary, to share it with us.

Water supply in to the municipalities is a public service, but that is not a reason to manage the service from the business efficiency point of view. What we mean to say is that the price of the service should be such, to allow covering the costs and, in consequence, to avoid all subsidiary policies and prices below costs, only leading to deficitary services.

We can state that, seeking a proper management, there is an increasing number of experts, authorities, management companies, etc., that are showing a strong interest to define performance indicators to be used measuring the efficiency in the supplies. Even to obtain some reference values that may be of use as a measure oh how healthy the supply is.

References in the paper show the definitions and reasons for the use of performance indicators given by the authors.

In this paper, and taking as a starting point the proposals of the mentioned references and by adding our personal experience we are going to define a set of performance indicators, that in our opinion, may be used for the diagnosis and improvement of a water supply system. We shall also include the values for some real supply systems that will allow us to make comparisons and if possible, reach some conclusions.



*E. Cabrera and J. García-Serra (eds.), Drought Management Planning in Water Supply Systems,* 179–192.

## 2. Performance Indicators Evolution

Water supply systems managers have traditionally relied on performance indicators that we could describe as *macroeconomical*, to get global information on the supply and support the management process.

Among these ratios the following can be highlighted: Annual production volume; Electrical power consumption in Wh/m3; Annual billing volume; Number of customers; Distribution system length; Number of employees; Total annual income; Annual cost of the service, distributed into personnel, Electric power, Operation and maintenance, Management,....

With the new management tools, based on computers, such as cost accountancy and budget accountancy, altogether with the enormous advance in computer science, most supply systems have already developed software that allows to increase the degree of detail in expenses and income analysis and, as a consequence, to deeply study and define some other ratios that allow to have detailed and partial information on the managed service.

However, the existence of generic ratios or performance indicators used in most supplies, is still quite far from becoming true. Our intention with this paper is to outline a number of performance indicators, that may prove to be helpful in the management of a supply system as they become used and perfected.

## 3. Classification and definition of the performance indicators

One of the first difficulties to be encountered is to properly define and classify the performance indicators.

Our proposal is to do so in 5 main groups:

- structural
- quality of service
- operational
- personnel and productivity
- economical

In the first group we will include those indicators that provide an idea of the size, extension and coverage of the supply.

In the second one the indicators that show the degree of fulfilling a service are covered.

In the third group the indicators that show the operational behaviour of the supply are included. These indicators cover every day matters.

The indicators concerning the number of employees and their productivity can be found in the fourth group.

Finally, the economical indicators offer an economic measure of the supply.

These are the indicators contained in each group and their definitions. It is important to outline that in order to make comparisons between different supply systems it is important to use a single definition for a certain indicator, and such definition must be clear and lead to a single interpretation.

### 3.1. STRUCTURAL INDICATORS

The following indicators are included in this group:

#### *3.1.1. Annually produced water*

It represents the amount of water from the production centres (treatment plants or perforations) that is injected into the network.

#### *3.1.2. Capacity of production*

It is equal to the annually produced water over the nominal capacity of the installation, expressed as a percentage.

#### *3.1.3. Water abstraction (surface/ground)*

Fractions of the total amount of water that surface and ground water abstractions represent, expressed as a percentage.

#### *3.1.4. Total length of the distribution system*

It shows the length of the pipes in the distribution system, in Km, being the distribution system the set of pipes that leaving the production centres cover the geographical area of the supply system.

- Pipe distribution by materials: where it shall be specified the length in Km for each material, as an absolute value or a percentage of the total length of the distribution system.

As additional information it is also rather interesting to have statistics on the distribution by pipe materials and diameters.

#### *3.1.5. Population supplied*

It refers to the population of the geographical area of the supply system. Official data are preferred on the last population census. Of a special interest are the areas in which the population has a strong seasonal behaviour (touristic areas), where the actual population may be notably increased over certain periods during the year.

#### *3.1.6. Number of customers*

It shows the number of consumers with a meter installed. It is important to specify if such meters are installed in every household or belong to a certain community.

- Number of domestic customers, expressed as a percentage of the total number of customers

- Number of industrial customers, expressed as a percentage of the total number of customers
- Distribution of customers by meter diameters.

*3.1.7. Number of service pipes*

It is understood by service pipe, the one that links the distribution pipe with the distribution network in every building. Generally every service pipe will provide service for more than a customer.

*3.1.8. Storage volume in regulation tanks*

Total volume of storage in regulations tanks, if applicable. This ratio offers a measure of the system's security over a failure in the supply from the production centres.

*3.1.9. Customers density*

It is a measure of the degree of concentration of the customers of the service. It may be obtained by dividing the total number of customers and the total length of the network.

*3.1.10. Service pipes density*

Shows the number of services pipe per Km of network length.

*3.1.11. Allocated production per customer*

May be obtained by dividing the average daily production volume over the total number of customers. It usually is expressed in (l per capita/day)

*3.1.12. Produced volume per Km of network*

It is equal to the total volume produced in a year in $m^3$ over the total length of the network in Km.

*3.1.13. Increase of customers*

Equals to the average increase over the last five years expressed as a percentage of the number of customers. This indicator is important since it will condition future growth expectations for the supply and also because it is essential to calculate the income budget for the supply.

### 3.2. QUALITY INDICATORS

In this group we can find the following indicators:

*3.2.1. Quality of water*

Under this point the results of the physical, chemical and bacteriological tests will be reflected. The tests should be carried out accordingly with each country's legislation.

*3.2.2. Number of tests per year*

- N1- number of minimum tests
- N2- number of normal tests

- N3- number of complete tests

*3.2.3. Number of anomalous tests*

Adds up for the samples that have resulted in an anomalous result when tested. Statistics on the anomalous parameter should be tracked.

*3.2.4. Percentage of anomalies in tests*

Ratio concerning the number of failed tests and the total number of tests carried out expressed as a percentage.

*3.2.5. Service complaints*

Number of formal complaints received during a year. Statistics will be done on the complaints and the groups they come from.

*3.2.6. Density of complaints*

Shows the actual number of complaints over the total number of customers, expressed as a percentage.

*3.2.7. Residual pressure in the distribution network*

Residual pressure at the service pipe, existing in normal operating conditions in the network.

*3.2.8. Service Interruptions*

Defined with duration and reach (number of customers affected). It must show the number of hours without service, and detailed statistics on the causes and the effects.

*3.2.9. Continuity of the supply*

Ratio between the number of hours without interruptions and the annual total of service hours.

### 3.3. OPERATIONAL INDICATORS

*3.3.1. Billing period*

Number of months

*3.3.2. Number of annual bills*

Number of bills issued in a natural year. From the management point of view it is important to carry out statistics which show, for every month of the year, the number of bills issued up to that month and the number issued in the previous 12 months.

*3.3.3. Annually metered volume*

Volume of water registered by domestic meters in $m^3$.

*3.3.4. Annually billed volume (ABV)*

Volume of water billed to every customer supplied. If no billing minimum is considered this volume will be equal to the metered volume.

The ABV must include the $m^3$ supplied at no cost (as if the price was 0), as well as those cases in which the billing is done by estimation (billing volumes assigned per period) and that usually correspond to non-metered supply for irrigation, free discharge to the sewer system, and other uses defined by the supply.

The ABV shall be calculated by adding the billed volumes for every billing period. In order to track it, detailed statistics must be carried out on the number of bills issued, the amount of water billed, the average number of days billed and the average allocated volume per bill in l/(bill*day).

*3.3.5. Number of non working meters (NWM), (%)*

Number of non working (stopped) meters at the end of a financial year. It shall be expressed as a percentage of the number of customers of the supply.

*3.3.6. Number of replaced meters (NRM)*

Number of meters replaced during the financial year, whether they may have been replaced as non working meters, or as a part of a replacement program designed to maintain the average age of the meters in the supply within reasonable limits. Just as an example, in Spain the tendency is to limit the meter's age to 10-12 years (in some areas it is limited to a shorter period by law). It is expressed as a percentage over the total number of customers.

*3.3.7. Global efficiency of the distribution network*

Understood as the ratio of the billing volume and the total produced volume.

*3.3.8. Billing distribution by customer*

- billing to domestic customers (%)
- billing to industrial customers (%)

Ratio between annual billed volumes to domestic and industrial users and the total annual billed volume.

*3.3.9. Allocation per customer*

Ratio between the total annual billed volume and the number of customers, expressed in l/(customer*day)

*3.3.10. Unacounted for water (%)*

Difference of the produced and billed volumes over the total produced volume, expressed as a percentage

*3.3.11. Unacounted for water per Km of network*

Ratio between the unacounted for water and the length of the distribution network in Km ($m^3$/(Km*year)).

This indicator is most representative to show the efficiency of the distribution network, and it is directly related with leaks detection and network renewal policies developed in the supply system.

*3.3.12. Total number of leaks*

Shows the total number of leaks repaired during a year in the supply. To measure it, statistics will be performed according to this classification:

- leaks in pipes
- leaks in service pipes
- leaks in accessories (valves, drains, etc.)

*3.3.13. Density of leaks*

Total number of leaks over the distribution network length in Km.

*3.3.14. Length of distribution network inspected*

Shows the number of Km of network that have been inspected during a financial year, with correlation equipment, etc., aimed to detect leaks before they naturally appear. It shall be expressed as a percentage of the total network length.

*3.3.15. Leaks detected through inspection*

Number of leaks detected and repaired within the inspection program. It shall be expressed as a percentage of the total number of leaks in the network.

*3.3.16. Annual increase in network length*

Km of newly installed pipes in the network during the financial year. It shall be expressed as a percentage of the total length of the network.

*3.3.17. Number of Km of replacement pipes*

Total length of the pipes that have been installed to replace those damaged. It shall be expressed as a percentage of the total length of the network.

*3.3.18. Specific energy consumed*

Ratio of energy consumed by the production centres and the total produced volume.

## 3.4. PERSONNEL AND PRODUCTIVITY INDICATORS

The personnel and productivity indicators may give us an idea of the work load and the efficiency of the different supply areas: administrative, planning, operation and maintenance. The following indicators are proposed for this group:

*3.4.1. Total number of employees*

Adds up for the total number of employees within the supply, regardless of the kind of activity they carry out. The following unitary indicators are obtained:

- Employees per million of $m^3$ produced
- Employees per thousand consumers supplied
- Employees per Km of distribution network
- Employees per thousand service pipes

*3.4.2. Distribution of the personnel by activity*
Covers the distribution of the personnel, expressed as a percentage of the total number of employees according to the following classification:

*Administrative (only customer service):* Percentage of the personnel in charge of attending customer service

*Operation and maintenance:* Percentage of the personnel in charge of operation and maintenance tasks both corrective and preventive.

*Planning and construction:* Percentage of the personnel in charge of planning and construction of new infrastructures

*Management and administration:* Percentage of the personnel in charge of administrative tasks (accountancy, finances, personnel, computing and billing).

*3.4.3. Annual absenteeism rate*
Ratio of the annual number of hours of absenteeism and the number of working hours in a year.

*3.4.4. Extra hours percentage*
Ratio of the annual number of extra hours and the number of regular working hours in a year.

### 3.5 ECONOMICAL INDICATORS

These indicators are defined to quantify the economical aspects of the supply and its efficiency. In this group we may define the following:

*3.5.1. Total cost of the service, in $\$/m^3$*
Ratio between the Total Annual Cost of the Service and the annual Production Volume.

*3.5.2. Total revenues of the service, in $\$/m^3$*
Ratio between the total revenues for the service in a year and the Annual Production Volume.

*3.5.3. Coverage degree of the service*
Expressed as a percentage of the annual revenues for the service and the total cost of the service.

*3.5.4. Cost Structure for the Service*

Covers the cost structure for the service according to the following classification:

*Personnel costs:* Including all personnel costs, disregarding the activity they develop

*Operation and Maintenance costs:* Including all external resources used for the operation and maintenance of the supply installations, like: electrical power , materials, and external contractors.

*Water import costs:* If the supply does not have its own production means and has to purchase water to some other entity, which is metered at the distribution network's entry, it is convenient to split this concept from the operation and maintenance costs, since it represents an important fraction of the total service costs.

*Meter reading and billing costs:* Includes all external resources used to read meters, issuing and printing bills and receipts, bank commissions, bills delivery, etc...

*Capital costs:* Includes the costs of the company's capital offering the service. They are often considered as a percentage of the capital invested in the supply or as a percentage of the total income for the service.

*Administration costs:* Includes costs of administering the service like: rentals (buildings and vehicles), insurance, taxes on the service and similar ones.

*Depreciation costs of the installations:* Includes the annual depreciation cost, both technical and financial, of the investments made during the financial year.

Every one of these costs will be expressed in currency units, as unitary cost per Km of distribution network for all the listed costs except the meter reading and billing cost, which shall be expressed in relation with the total number of customers.

*3.5.5. Operation and maintenance costs with relation to the fixed assets value in the service*

Shows how important are operation and maintenance costs with relation to the fixed assets of the service installations value .

*3.5.6. Annual investment in the distribution network*

Represents the annual amount invested in the supply, growth and replacement of the distribution network. There will consequently be two groups:

- Annual investment in the network's growth
- Annual investment in network's replacement

*3.5.7. Average price of water ($/billed $m^3$)*

Ratio between the annual revenues for the sale of water, and the annual billed volume of water. The rate structure of the service must be indicated as well.

*3.5.8. Bill charging efficiency*
Ratio between the total charged amount ($) and the total annual billed amount, expressed as a percentage.

## 4. Application examples

As a summary to this paper and to let them be used as examples, some data tables are included containing the values of the presented performance indicators for five different water supply systems.

*Table 1. Comparison of quality performance indicators for 5 different supplies.*

| COD | NAME | SUP 1 | SUP 2 | SUP 3 | SUP 4 | SUP 5 | UNITS |
|---|---|---|---|---|---|---|---|
| 3.2.1 | Water Quality | | | | | | |
| 3.2.2 | Number of tests per year | 417 | 423 | 445 | 510 | 2446 | |
| | Classification of tests | | | | | | |
| | Number of anomalous tests | 415 | 7 | 24 | 66 | 378 | |
| | Number of minimum tests | 1 | 415 | 420 | 439 | 2057 | |
| | Number of complete tests | 1 | 1 | 1 | 5 | 11 | |
| 3.2.3 | Number of anomalous tests | 0 | 0 | 0 | 0 | 8 | |
| 3.2.4 | Percentage of anomalies in tests | 0.0% | 0.0% | 0.0% | 0.0% | 0.3% | |
| 3.2.5 | Service complaints | 0 | 0 | 40 | 50 | 50 | Complaints |
| 3.2.6 | Density of complaints | 0.0% | 0.0% | 0.8% | 0.5% | 0.2% | |
| 3.2.7 | Residual pressure in the distribution network | 24 | 23 | 25 | 25 | 30 | mwc |
| 3.2.8 | Service interruptions | 0 | 8 | 0 | 10 | 0 | |
| 3.2.9 | Continuity of the supply | 100% | 100% | 100% | 99.9% | 100.0% | |

*Table 2. Comparison of structural performance indicators for 5 different supplies.*

| COD | NAME | SUP 1 | SUP 2 | SUP 3 | SUP 4 | SUP 5 | UNITS |
|---|---|---|---|---|---|---|---|
| 3.1.1 | Annually produced water (APW) | 2,354.0 | 2,581.0 | 1,120.0 | 2,142.0 | 8,869.6 | Thousands $m^3$ |
| 3.1.2 | Capacity of production | 100% | 100% | 28% | 45% | 44% | |
| 3.1.3 | Water abstraction | | | | | | |
| | Surface | 1,490.0 | 2,091.0 | 0.0 | 0.0 | 0.0 | Thousands $m^3$ |
| | Surface % s/APW | 63.30% | 81.02% | 0.0 | 0.0 | 0.0 | |
| | Ground | 864.0 | 490.0 | 1,120.0 | 2,142.0 | 8,869.6 | Thousands $m^3$ |
| | Ground s/APW | 36.70% | 18.98% | 100.00% | 100.00% | 100.00% | |
| 3.1.4 | Total length of the distribution network | 65.2 | 67.5 | 35.8 | 63.2 | 162.0 | Km |
| | Pipe distribution by materials | | | | | | |
| | Asbestos cement | 92.1% | 94.1% | 97.0% | 94.3% | 85.0% | |
| | Cast iron | 4.9% | 0.0% | | 0.0% | 12.0% | |
| | Polyethylene | 3.1% | 5.9% | 3.0% | 5.7% | 3.0% | |
| 3.1.5 | Population supplied | 24,500 | 22,450 | 12,650 | 27,100 | 55,000 | Inhabitants |
| 3.1.6 | Number of customers | 10,743 | 9,001 | 5,133 | 10,420 | 30,001 | Customers |
| | Number of domestic customers | 96.9% | 95.7% | 99.4% | 97.5% | 88.9% | |
| | Number of industrial customers | 3.1% | 4.3% | 0.6% | 2.5% | 11.1% | |
| | Distribution of customers by meter diameters | | | | | | |
| 3.1.7 | Number of service pipes | 3,350 | 3,210 | 1,750 | 3,750 | 7,800 | Service Pipes |
| 3.1.8 | Storage volume in regulation tanks | 5,000 | 700 | 520 | 0 | 2,300 | $m^3$ |
| 3.1.9 | Customers density | 164.9 | 133.3 | 143.4 | 164.9 | 185.2 | Cust./Km |
| 3.1.10 | Service pipes density | 51.4 | 47.6 | 48.9 | 59.3 | 48.1 | S.P./Km |
| 3.1.11 | Allocated production per customer | 600.3 | 785.6 | 597.8 | 563.2 | 810.0 | l/(cust. *day) |
| 3.1.12 | Produced volume per Km of network | 36.1 | 38.2 | 31.3 | 33.9 | 54.8 | Thousands $m^3$/(Km*yr) |
| 3.1.13 | Increase of customers | 1.5% | 1.2% | 0.95% | 1.0% | 1.2% | |

*Table 3. Comparison of operational performance indicators for 5 different supplies.*

| COD | NAME | SUP 1 | SUP 2 | SUP 3 | SUP 4 | SUP 5 | UNITS |
|---|---|---|---|---|---|---|---|
| 3.3.1 | Billing Period | 3 | 3 | 3 | 3 | 2 | Months |
| 3.3.2 | Number of annual bills | 42,972 | 36,004 | 20,532 | 41,680 | 180,006 | Bills |
| 3.3.3 | Annually metered volume | 1,571.0 | 1,455.0 | 705.0 | 853.0 | 6,019 | Thousands $m^3$ |
| 3.3.4 | Annually billed volume (ABV) | 1,571.0 | 1,455.0 | 784.0 | 1,317.0 | 6,019 | Thousands $m^3$ |
| 3.3..5 | Number of non working meters | 75 | 0 | 0 | 195 | 1,525 | |
| 3.3.6 | Number of replaced meters | 292 | 233 | 48 | 790 | 2,480 | |
| 3.3.7 | Global efficiency of the distribution network | 66.7% | 56.4% | 70.0% | 61.5% | 67.9% | |
| 3.3.8 | Billing distribution by customer | | | | | | |
| | Industrial Customers | 17.7% | 9.2% | 7.7% | 9.0% | 24.8% | % s/ABV |
| | Domestic Customers | 82.3% | 90.8% | 92.3% | 91.0% | 75.2% | % s/ABV |
| 3.3.9 | Allocation per customer | 400.6 | 442.9 | 418.5 | 346.3 | 550 | l/(cust.*day) |
| 3.3.10 | Unaccounted for water | 783.0 | 1,126.0 | 336.0 | 825.0 | 2,851 | Thousands $m^3$ |
| 3.3.11 | Unaccounted for water per Km of network | 12,018.4 | 16,681.5 | 9,385.5 | 13,053.8 | 17,596 | $m^3$/(Km net *year) |
| | | 1.4 | 1.9 | 1.1 | 1.5 | 2.0 | $m^3$/h/Km net |
| 3.3.12 | Total number of leaks | 374 | 407 | 85 | 260 | 677 | |
| | leaks in pipes | 122 | 120 | 35 | 60 | 109 | |
| | leaks in service pipes | 211 | 176 | 38 | 186 | 289 | |
| | leaks in accesories | 41 | 111 | 12 | 14 | 276 | |
| 3.3.13 | Density of leaks | 1.9 | 1.8 | 1.0 | 0.9 | 0.7 | |
| 3.3.14 | Length of distribution network inspected | 32.5 | 29.9 | 35.8 | 63.2 | 25.0 | |
| 3.3.15 | Leaks detected through inspection | 75 | 62 | 22 | 94 | 18 | |
| 3.3.16 | Annual increase in network length | 1.5 | 1.3 | 0.8 | 1.2 | 2.0 | |
| 3.3.17 | Number of Km of replacement pipes | 6.8 | 1.2 | 0.2 | 0.6 | 0.0 | |
| 3.3.18 | Specific Energy consumed | 632.0 | 350.0 | 351.0 | 347.5 | 472.0 | Wh/$m^3$ |

*Table 4. Comparison of economical performance indicators for 5 different supplies.*

| COD | NAME | SUP 1 | SUP 2 | SUP 3 | SUP 4 | SUP 5 | UNITS |
|---|---|---|---|---|---|---|---|
| 3.5.1 | Total cost of the service | 119,008.1 | 103,156.6 | 35,625.8 | 76,135.8 | 283,302.0 | Thousands $ |
| | Unitary cost for produced $m^3$ | 50.6 | 40.0 | 31.8 | 35.5 | 31.9 | $/$m^3$ Prod |
| 3.5.2 | Total revenues of the service | 119,008.1 | 87,706.6 | 35,764.1 | 72,146.0 | 287,135.0 | Thousands $ |
| | Unitary revenue for produced $m^3$ | 50.6 | 34.0 | 31.9 | 33.7 | 32.4 | $/$m^3$ Prod |
| 3.5.3 | Coverage degree of the service | 100% | 85% | 100% | 95% | 101% | |
| 3.5.4 | Cost structure of the service | | | | | | |
| | Personnel costs | 16.9% | 14.5% | 40.0% | 34.3% | 27.0% | % w/r Total cost |
| | Average personnel costs | 2,876.6 | 2,995.0 | 3,559.0 | 5,227.6 | 5,099.3 | Thosands $/Empl. |
| | Unitary personnel costs | 309.1 | 221.9 | 397.7 | 413.6 | 472.2 | Thousands $/Km net |
| | Operation and maintenance costs | 20.8% | 12.0% | 38.2% | 37.8% | 35.6% | % w/r Total cost |
| | Unitary operation and maintenance costs | 380.0 | 183.6 | 379.8 | 455.9 | 622.0 | Thosands of $/Km net |
| | Water import costs | 21.2% | 52.3% | 0.0% | 0.0% | 0.0% | % w/r Total cost |
| | Unitary water import costs | 388.1 | 799.8 | 0.0 | 0.0 | 0.0 | Thousands $ /Km net |
| | Meter reading and billing costs | 1.5% | 3.2% | 5.5% | 4.7% | 7.3% | % w/r Total cost |
| | Unitary meter reading and billing costs | 40.6 | 90.4 | 95.0 | 85.2 | 114.6 | Pts/Bill |
| | Capital costs | 2.6% | 3.8% | 6.0% | 6.0% | 6.5% | % s/ Total cost |
| | Unitary capital costs | 48.0 | 58.8 | 59.9 | 71.9 | 114.1 | Thousands Pts/Km |
| | Administration costs | 4.4% | 5.7% | 6.0% | 7.7% | 6.6% | % s/ Total cost |
| | Unitary administration costs | 80.2 | 87.8 | 59.5 | 93.3 | 114.8 | Miles Pts/Km |
| | Depreciation costs of the installations | 32.5% | 8.4% | 4.4% | 9.4% | 17.1% | % s/ Total cost |
| | Unitary depreciation costs of the installations | 594.5 | 128.1 | 43.8 | 113.8 | 298.4 | Thousands $/Km net |
| 3.5.5 | Operation and maintenance costs w/r fixed assets value | | | | | | |
| 3.5.6 | Annual investment in the distribution network | 36,000.0 | 12,000.0 | 1,000.0 | 6,300.0 | 5,000.0 | |
| 3.5.7 | Average price of water | 75.8 | 60.3 | 45.6 | 54.8 | 47.7 | $/billed $m^3$ |
| 3.5.8 | Billing charging efficiency | 95.5% | 91.7% | 98.2% | 97.0% | 97.0% | |

*Table 5. Comparison of personnel and productivity performance indicators for 5 different supplies.*

| COD | NAME | SUP 1 | SUP 2 | SUP 3 | SUP 4 | SUP 5 | UNITS |
|---|---|---|---|---|---|---|---|
| 3.4.1 | Total number of employees | 7 | 5 | 4 | 5 | 15 | Employees |
| | per million of $m^3$ produced | 2.97 | 1.94 | 3.57 | 2.33 | 1.69 | |
| | per thousand inhabitants | 0.29 | 0.22 | 0.32 | 0.18 | 0.27 | |
| | per Km of distribution network | 0.11 | 0.07 | 0.11 | 0.08 | 0.09 | |
| | per thousand service pipes | 2.09 | 1.56 | 2.29 | 1.33 | 1.92 | |
| 3.4.2 | Distribution of the personnel by activity | | | | | | |
| | Administrative (only customer service) | 28.6% | 20.0% | 25.0% | 40.0% | 26.7% | |
| | Operation and maintenance | 57.1% | 20.0% | 50.0% | 40.0% | 60.0% | |
| | Planning and construction | 7.1% | 40.0% | 12.5% | 10.0% | 6.7% | |
| | Management and administration | 7.1% | 20.0% | 12.5% | 10.0% | 6.7% | |
| 3.4.3 | Annual absenteeism rate | 3.6% | 3.8% | 2.4% | 3.8% | 5.0% | |
| 3.4.4 | Extra hours percentage | 0.8% | 1.1% | 0.3% | 1.1% | 2.6% | |

## 5. References

THE MALASYAN WATER ASSOCIATION *Performance Indicators for Water Supply.* Workshop IWSA-AIDE. Lisboa 5-7.May.1.997.

HIRNER, W. *Technical, Operational an Economic Perfomance Indicators of Water Supply.* EWAG, Nuremberg. Germany. Workshop IWSA-AIDE, Lisboa 5-7 May 1.997.

# SUSTAINABLE WATER MANAGEMENT IN AN URBAN CONTEXT

WOLFGANG SCHILLING
Norwegian University of Science and Technology, Trondheim

ARISTOTELIS MANTOGLOU
Laboratory of Reclamation Works and Water Resources Management, Technical University of Athens, Greece

## Abstract

The problem of urban water systems is examined with emphasis on sustainable management of water supply. In case of drought, water is a scarce resource, and sustainable management is of paramount importance.

The availability of the existing freshwater resources within an urban area is constrained by the hydrophysical, technical, economic and institutional conditions of the system. These conditions determine the general availability of water in the area and have an effect on, both, the supply and the demand of fresh (drinking) water. In addition, they are affecting each other in a complex way such that the effect of introducing new technical, economic and/or institutional measures on water supply can not easily be assessed in advance.

A reasonable management aspiration, which has become very popular in recent years, is to operate this system in a *sustainable* way. Related to sustainability are the concepts of *dynamically balanced* or *healthily balanced* system. This essentially means that water supply satisfies the needs of the society of the urban region without jeopardizing the long-term quality of the water resources and the environment in general.

The goal of sustainable development addresses more the issue of the needs of the society rather than the desires of a few people or a single industry for increasing comfort. Relative to this concept is the definition of absolute and relative water scarcity. When water is *absolutely* scarce it limits the survival and/or development of an individual, a population, a society or an ecosystem. If it is *relatively* scarce, its limiting character can be overcome with technical, economical or institutional measures, usually at higher costs. Absolute scarcity of water is very rare, and most situations involve relative scarcity of various degrees of severity. Here, the challenge is to find the right balance of technical, economic and institutional measures that will guarantee satisfaction of the needs of society with the least consumption of the available water and environmental resources.

Various technical, economic and institutional measures are described in the paper. In addition, case descriptions from regions with, both, water abundance and scarcity illustrate how these measures were applied in practice and how they affected the performance of the respective water supply systems.

## 1. Introduction

The vast majority of the population in Western societies lives and works in towns and cities, which mostly are located in the vicinity of freshwater resources (lakes, rivers). In the context of this paper an urban water system is defined *geographically* as the urban



*E. Cabrera and J. García-Serra (eds.), Drought Management Planning in Water Supply Systems,* 193–215.

area including its surrounding areas, where water imports and ecological effects of pollution discharges are involved. *Hydrologically* it comprises all freshwater resources of the area (i.e. surface water and groundwater, clean and polluted). *Physically* it is the natural and technical system that collects, stores, transports, treats and distributes water and wastewater. *Institutionally*, it includes the organizations dealing with water management (general water resources, water supply, wastewater) as well as those who are served or otherwise affected (i.e. customers).

The current urban water system typically requires a natural surrounding from which large volumes of fresh water are imported and large volumes of polluted wastewater are discharged. In case of rapid urbanization deleterious effects on the water environment can be observed, while the costs of urban water infrastructure is rising from "high" to "very high". Due to this process of urbanization the map of environmental degradation shows considerable geographical variability since urban areas are major concentrations of both water needs as well as production of wastes. The greater density often brings an even greater concentration of pollutants with major consequences to the local as well as the wider environment. Infiltration of rainfall to the ground is often reduced and runoff is increased. Fig. 1 describes the adverse effects that ultimately occur:

- decreased freshwater resources
- decreased quality of freshwater
- increased frequency and volume of floods
- decreased natural area.

Today's technology of urban water management is a major achievement of modern societies and is recognized as one of the prime factors for human welfare by maintaining a high hygienic standard and reducing the risk of urban flooding. However this service leads to pollution and / or reduction of limited natural resources like freshwater and natural land. The natural resources such as the freshwater are quite limiting in areas of fast urban development as well as near sensitive coastal regions. The high increase of the water demand due to urbanization as well as due to tourist development leads to problems of water scarcity and reduction of quality of water resources in such regions. Quite often the socioeconomic pressures for development do not consider the need of sustainability which leads to bad management of water resources. This approach leads to the waste of natural resources near urban regions, particularly of land and water and proves to be not sustainable. The high service level of modern societies it is not sustainable on the long run.

The anticipated climatic change and the associated droughts paint a bleak picture for the future of many modern cities relative to water supply. The solution to these problems requires both, structural measures and the participation and cooperation of the society. This paper discusses how the goal of sustainability can be achieved in urban water management. It is emphasized that reduction of wasteful consumption is needed since the high service level offered by modern societies is not sustainable in the long run. The problem of water supply in Athens which was recently stricken by a severe drought and the measures taken for the alleviation of the problems are presented as a characteristic example. Furthermore, it is shown that even in regions with abundance of freshwater such as Northern Europe water shortages occur, and special measures are taken to prevent consequences.

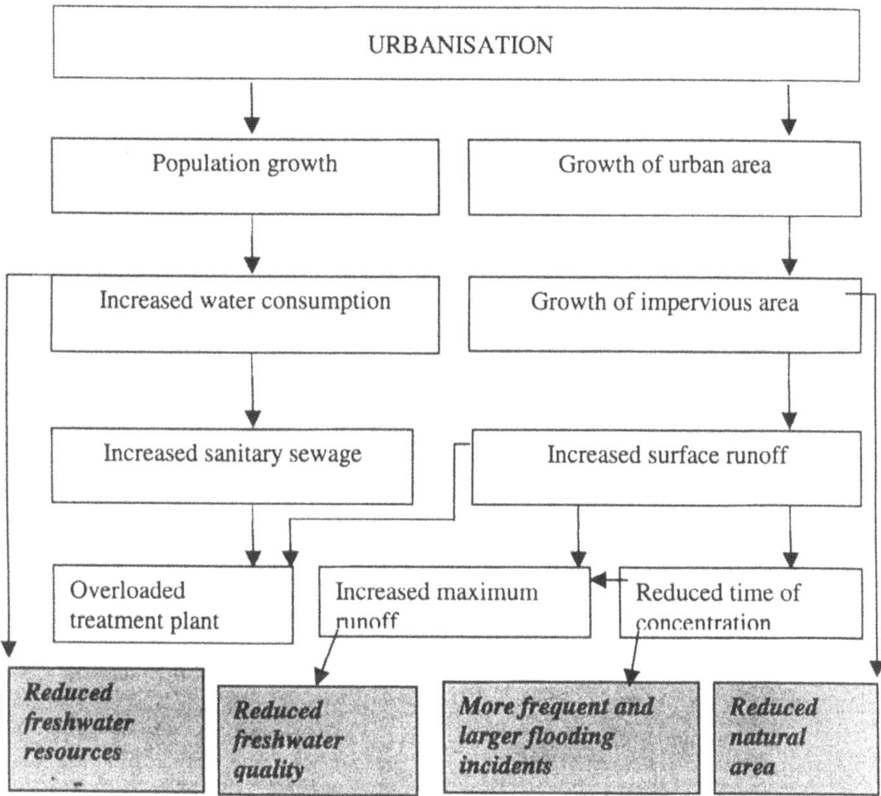


*Fig. 1 Urbanization and its detrimental effects on regional freshwater resources.*

## 2. Sustainable urban water resources management

Water is a resource that cannot be exhausted since it is naturally self-regenerating. However, the availability of freshwater resources varies in space and time. Changes in water consumption, pollution, and urban growth as well as changes in the global environmental system can affect the amount of good quality water in a region. Hence, regional water resources can be degraded, and they are limited to the extent of their use.

There is a tendency in the current economic development patterns in many countries to increase water use and therefore reduce the availability of water resources. The current mode of operation of modern societies depends on increased levels of consumption of goods and services and most people tend to believe that welfare means to consume more in general. This is crucial since a sustainable situation is rarely tenable under such conditions of increased consumption. This growth syndrome has many consequences for water resources and the environment in general. However, as will be illustrated with some examples below, economic development does not necessarily mean increased consumption of water.

Sometimes the purpose of public or private companies that manage the water supply system is to make profit from the sale of water. Consequently, they promote increase of consumption without much consideration of environmental sustainability. Although public agencies do not necessarily need to make profits, they are largely financed through the sale of bonds and thus have to make equivalent profits in order to cover operating costs and ensure returns to their investors. Thus, continued expansion of water demand and delivery is typical, and quite often efficiency in water supply and water use is not a major policy concern. With the growth and expansion of urban areas the demand for water grows, and this water must be increasingly imported from outside boundaries. The implications for the environment are profound, since as discussed the goal of the profit makers is to accelerate the consumption of the resource.

There are limits to such consumption, because water is not only a scarce resource globally and even scarcer locally in many parts of the world, but it is also a necessity for human life. Thus, eventually choices will have to be made on how to use a limited amount of water available in an area. Making those choices in a specific area could be delayed by importing water from neighboring regions, as has increasingly been done during the last centuries of rapid urbanization, but paying for those investments and their upkeep means that water becomes more expensive, and in many cases natural areas are spoiled. An example is the exploitation of groundwater aquifers south of Hamburg / Germany where vegetation was affected due to water table drawdown. Another example is the water supply resource of Trondheim / Norway, Lake Jonsvatn, where a tunnel connection to a larger lake was constructed and, through the tunnel, a small crawfish species entered the lake and disturbed the ecological balance so much that some game fish were literally exterminated.The relation between urban design and degradation of water resources quality as well as the associated dangers to public health are often not considered in engineering studies. Quite often the goals of society (as expressed by politics) and those of the water companies are of short term and of an economic nature. However, in order to attain sustainability, the goals must be long term and should be of a broad nature and must include social as well as environmental considerations. It must be recognized that the aggravation of the effects of droughts on water supply is at least partially due to bad practices regarding urban development as well as from the lack of cooperation between water users, social groups and water authorities regarding wise management of natural resources.

Efficiency can be gained by ensuring as little waste as possible (e.g. through proper pricing policies) and by recycling the resource, which means using it more than one time during one sequence of the hydrological cycle. As will be discussed with the example of Athens after the severe drought of 1989-1995, the results are better when there is an integrated effort between structural changes in the system of management as well as changes in the behavior of the users of the resources. The example indicates that structural changes are not by themselves sufficient, and the cooperation of the society is needed for effective alleviation of problems caused by severe droughts.

An urban region is a complex system with a multiplicity of relations of various factors such as water use, landscape management, urban development, social factors, economy, etc. In a complex system such as this it is not possible to select one factor, for example water supply, and look at it in isolation from the others since all the factors interact with each other in an intricate way. Water supply depends on factors of economic, social and environmental nature that all must be examined simultaneously.

From the considerations above it can be seen that sustainable water management in the urban context must refer to the urban area itself as well as its natural surroundings, as far as water imports and wastewater exports affect them. The challenge is to harmonize the dynamics of a system that consists of the three entities:

- The physical system that governs the freshwater resources and its balance
- The water services institution and its operating objectives
- The regional society and its political aspirations.

Sustainability can only be reached if the interests of these three entities are balanced over a very long time horizon into the future (i.e. several generations).

## 3. Measures to reach sustainability

Concrete measures to promote sustainable water management can be grouped into technical, economic, and institutional. Technical measures increase the availability of clean water or improve the efficiency of the existing system. It is worthwhile to consider the whole water supply system for potential improvements. Among the most prominent are:

- Use of yet untapped sources and/or conjunctive use of several sources
- Long distance transfer
- Systematic leakage control in the distribution system
- Installation of individual water meters
- Installation of water saving equipment in households

Economic measures include:

- Relate water fee to consumption (rather than flat rates per capita or per unit area)
- Surcharge fees for over-use of water
- Pricing policy that refers to the important aspects, namely that water is a necessity for life, that water supply systems are costly, and that water might be scarce (the IWSA water tariff policy gives a reasonable guideline)
- Economic incentives for customers to install rainwater cisterns or re-use systems

Institutional measures are:

- Protection of freshwater resources from detrimental use
- Enforcement of pollutant discharge reduction
- Discussion not only with "shareholders" but with all "stakeholders"
- Approval of new development under, both, economic and environmental considerations
- Information of customers on water saving technology

In a European perspective it is interesting to study the relation between the gross water resources per capita versus the average water price per cubic meter. The following table shows that water prices tend to decrease with increasing availability of water, although there are a number of exceptions.

Gross freshwater resources per capita and year versus average water price per cubicmeter in various countries (1994).

| Country | Freshwater resources (1000 cbm/cap.yr) | Average water fee (US $/cbm) |
|---|---|---|
| Norway | 122 | 0.34 |
| Finland | 22 | 0.75 |
| Sweden | 23 | 0.58 |
| Germany | 1.5 | 1.62 |
| France | 3.5 | 1.16 |
| Belgium | 1.2 | 1.28 |
| Netherlands | 1 | 1.21 |
| Ireland | 13 | 0.55 |
| Great Britain | 2.5 | 0.75 |
| Italy | 4.5 | 0.63 |
| Canada | 150 | 0.30 |
| Australia | 100 | 0.75 |

Operating a complex water supply system in a sustainable way requires the efforts of people from various disciplines and decision bodies. For example strategic management issues are examined at a political level, and the decision-makers are the appropriate regional political bodies. Water resources managers and engineers help in making appropriate strategic decisions by providing the necessary information to the decision-makers regarding the consequences of technical, economic and institutional measures. The actual execution of the water supply service requires the work of water resources managers, engineers, and other staff who run the various sub-systems (e.g. waterworks, distribution networks) based on the decisions derived by the strategic decision team.

The issue of finding a balanced, integrated and effective strategy to the management of water supply in urban regions can be approached from two different viewpoints. The first viewpoint is based on using practical knowledge, obtained from experience and intuition regarding the behavior of such systems. The second viewpoint is based on the development of quantitative models based on mathematical and computer simulations of the physical/environmental and the social/economic systems. We believe that both approaches must be used in an integrated way if effective decisions are to be made.

The structural management decisions obtained using such integrated approach must be possible to implement in practice. This requires the application of measures of enforcement as well as education of the users to cooperate for reducing consumption and using water resources more effectively. Education of the users can help achieving significant water savings, as seen from the example of Copenhagen where the use of water dropped from 168 l/p.d in 1989 to 133 l/p.d in 1996. Other cases from North

European cities showing similar patterns are described below. The case of Athens is discussed in some detail below. There, a number of measures including information of the public assisted reduction of water consumption alleviated the fear of a big metropolitan city running completely out of water during a severe drought period that lasted from 1989-1995.

## 4. European cases

This section discusses some cases from South as well as North European countries. As will be seen there seems to be a general effort towards reducing consumption in both southern and northern countries. While reduction of consumption in the south is driven mainly due to limited water supply and problems of droughts, water reduction in the north seems to be driven by issues of water quality and protection of the environment.

### 4.1 WATER SUPPLY MANAGEMENT IN ATHENS DURING A DROUGHT PERIOD

Athens is a metropolitan city of approximately 3.5 million inhabitants. Due to the expansion of the city and the increase of the material standard of living in recent years, there has been a significant increase on the water consumption during the period prior to 1989. However, a severe drought period that lasted from 1989 until 1995 brought into mind that it is not possible to increase water consumption at the same rate, and it was necessary to make an effort to reduce consumption. A number of measures were taken to alleviate the problems caused by the drought. These measures included information of the public, flexible water prices based on consumption, prohibiting using water from the mains for garden irrigation, swimming pools etc., drilling new groundwater wells, planning for other short term (e.g. carrying water by tankers, etc.) and long term solutions (Euinos project). These measures assisted in alleviating the problems caused by the drought and, with the cooperation of natural forces, Athens now enjoys plenty of water and is in a much safer condition.

Note that for the period prior to 1989 the annual increase of water consumption was of the order of 10% or more, but since 1990 the consumption has been significantly reduced. The consumption today is quite lower than that of 1989 even though the city has expanded since then, and the reservoirs are now quite full.

In this discussion, the outline of the Athens water supply system is presented first, then the hydrologic properties of the drought of 1989-95 are described, and finally the measures taken as well as the results achieved are presented.

#### *4.1.1 The Athens Water Supply System Today*

The water supply sources of the metropolitan Athens area are presented in Fig. 2 and are summarized below (Passios, 1997; personal communication).

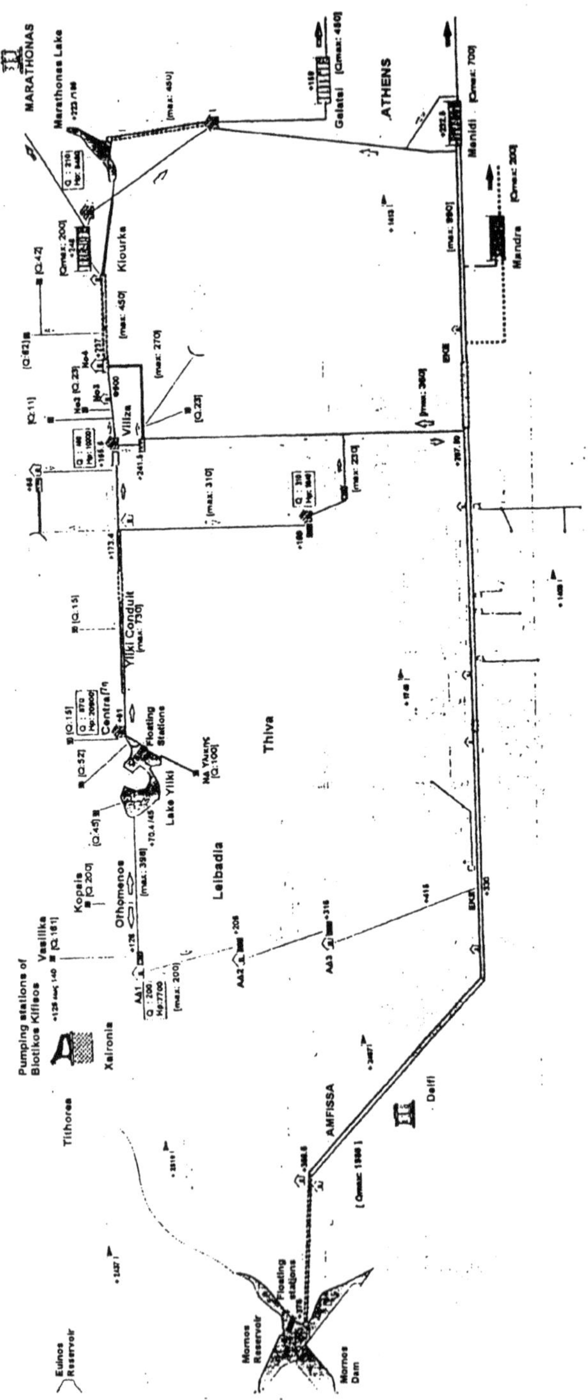


*Fig. 2. Water supply system of Metropolitan Athens area (map not in scale)*

A. Mornos and Euinos Reservoirs

The Mornos reservoir was constructed in 1980 and constitutes the most important water supply source available. The Euinos dam has been recently completed and is now in operation. Before 1989/90 it was believed that the system of the two reservoirs would offer 450-500-million $m^3$/ year. However, after the extended drought period that lasted from 1989 until 1995, the estimation was reduced to a safer figure of approximately 380 million $m^3$/year. This is a conservative estimate and should be sustained even during periods of extended drought. Under more normal conditions the system of two reservoirs is estimated to supply approximately 450 million $m^3$/year. This volume is sufficient to satisfy the needs of the city safely for at least the next ten years.

B. Yliki lake

Lake Yliki is a surface water resource that is in close dynamic interaction with the groundwater resources of the region. Because of this interaction, hydrologic simulation is extremely difficult and it is thus very difficult to estimate the safe long-term yield of this system. However, during the long drought period of 1989-95 the water company was forced to pump water at the maximum possible rate and hence made a large scale experiment on the long term supply of the system. Based on this experiment, it was possible to estimate the supply of the lake at 140 million $m^3$/year. This estimated volume considers the losses of water from the lake to groundwater aquifers and possibly to the sea. Based on field studies it was concluded that it is extremely difficult to eliminate these water loses by waterproofing.

C. Groundwater Wells

The Water Company owns many well fields. The groundwater supply from the well site of Northeast Parnitha mountain is evaluated at 50 million $m^3$/ year while at the site of Yliki the long term supply is evaluated at 20 million $m^3$/ year. The calculations of the groundwater supply of the Biotikos Kifisos aquifer vary. An optimistic estimate is 370 million $m^3$/year (this includes water used for irrigation) but a number of hydrologists believe that groundwater flow in this region is towards the lake Yliki and thus water pumped from Biotikos Kifisos aquifer appears as a deficit at Yliki lake. Based on this hypothesis, the net aquifer supply is estimated at 55 million $m^3$/year. Furthermore, even this value is quite uncertain and the Water Company has decided not to consider this supply in their long-term plans.

The safe supply of the various sources is summarized in the following table.

| **Athens Water Resources** | **Long term safe supply (million $m^3$/year)** |
|---|---|
| Mornos & Euinos reservoirs | 380 |
| Yliki lake | 140 |
| Groundwater resources | 70 to 125 |
| Total supply (in the source) | 590 to 645 |
| Estimation of loses 10% | (-60) to (-65) |
| **Total water supply (distributed)** | **530 to 580** |

Based on the above data the available long-term safe supply is between 530 to 580 million $m^3$/year, which should be sufficient to satisfy the needs of the city for at least the next 10 years. Under normal conditions (no droughts) the system may cover the needs of the city from the lower cost (based on natural flow) system of Mornos & Euinos. Lake Yliki and the groundwater aquifers play an important role in minimizing risk and satisfying the needs during drought periods and should therefore be maintained in a good operating condition.

Fig. 2 also displays the water conduits used for transportation of water to the city of Athens. Flow from the reservoirs of Mornos and Euinos to the city is natural, using an open channel of 200-km length. Lake Yliki is connected to Athens through a channel of about 60-km length, which includes some pumping stations. Water from lake Yliki and groundwater from NE Parnitha flows into the Yliki channel while water pumped at Biotikos Kifisos can be pumped either to Mornos or Yliki channels. The Yliki channel ends at Marathonas reservoir, which offers additional space for untreated water storage. It is estimated based on a hydraulic model, that the total carrying capacity of the water transport system from the reservoirs to the treatment plants under normal operating conditions is 526 million $m^3$/ year. This capacity should be able to cover the city water needs for at least the next 10 years. The system of conduits that connects Mornos reservoir to the city may transfer 460 million $m^3$/year, which is the anticipated supply of both Mornos and Euinos reservoirs under normal conditions. Thus, the safety of the system depends on the uninterrupted operation of the Mornos conduit.

When water reaches the city it is treated in one of four water purification units as shown in Fig. 2. The capacity of these units is 533 million $m^3$/year. The quality of the pretreated water is generally considered to be very good. The overall quality of the water offered by the Water Company is in accordance with the rules set by EU. The only problem seems to be a lack of good protection of the open water channels carrying water to Athens from pollution along their length.

The water distribution system of the city of Athens has expanded based on immediate needs and not on rational planning and design, and is therefore quite complex and in some cases not very effective. Some details of the distribution system are found in the following table

| **Population** | 1990 (estimated) | **3.400.000** |
|---|---|---|
| **Annual consumption** | 1995 (million $m^3$) | **285** |
| **Number of connections** | households | **1.580.000** |
| | industrial | **20.000** |
| | rate on increase per year | **30.000** |
| **Length of pipes** | main pipes (400-1800 mm) | **1.500** |
| | secondary pipes (100-400mm) | **5.500** |
| **Storage tanks** | number | **40** |
| | capacity $m^3$ | **190.000** |
| **pumping stations** | number | **70** |
| **elevation of main pipes** | m | **159-248** |
| **elev. distribution system** | m | **0-600** |
| **water losses** | % of treated water | **10-15%** |
| | $m^3$/day/km | **9** |

The area covered by the water company has an elevation from 0-600 m, and is divided in zones of a height difference of approximately 30-m. There are three very large zones, which cover about 59% of all connections at the center of Athens and these are the zones of +90, +125 and +159 m. The water company is obliged to offer water without interruption, except in cases of technical problems, with a pressure between 1 to 12 bar and the company's goal is to offer pressures of 6 bar. It is noted that citizen complains have been reduced in recent years and are now practically down to zero. Most complains were made during the drought of 1989 when the water distribution system was not able to cover the needs.

*4.1.2 Characteristics the 1988-1995 Drought*

A. Hydrological Characteristics of the Drought

The reservoirs at Mornos and Yliki are replenished with water from the rivers of Mornos and Biotikos Kifisos respectively. The water input to Yliki comes mainly from Biotikos Kifisos having a catchment of 2010 $km^2$, which through a channel discharges to the lake. The flow rates of Biotikos Kifisos are measured since 1906. The catchment of Mornos river has a size of 557 $km^2$ at the site of the reservoir. Discharge data of Mornos river are measured since 1951 at various locations along the river.

Figure 3 plots rainfall and discharge data of B. Kifisos since 1920 and the data indicate a long-term trend towards lower values in recent years (Nalmpantis et. al. 1994). The data at the Mornos catchment are of shorter length and do not indicate a long-term trend.

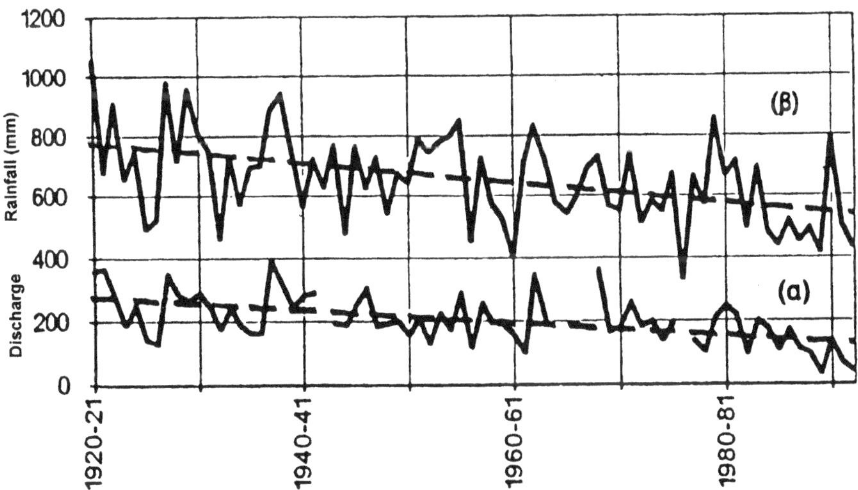


*Fig 3. Rainfall and discharge of Biotikos Kifisos River (Nalmpantis, et. al. 1994)*

During the period from 1988 until 1995 there was a severe drought in Greece which resulted in reduced discharge of the two rivers supplying the reservoirs. The flow rates of the two rivers Mornos and Biotikos Kifisos during this period were reduced to very low levels compared with previous years. In fact the discharges of both rivers during the years 1989-90, 1991-92 and 1992-93 were the lowest observed during a long historic period (Nalmpantis, et al 1994).

It is of interest to investigate the water discharge of the rivers during the drought period. Fig. 4 shows the discharge of Mornos and B. Kifisos for years 1987-1993 (Nalmpantis, et. al, 1994). The discrete horizontal line represents the average values for the period before 1987-88 while the continuous horizontal line represents the average value for the period 1987-1993. The data display a dramatic decrease of discharges for both rivers, which tend to be about half of the average discharges in previous years. This indicates the seriousness of the drought during this period.

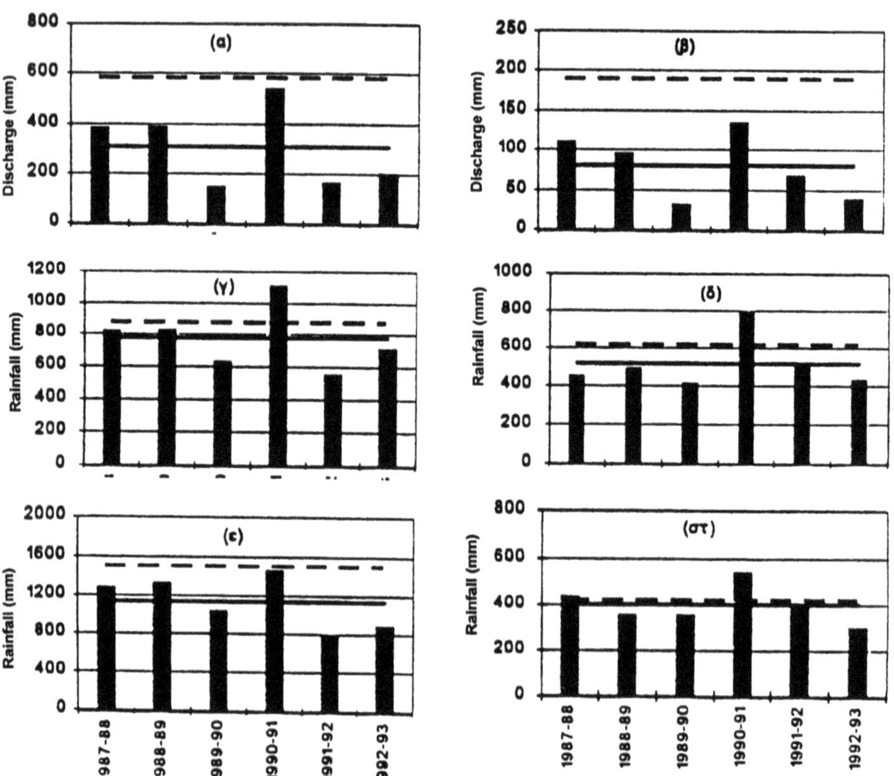


*Fig 4 Rainfall and discharge of rivers during the drought period (a) discharge of Mornos river (b) discharge of B. Kifisos river., Other data refer to rainfall measured in various stations in the region (Nalmpantis, et. al. 1994)*

B. Water Consumption and Water Reserves During the Drought

Figure 5 plots the annual water consumption in the city of Athens as a function of time in the years from 1977 until 1996 (Passios, 1997; personal communication). The graph indicates a very quick increase of consumption from 1983 until 1989, which is the year when the aftereffects of the drought were first felt. The consumption increased at an annual rate of about 5-10%, which was due to expansion of the water distribution system as well as due to increased consumption of water by individual users.

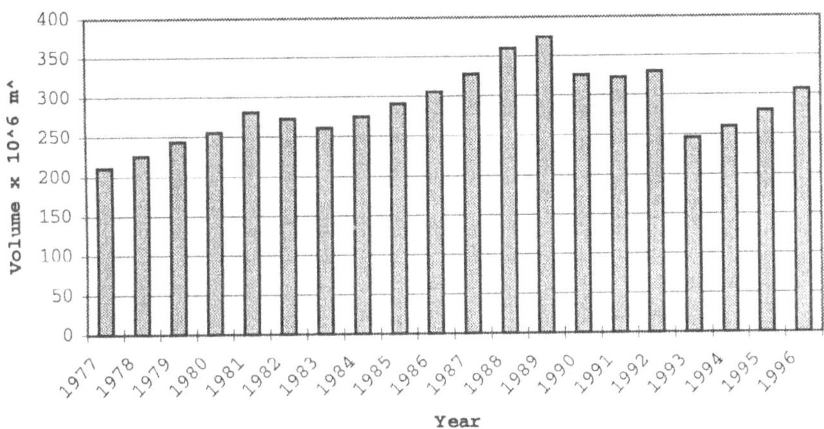


*Fig. 5 Annual water consumption in the city of Athens in $10^6$ $m^3$*

The water reserves in the reservoirs of Mornos, Yliki and Marathonas are plotted in Fig. 6 for the period between 1987 and 1997, (Passios, 1997; personal communication).

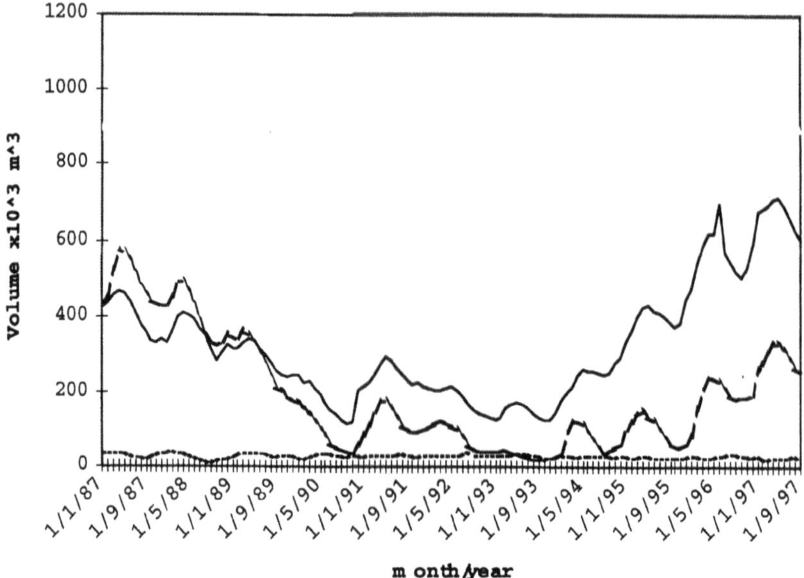


*Fig 6. Volume of water in storage in the reservoirs. 1. Lower curve is the storage of Marathonas reservoir, 2. Next curve represents storage of Yliki lake, 3. Following curve represents storage of Mornos reservoir and the top curve represents the cumulative storage of all reservoirs.*

As shown from the above figure, the reserves of all reservoirs went to minimum levels for the first time at the end of summer of 1990 after the severe drought of 1989. Comparison with the consumption data indicates that consumption was reduced during this period and this reduced consumption was kept more or less constant until 1993 when another drought stroke. Due to the severity of the new drought additional measures were enforced that led to reduction of consumption to even lower levels as indicated by Fig. 5. It is stressed here that if the savings in the consumption of 1993 did not take place, the reservoirs would have practically dried out. The following section discusses in some detail the measures taken to alleviate the drought problems in Athens during this long drought period.

#### *4.1.3 Immediate Measures for Problem Alleviation*

According to Christoulas (1994), (Professor Christoulas was the general director of the water company during the drought period), after a very dry winter season, at the end of spring of 1989 the system was already very susceptible even to a mild drought. This danger was recognized and was addressed publicly in 1989. The eventual water consumption in 1989 was 376 million $m^3$ which led to an estimated consumption of 400 million $m^3$ per year in the following years. At the beginning of hydrologic year 1989-1990 (1/10/89) and based on the existing pumping stations installed in the reservoirs, the useful water storage of all reservoirs was estimated around 240 million $m^3$. This painted a very bleak picture for the future particularly if the drought period would last for more than 2 years.

The first measure was to increase the availability of water by installing new floating pumping stations in the reservoirs, which added an additional volume of 200 million $m^3$ to the useful reservoir storage. Then a series of groundwater wells were drilled which started to operate in the spring of 1990. Due to a long election period and the associated political risks, the authorities were reluctant to increase water prices in order to conserve water. The water company made an effort to alert consumers but this did not produce any measurable results.

The hydrological conditions in the following years were much worse than anticipated as illustrated by Fig. 3, Fig. 4 and Fig. 6. After a very dry winter 1989-90 and due to a high consumption rate, the useful water reserves in the spring of 1990 were only 313 million $m^3$, including the extra volumes added by the new floating pumping stations. The data of the first months of 1990 showed a consumption increase of 4,5% relative to the 1989 data which indicated a serious danger that the capital would run out of water before next fall when the rainy season would start. The only solution that could offer immediate results was to reduce the rate of consumption. This objective was accomplished using a pricing policy based on new tariffs with escalating prices for increased consumption, parallel to an information and advising campaign. Suppression measures such as fines or other punishments for over-consumption were not exercised. Notice that before these measures took effect, water prices were quite low at a level lower than the cost.

The underlying notions behind the new pricing measures were:

a) high increases, particularly for high consumption rates, so that even consumers which are well off economically would have a serious incentive to reduce consumption for swimming pool, garden or other extravagant uses,
b) highly scaled pricing even at the lowest consumption levels, so that even smaller users to consider the non-proportional increase of the water bill with increased consumption,
c) special price reductions for families with 3 or more children,
d) threefold increase of water for industrial uses and special reductions for industries using water recycling practices.

The purpose of the public campaign was to inform the users about the real danger of drought, persuade them of the need to save water and inform them about ways to do so. The campaign was designed using various means such as newspapers, radio, TV, large posters in the city, information leaflets included in the bills, etc. In parallel to the above measures there was a serious effort of the water company to quickly fix leakage problems and reduce the water loses in the distribution network.

The above measures led to a reduction of consumption in the range of 20% for the period between 1/5-30/10/90 relative to the 1989 consumption for the same period, and helped alleviating the danger during the dry season of 1990.

The following winter 1990-91 was quite rich in rainfalls and thus the fear of a persistent drought was eased. The government reduced the prices of the first two levels on 1/7/91 and due to economic and psychological reasons there was a subsequent increase of consumption in years 1991 and 1992 as shown in Fig. 5.

Meanwhile the hydrologic situation developed very poorly in the following two years 1991-92 and 1992-93 (see Fig. 3, Fig. 4 and Fig. 6) which created fear of a persistent drought. The groundwater wells started to offer significant amounts of water, but these seemed to be insufficient to cover the needs of the city. This condition required the application and strong enforcement of new measures for water savings in 1993 as follows:

a) it was forbidden to use water from the mains for swimming pools, garden irrigation, car and pavement washing, etc.,
b) setting a maximum limit of consumption at a level of 70% compared with the consumption of the year before, and
c) possibility of quite high fines for not obeying the rules. At the same time there was a systematic program of public information.

The results of the new measures were quite spectacular and as it turned out saved the city from disaster. The eventual consumption in 1993 did reduce to 246 million $m^3$ whereas the consumption of 1992 was 330 million $m^3$ and that of 1989 was 376 million $m^3$. Comparison of these figures with the available water stored in the reservoirs indicates that without the emergency measures the city would be in a state of dismay during the summer of 1993. The available water level on 13/11/93 was just 113 million $m^3$, which indicates that without the emergency measures Athens would completely run out of water. Of course the result would be the same if the program of groundwater utilization, which offered at least 200 million $m^3$ until 13/11/93, was not established.

During this period the water company drilled many new wells and energized old pumping stations. In addition new licenses were given to municipalities to drill wells in the Athens area for pumping water for irrigation of public gardens. During this time many private users also drilled wells in their gardens without permits.

#### *4.1.4 Other Proposals for Drought Alleviation*

There were several alternative proposals for alleviation of the problems of drought in the years 1989-1995 and two proposals were finally selected and would have been applied if the situation had gotten worst. One proposal was to use water from a large lake in the western part of Greece (lake Trihonis) and pump water to the reservoir of Mornos at a rate of 600000 $m^3$/day. Water deficit of the lake would be replenished with transportation of water through channels from Ahelooos river. This solution would disturb the environmental balance of the region and luckily did not have to be applied (Hatjimpiros and Papagrigoriou, 1994).

A second proposal was to carry water to the capital by tankers from Acheloos river, which is at the western part of Greece, at a rate of 200000 $m^3$/day. This solution would have been quite costly and there was a problem with cleaning the containers of the tankers. The above-discussed measures would have been short-term solutions and luckily did not have to be exercised since the measures of reducing consumption already managed to overcome the problem. Parallel to these short-term measures the construction of Euinos dam near Mornos went on and is now offering new reserves of water to the capital.

## 4.2 NORTH EUROPEAN CASES

In Northern Europe freshwater is considered to be an abundant resource. Fig. 7 shows the freshwater resources per capita and year for a number of European countries. Although the total resources per capita are more than sufficient in all countries, if compared to basic per capita water needs, it is obvious that especially the Scandinavian countries are in an extremely comfortable situation.

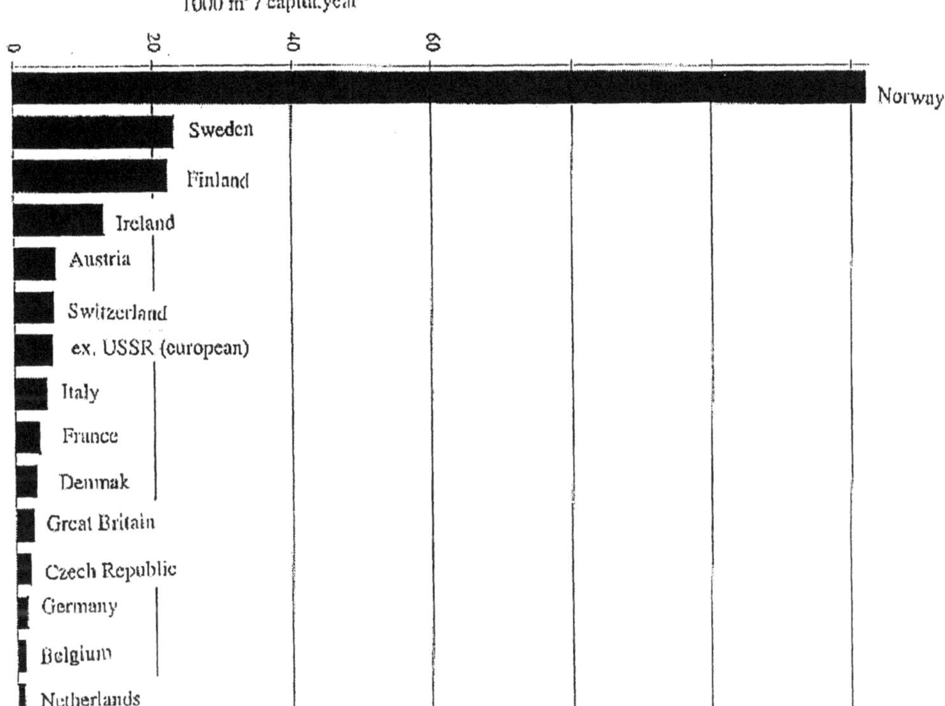


*Fig. 7 Total freshwater resources per capita and year for some European countries*

### *4.2.1 Hamburg / Germany*

Hamburg is the second largest German city located in the North of the country. The drinking water resources are largely groundwater obtained from various regional aquifers. It has been politically difficult to develop new water sources because of environmental concerns. Since at least two decades it is clear that the city has to live with the recourses that are currently available. Therefore the water company started extensive campaigns to encourage consumers to avoid unnecessary consumption. Water meters have been installed for each lot already a long time ago, and since 1987 law requires in addition, that water meters have to be installed for each apartment in houses with more than one apartment. Pricing of water service is largely based on quantity. Together with tight leakage control this resulted in a decreased consumption since approximately 1975 (Fig.8).

**Entwicklung des Wasserverbrauchs im HWW-Versorgungsgebiet**

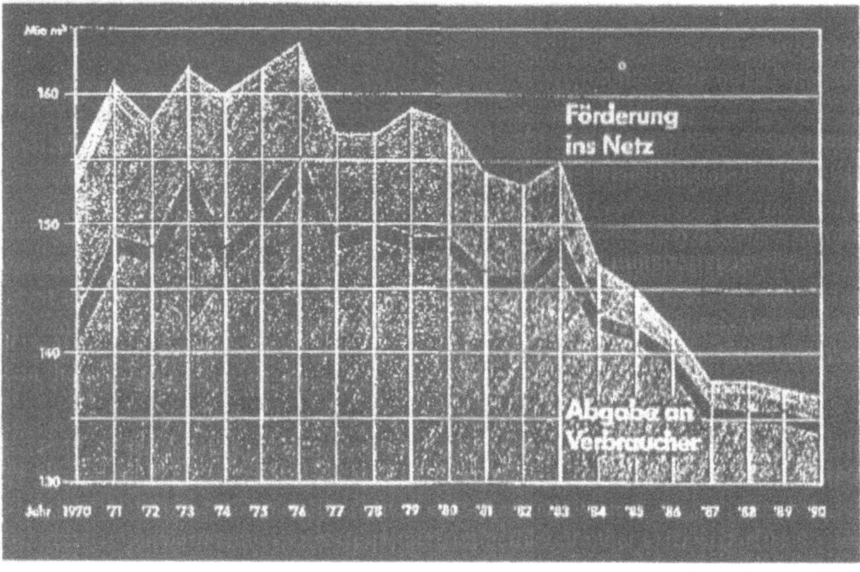


*Fig. 8. Water consumption in Hamburg / Germany 1970-90 (upper curve: total production, lower curve: total consumption), copied with permission of Hamburger Wasserwerke GmbH from a public brochure*

More recently, the water company embarked on programs encouraging customers to use rainwater for toilet flushing and garden irrigation. Several thousands of these installations were partially sponsored with direct cash subsidies and technical consultation. Fig. 9 shows a copy of an information leaflet that was distributed to the customers.

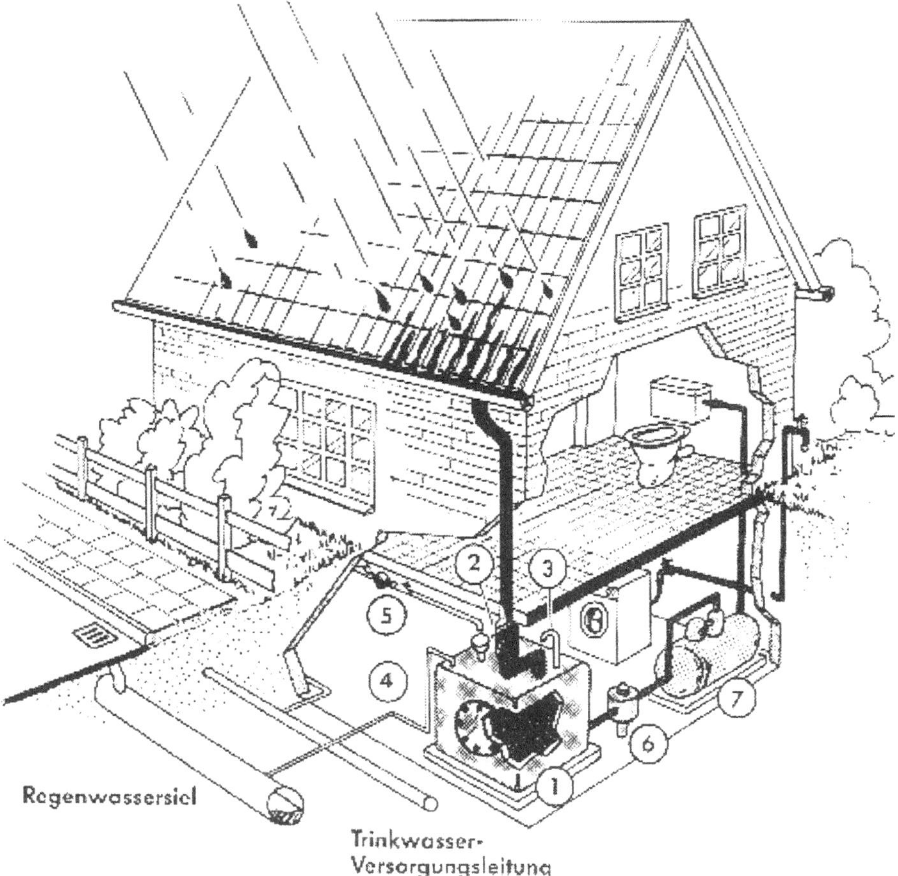


*Fig. 9. Subsidized roof runoff usage system: (1) tank, (2) coarse filter, (3) vent, (4) overflow to stormwater sewer, (5) drinking water back-up, (6) fine filter, (7) booster pump, copied with permission of Hamburger Wasserwerke GmbH from a public brochure*

In parallel the customers are informed about the economic savings if water-saving appurtenances are installed. Fig.10 shows another brochure that shows an estimation of the amount of money to be saved by installing flow-throttles in water taps, stop-handles at toilet-flushers, or more modern kitchen whiteware.

## Rechenbeispiele

| Entnahme-stelle | Häufigkeit | Liter | Liter/Jahr | Sparmaßnahme und Kosten | Neuer Verbrauch mit Sparmaßnahme Liter/Jahr | Mögliche Einsparung im Jahr | | |
|---|---|---|---|---|---|---|---|---|
| | | | | | | In Liter pro Person (1000 l = 1 m³) | in DM | |
| | | | | | | | 1 Person | 3 Personen |
|  | 2 Minuten täglich | 30 | 10.950 | Durchflußmengenbegrenzer (6 Liter/min), ca. DM 10,- | 4.380 | 6.570 | 49 | 147 |
|  | 1,5 Minuten täglich | 23 | 8.395 | Durchflußmengenbegrenzer (6 Liter/min), ca. DM 10,- | 3.285 | 5.110 | 38 | 115 |
| | 5x täglich | 50 | 18.250 | Stoptaste, (1x Normal- und 4x Sparspülung) ca. DM 15,- | 9.490 | 8.760 | 66 | 197 |
| 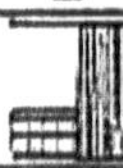 | 1x wöchentlich | 140 | 7.280 | 3-Minuten-Dusche mit Durchflußmengenbegrenzer, (12 Liter/min), statt Vollbad ca. DM 10,- | 1.872 | 5.408 | 40 | 121 |
|  | 2 Minuten täglich | 40 | 14.600 | Durchflußmengenbegrenzer (12 Liter/min), ca. DM 10,- | 8.750 | 5.840 | 44 | 131 |
|  | 1x wöchentlich | 140* | 7.280 | Neuer Waschvollautomat (58 - 75 Liter/Waschgang), mit mind. 800 Schleudertouren, ca. ab DM 1.200,- | 3.380 | 3.900 | 29 | 88 |
|  | 1x wöchentlich | 60* | 3.120 | Neuer Geschirrspüler (22 Liter/Spülgang), ca. ab DM 1.000,- | 1.144 | 1.976 | 15 | 44 |

*Geräte bis Baujahr 1985

Die Rechenbeispiele beziehen sich auf einen Wasserpreis von DM 2,67/m3 plus MwSt. und Sielbenutzungsgebühren von DM 4,70 m3

*Fig. 10. Potential economic savings after application of water saving equipment in households based on a water fee of 2,67 DM/m$^3$ and a sewer fee of 4,70 DM/m$^3$ (columns from left to right: type of equipment, frequency of use, liters per use, liters per year, investment cost, new reduced consumption per year, water savings per year and person, savings per person, savings per 3 persons), copied with permission of Hamburger Wasserwerke GmbH from a public brochure*

### *4.2.2 Zurich / Switzerland*

Zurich, the largest Swiss city is located at Lake Zurich and the lake effluent, the river Limmat. The hydrologically available freshwater resources are far larger than the demand. Nevertheless the water company applies a water pricing system that discourages customers from excessive water use:

- A flat connection fee based on the size of the water meter and the value of the house
- An annual connection fee based on the size of the water meter and house value
- A consumption fee per m$^3$
- A doubled consumption fee per m$^3$ exceeding a limit (e.g. 400 m$^3$ per year for a single-family residential house)

Although most customers have only a vague feeling for the quantities of their own water consumption it is obvious that the mere existence of the term "over-consumption rates" leads to more cautious use of water. The effect of these measures could be observed since ca. 1975 when water consumption peaked and, since then, declined steadily (ca. 12 % between 1975 and 1992).

*4.2.3 Oslo / Norway*

From Fig. 7 above it is obvious that Norway has enormous freshwater resources. Yet, water shortages are known even in Norway. The last severe episode lies only 2 years back when Oslo needed to prohibit car washing and garden watering over a period of several months. At the same time the leakage losses in the distribution system in Oslo are today larger than the actual household consumption. The leakage rate is 35 % of the total water production. This is typical for larger Norwegian cities, and is much higher than most European cities. In smaller towns leakage rates are even higher. The Oslo Waterworks are convinced that a 50% leakage reduction should be sufficient to cover increasing needs due to population growth.

One obvious explanation for the surprisingly high leakage rates in Norway is the abundance of water. It appears that at least in North European cities water losses are closely related to the available resources. From the figures above it is obvious that losses are smaller in Zurich and much smaller in Hamburg.

Another reason for the high losses in Norway is the low alkalinity of most Norwegian drinking water that makes it particularly aggressive against pipe materials. Norwegian waterworks have to fight an intensive battle against pipe corrosion and, with the current low renewal and rehabilitation rates of pipes, they are likely to loose that battle. Hence, the question for sustainable water management in Norway is focusing more on the economic than on the hydrologic consequences of water management. Also the reduction of leakage is largely an economic optimization problem rather than a physical necessity. The example of the water shortages in Oslo, however, shows that there are exceptions from this rule.

## 5. Discussion and conclusion

In order to make effective decisions regarding sustainable urban water use, first the strategic objectives must be chosen. They should be based on a regional development balanced with the protection of the environment and the character of the regions. In order to succeed in achieving sustainability we need to change our current attitudes and, where water is or will be scarce must reduce our current level of water consumption. At least for domestic and industrial use the examples above show that this is feasible without obstructing other development objectives. We should aim in satisfying basic human needs and not extravagant desires, and objectives that can be explained to and understood by most people.

In order to reach sustainability of urban water supply, greater efficiently in the design and operation of urban water, wastewater and drainage systems is necessary. However, it is not a sufficient condition! Management of a water supply system must consider in general the complete urban water system (i.e. it should include wastewater production, treatment and disposal). At the same time ecologically sustainable patterns of urban development and renewal must be applied. A mix of technical, economic and institutional measures is required. We need to bring far more technical and policy flexibility into urban water management than currently exists, with improvements in the efficiency and ecological impacts of the water sector.

Most situations of non-sustainability or even acute water emergency seem to occur because water managers "are caught by surprise". In other words, there is a lack of long term planning that includes the complex nature of urban water supply management, i.e. its impacts and interactions with regional development.

It is obvious that engineering measures to increase the technical efficiency of a water supply system cannot be the final answer to the quest of sustainability. The availability and price of water has an important impact on the regional development and vice versa. Much of this interdependency is still far from being understood and more work and research is necessary in the following areas:

- Economics of urban water management and pollution control, with particular emphasis on economies/diseconomies of scale in water infrastructure arrangement, the economics of water pollution alleviation, the choice of policy instruments, and institutional development for improved integrated catchment management in urban regions.
- Regional case studies in collaboration with urban water management agencies, particularly emphasizing issues of water design and management for new urban developments.
- Policy and planning of urban water management at state and local government levels and fostering international research links in integrated urban water management.
- Studies of the community's knowledge, attitudes and behavior in water consumption and pollution, and development of programs for community participation and cooperation.
- Decision support systems for the planning and management of urban water supply considering the above factors.

It must be noted that understanding the nature of environmental problems and how they might be solved requires much more than a scientific appreciation of environmental processes. It demands an understanding of how societies work, and how collective action within those societies is both organized and constrained. Only with that understanding is it possible to discuss how change might be achieved.

Even when the supply of water seems to be sufficient, it should be recognized that it is often better to leave the water to run free in the streams and be «lost» to the sea rather than using it for wasteful consumption in the cities and, thus, create in addition wastewater that needs to be disposed of with potential dangers to the environment. When we consider the ecological dimensions of the water supply projects we may realize that this «lost» water is often badly needed for the support and welfare of life downstream.

The dominant ideology of our days is based on a strong belief in human ability to dominate nature, through science and technology. Whether that vision is correct is better not put to test, since if it fails it might be too late for alternatives.

**Bibliography**

Christoulas D.G., "Necessity and possibilities for saving drinking water", presented in the meeting: "The water supply problem of Athens", National Technical University of Athens, April, 1994

Gaut, A.; Jenssen, P.; Lindholm, O.; Pedersen, N.E.; Schilling, W.; Ødegaard, H. "New sustainable water- and wastewater technology"(in Norwegian), preliminary report, Jordforsk, Ås/Norway, September 1997.

Hamburger Wasserwerke GmbH (Hamburg Water Works), General information and selected reports, in German.

Hatjimpiros K. and S. Papagrigoriou, "Environmental degradation from the water supply projects of Athens and the need of solutions through management" presented in the meeting: "The water supply problem of Athens", National Technical University of Athens, April, 1994

Henze, M.; Somlyody, L.; Schilling, W.; Tyson, J. (eds.), "Sustainable Sanitation", Water Science and Technology, Vol. 35, No. 9, ISBN 0 08 043290 5, 212 pages, 1997.

Napmpantis I., N. Mamasis, D. Koutsoyiannis, E. Mpaltas, E. Aftias, M. Mimikou and U. Xanthopoulos, "Hydrologic Characteristics of the drought" presented in the meeting: "The water supply problem of Athens", National Technical University of Athens, April, 1994

Passios G., Director of water supply of Athens Water Company, 1997, Personal communication

Oslo Water- and Sewerage Works Department "Master Plan Water Supply" (in Norwegian), Draft 15. Sept., Oslo, 1997.

Näf, A.; Naef, H, "Zurich's Water Supply" (in German), gwa - Gas Wasser Abwasser, 634 – 661, 1993.

Schilling, W., Mantoglou, A., et al. "Sustainable Sanitation Concepts for the 21st Century - A Decision Support System for Sustainable Water Management Policies for Sensitive Regions", Project proposal to the EU Commission, 1996.

Schilling, W., "Sustainable Management of Urban Water Systems in the 21th Century", Summary Presentation, Engineering Foundation Conference "Stormwater management - Creating sustainable urban water resources for the 21. Century", Malmö, Sweden, 7 - 12 September 1997.

**Addresses of the authors:**

Wolfgang Schilling
Department of Hydraulic and Environmental Engineering, Norwegian University of Science and Technology, N - 7034 Trondheim, Norway

Aristotelis Mantoglou
Laboratory of Reclamation Works and Water Resources Management
Dept. of Rural and Surveying Engineering, National Technical University of Athens
9, Iroon Polytechneiou, GR 15773 Zografos, Greece

# MANAGING WATER QUALITY AND QUANTITY UNDER DROUGHT CONDITIONS

Robert M. Clark
and
Jill Neal
Water Supply & Water Resources Division
U.S. Environmental Protection Agency
Cincinnati, Ohio 45268

and

Virendra Sethi
Oak Ridge Post-doctoral Appointment
Water Supply & Water Resources Division
U.S. Environmental Protection Agency
Cincinnati, Ohio 45268

## INTRODUCTION

It is common practice to select the most pristine available water source when choosing a drinking water supply. However in many parts of the world and some parts of the United States both the quality and quantity of available source water is declining. This trend together with increasing populations in urban areas is making it increasingly difficult for drinking water utilities to find adequate water sources. Compounding this problem is the tightening of drinking water standards throughout the world. One consequence of this trend is the need to explore various options, including mixing and blending of water from several sources or the installation of advanced treatment in order to provide water of acceptable quality to consumers.

This paper explores the problems of maintaining water quality under drought conditions in the context of a case study based on the North Marin Water District (NMWD) in Northern California. The NMWD serves a suburban population of 53,000 people who live in or near Novato, California. NMWD uses two sources of water: Stafford Lake and the North Marin Aqueduct. The North Marin Aqueduct is in use year round and Stafford Lake is in use during the warm summer months when precipitation is virtually non-existent and when demand is high. Novato, the largest population center in the NMWD service area, is located in a warm inland costal valley with a mean annual rainfall of 69.1 cm. There is virtually no precipitation from May through September. Eighty five percent of total water use is residential and the service area contains 13,200 single family detached homes, which account for 65% of all water use.



*E. Cabrera and J. García-Serra (eds.), Drought Management Planning in Water Supply Systems,* 216–241.

The water qualities of the two sources are very different. Stafford Lake water has a high humic content and is treated using conventional treatment and a pre-chlorination dose of between 5.5 and 6.0 mg/L. The treated water has a residual of 0.5 mg/L when it leaves the treatment plant clear well. Total trihalomethane formation potential (TTHMFP) in Stafford Lake is very high. The source of the North Marin Aqueduct is a Rainey Well Field along the Russian River. While technically ground water, the source is likely to contain a high proportion of naturally filtered water. The aqueduct water is only disinfected and is very low in precursor material with a correspondingly low TTHMFP. Both sources carry a residual chlorine level of approximately 0.5 mg/L when the water enters the system. The quality and blending issues associated with these sources and their corresponding impact on the NMWD water supply will be discussed later in this paper.

In addition to the problem of inadequate supply, the NMWD is facing increasingly stringent drinking water standards. The Safe Drinking Water Act and its Amendments are posing a major challenge to drinking water utilities in the United States. Among other requirements, the Act has established rules for filtration of surface water sources and maximum levels for total coliform and total trihalomethanes (TTHMs) in distribution systems. Potentially even more stringent regulations for many contaminants, including TTHMs and other other disinfection by-products are possible in the future (Clark and Feige, 1993). Utilities are being forced to find a balance between minimizing the formation of disinfection by-products while at the same time providing protection against microbial contamination. Utilities must also provide sufficient quantities of water to satisfy consumer demands and fire safety requirements. At times, quantity demands and quality requirements may conflict, and a utility may be required to use sources of marginal quality which makes achieving water quality goals difficult. This has been the situation in which the NMWD has found itself. Some of the current and future regulations facing drinking water utilities in the United States are summarized below.

## DRINKING WATER REGULATIONS IN THE UNITED STATES

The Safe Drinking Water Act (SDWA) mandates that EPA identify and regulate drinking water contaminants which may have any adverse human effects and which are known or anticipated to occur in public water systems. The SDWA also requires the use of filtration and/or disinfection for public water supplies that serve most of the U.S. population. Currently, several regulations attempt to control for DBPs and pathogens in public drinking water supplies. These are as follows:

- Interim total trihalomethane (TTHM) standard (promulgated 1979) - This standard pertains to all public water systems that disinfect and serve more then 10,000 people. Systems must achieve less than 0.100 mg/l of total triahlomethanes as an annual average based on quarterly measurements in the distribution system.

- Total coliform rule (promulgated 1989) - This standard pertains to all public water systems. Systems must demonstrate that the frequency of total coliform occurrence is below acceptable limits depending upon population served. Small systems which collect fewer than 5 samples per month must conduct periodic sanitary surveys.

- Surface Water treatment rule (promulgated 1989) - This standard pertains to all public water systems that use surface or ground water under the direct influence of surface water. Systems must achieve at least 3 and 4 log removal and /or inactivation for Giardia and viruses, respectively, and if filtration is not part of the treatment process, meet specific criteria for avoiding filtration.

- Information collection rule (promulgated 1997) - Large public systems are required to collect approximately $130 million worth of occurrence and treatment information concerning pathogens and DBPs. Information collected under this rule will be used with research to support the development of the interim and long-term enhanced SWTR, and Stage 2 DBP rule, which will be discussed below.

There are many comprehensive regulations being considered for promulgation. The regulations that are in development are as follows:

- Stage 1 Disinfectant/Disinfection By-Product (D/DBP) rule (proposed 7/94) - This rule would pertain to all public water systems that disinfect and is intended to reduce risks from disinfectants and DBPs. Systems would be required to achieve new limits for TTHMs, (a level of 0.080 mg/L has been proposed) the sum concentration for five haloacetic acids, bromate, chlorite, chlorine, chlorine dioxide, and chloramines. Systems using settling and filtration would be required to achieve percent reductions of total organic carbon, depending upon source water quality, prior to disinfection.

- Interim enhanced surface water treatment rule (IESWTR) (proposed 7/94) - This rule would pertain to all public water systems using surface water or ground water under the direct influence of surface water that serve populations of 10,000 or greater. The purpose of this rule is to enhance protection from pathogens, including Cryptosporidium, and to prevent increases in microbial risk while large systems comply with the Stage 1 D/DBP rule. Several regulatory options have been proposed, including systems being required to achieve a) proportionally higher levels of pathogen removal depending upon pathogen measurements in the source water, and b) fixed level removal requirements independent of pathogen measurements in the source water.

- Long-term enhanced surface water treatment rule (ESWTR) (not yet proposed) - This rule, which could include changes to the IESWTR, would pertain to all public water systems using surface water or ground water under the direct influence of surface water. The purpose of this rule is to enhance protection from pathogens, including Cryptosporidium, and to prevent increases in microbial risk for systems serving less than 10,000 people while they comply with the Stage 1 D/DBP rule.

- Stage 2 DBP rule (proposed in part 7/94 with the Stage 1 D/DBP rule; to be reproposed when more data becomes available) - This rule would pertain to all public water systems that disinfect and is intended to further reduce the levels of risk achieved under the Stage 1 rule. Only tentative limits were proposed for TTHMs ( possibly as low as 0.040 mg/L) and the sum concentration for five haloacetic acids.

- Ground water disinfection rule (GWDR) (Not yet proposed) - This rule would pertain to all public water system using ground water not under the direct influence of surface water. This rule would require all systems to disinfect except those meeting disinfection avoidance criteria. This rule is intended to enhance protection from pathogens as well to prevent increases in microbial risk while systems comply with the Stage 1 D/DBP rule.

## CURRENT WATER QUALITY SITUATION IN NMWD

Figure 1 shows the entire NMWD service area and Figure 2 is a schematic of the distribution system. As mentioned earlier, depending on the time of year and the time of day, water enters the system from one or both of the sources. The North Marin Aqueduct operates year-round, 24 hours a day. The Stafford Lake source operates only during the peak demand period from 6:00 a.m. to 10:00 p.m. and generally operates for a period of 16 hours per day. Table 1 summarizes NMWD water use characteristics.

In May of 1992 the USEPA conducted a water quality survey and laid the basis for an EPANET analysis to evaluate the potential for making changes in the network that might potentially improve water quality in the system(Clark et al. 1994). In 1996 the US EPA in conjunction with the American Water Works Association Research Foundation (AWWARF) conducted a similar study (Vasconcelos et al. 1996).

The mix of water from the two sources may range from 100% Stafford Lake to 100% North Marin Aqueduct water at a given node. Figure 3 shows the predicted percentage of water from Stafford Lake at the various sampling points utilized during the two day study. The consequences of these variable flow patterns are discussed in the following section.

The dynamic nature of the system leads to both variable flow and quality conditions utilized as the basis for this analysis. Consequently flow directions frequently change and reverse within a given portion of the network during a typical operating day. Figure 4 shows the changes in flow direction and the percentage of water from Stafford Lake penetrating the system during a typical operating day. In this figure, N5 is the designation for Stafford Lake water. As can be seen, the system experiences dramatic flow reversals, depending on which sources are operating.

## US EPA WATER QUALITY STUDY

In order to characterize the NMWD water quality, the Water Supply and Water Resources Division designed a sampling protocol and sent a team of investigators to work with NMWD during the period 27-29 May 1992. Table 2 summarizes the results from this study(Clark et al. 1994). As can be seen from Table 2 water quality can change dramatically over time in the system (Clark et al. 1994). For example, at the "eighth street" sampling point, $CHCl_3$ levels vary from 38.4 μg/L to 129.9 μg/L over the two day period. This variability is due to the penetration of the water from the two different sources. As can also be seen from Table 2, a sample taken at "eighth street " may consist of water from the Aqueduct or Stafford Lake or a blend of both. However, over the sampling period, trihalomethane levels from the two sources were relatively constant. Given the extreme differences in water quality from the two sources and the variability in percentage source water penetrating within the network it is reasonable to assume that mixing and blending of water is the factor affecting water quality.

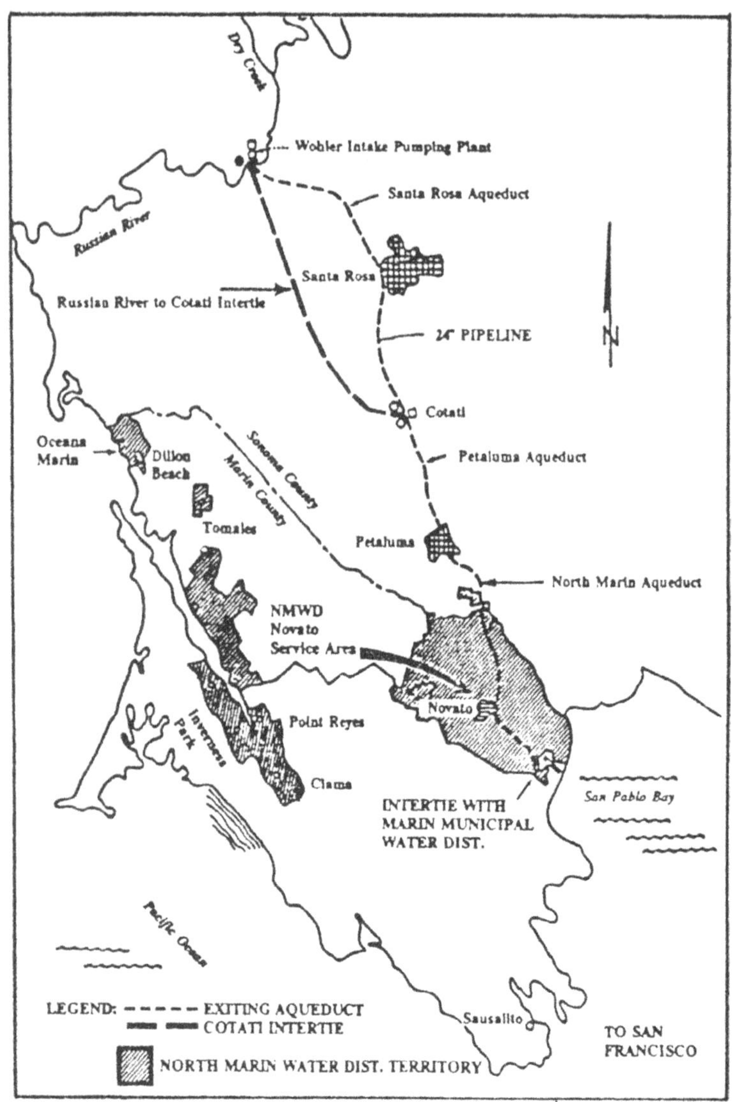


Figure 1. North Marin Water District (NMWD) Service Area

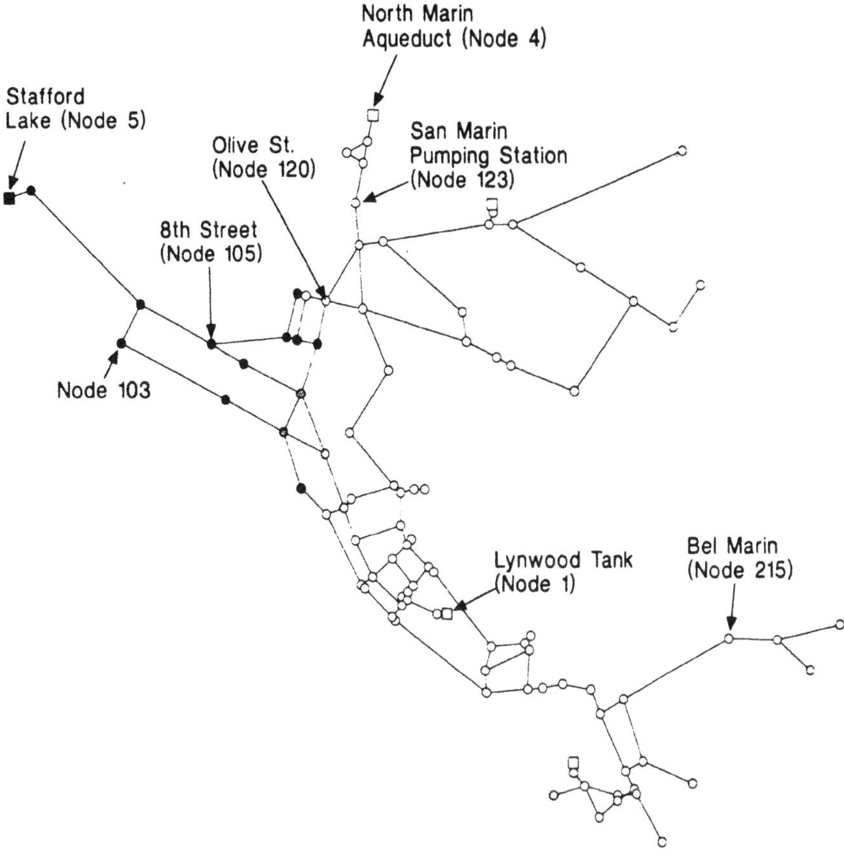


Figure 2. Schematic of North Marin Water District Distribution System

TABLE 1- NORTH MARIN WATER DISTRICT WATER USE CHARACTERISTICS

| ITEM | VALUE |
|---|---|
| Service Area | 259 $km^2$ |
| Principal service center | Greater Novato Area |
| Population | 53,000 |
| Character | Suburban (near San Francisco) |
| Normal Rainfall | 69.1 cm/year |
| Normal reference evapotransporation ($ET_0$) | 1,118.8 cm/year |
| Applied water requirement for cool season grasses | 70.6 cm/year |
| Surface Water Supply<br>North Marin Aqueduct<br>Stafford Lake | <br>83%<br>17% |
| Accounts Metered | 100% |
| Overall per capita use | 513.0 L/capita/day |
| Distribution of metered water use<br>Residential<br>Commercial | <br>84.7%<br>15.3% |
| Annual residential use:<br>Single family<br>Townhouse/Condo<br>Mobile Home<br>Apartment | <br>67.3% (547.2 L/capita/ day)<br>12.7% (437.0 L/capita/day)<br>16.6% (292.6 L/capita/day)<br>3.4% (300.2 L/capita/day) |
| Unaccounted-for water use and water loss | 5.7% |

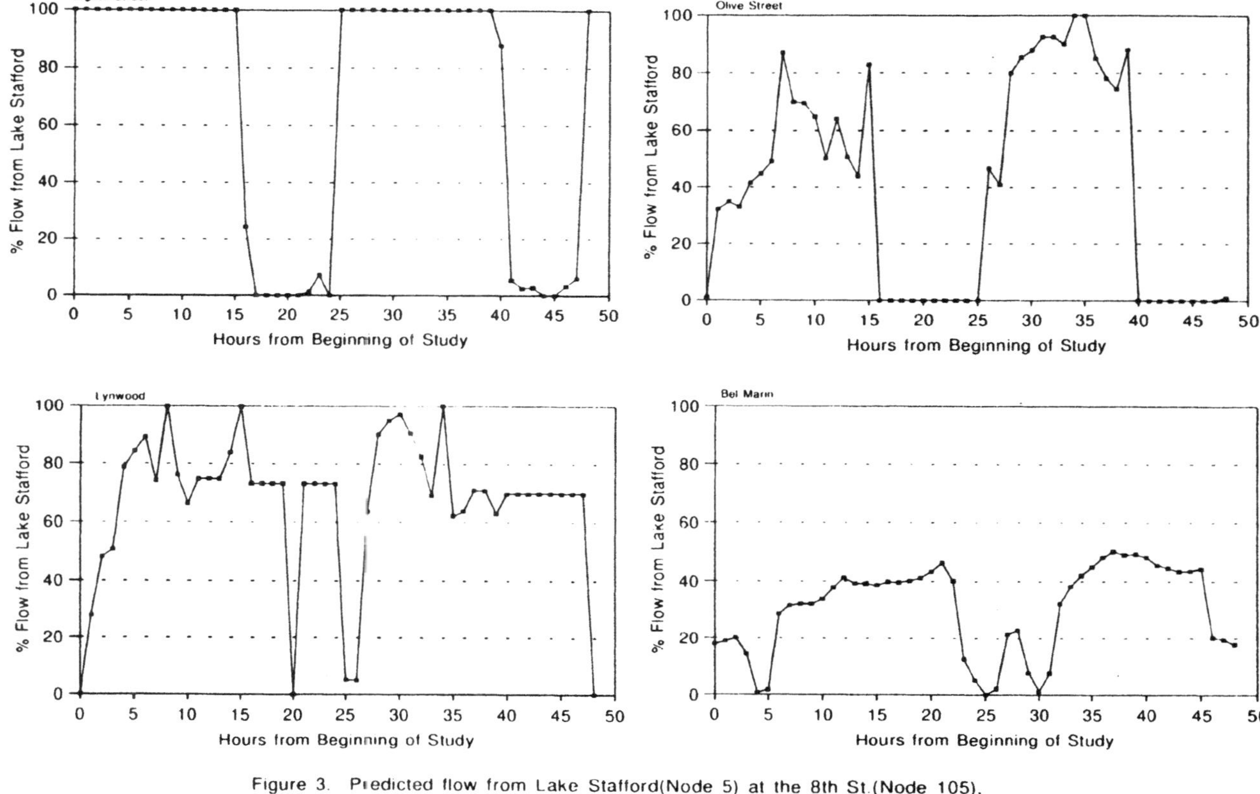


Figure 3. Predicted flow from Lake Stafford(Node 5) at the 8th St.(Node 105), Olive St.(Node 120), Lynwood Tank(Node 1), and Bel Marin(Node 215) sampling sites in percent

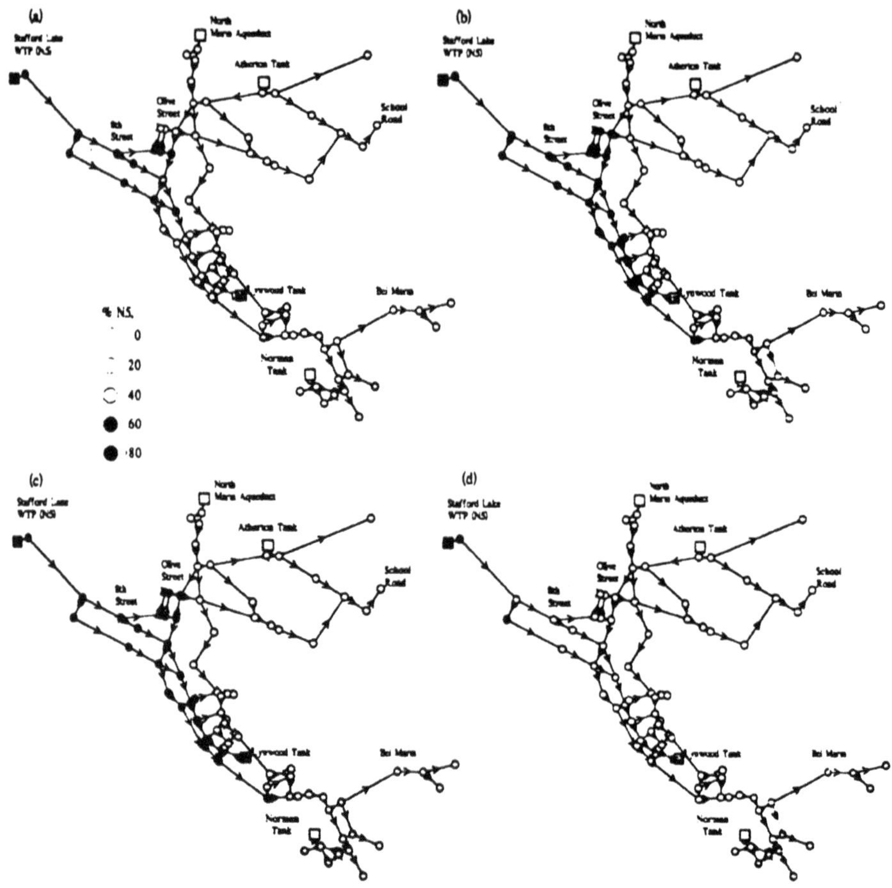


Figure 4. North Marin hydraulic calibrations at (a) 06.00, (b) 12.00, (c) 18.00 and (d) 24.00 hours

TABLE 2. WATER QUALITY CHARACTERISTICS OF
NORTH MARIN WATER DISTRICT
SOURCES AND SAMPLING POINTS

| | STATION | | | | | |
|---|---|---|---|---|---|---|
| | AQ | Bel Marin | Eighth St. | Lynwood | Olive | STP |
| $CHCl_3$[a] (μg/L) | 0.8-3.0 | 9.6-81.8 | 38.4-129.9 | 3.6-124.5 | 90.-134.7 | 108.0-140.5 |
| $CHBrCl_2$[a] (μg/L) | 3.1-6.4 | 6.3-22.0 | 12.4-30.7 | 4.2-28.9 | 5.2-30.2 | 25.8-32.1 |
| $CHBr_2Cl$[a] (μg/L) | 7.0-11.4 | 6.3-8.6 | 2.7-8.3 | 4.0-8.4 | 3.0-10.3 | 2.4-3.8 |
| $CHBr_3$[a] (μg/L) | 4.8-7.5 | 2.9-4.8 | <0.1-4.7 | <0.1-5.6 | <0.1-6.1 | <0.1 |
| UV[b] | 0.011-0.012 | 0.019-0.063 | 0.032-0.094 | 0.015-0.092 | 0.015-0.093 | 0.088-0.095 |
| DCAA[c] (p.p.b.) | <0.80 | <0.80-3.97 | <0.80-43.00 | <0.80-19.10 | <0.80-36.11 | <2.90-53.32 |
| TCAA[c] (p.p.b.) | <0.78 | 1.79-22.63 | 9.33-52.56 | 0.78-59,06 | 1.37-49.69 | 43.54-61.92 |
| BCAA[c] (p.p.b.) | <1.89-2.27 | <1.39-2.07 | <1.89-6.50 | <1.89-2.94 | <1.89-4.46 | 5.65-7.63 |
| DBAA[c] (p.p.b.) | <0.99-7.08 | <0.99 | <0.99 | <0.99 | <0.99 | <0.99 |
| Temp (°F) | 66-70 | 68-73 | 68-73 | 68-73 | 67-72 | 67-71 |
| $Cl_2$-F[d] | 0.66-0.45 | 0.02-0.22 | 0.05-0.61 | 0.00-0.55 | 0.02-0.41 | 0.40-0.58 |
| $Cl_2$-T[d] | 0.32-0.45 | 0.12-0.23 | 0.17-0.84 | 0.06-0.56 | 0.14-0.51 | 1.08 |
| pH[e] | 7.40 | - | - | 7.50-8.36 | - | 8.90-9.10 |
| Na[f] (μg/L) | 9.50-10.00 | 9.20-20.00 | 12.00-29.50 | 9.20-27.00 | 9.40-29.00 | 27.00-30.00 |

a EPA Method 551. Determination of Chlorination Disinfection By-Products and Chlorinated Solvents in Drinking Water by Liquid-Liquid Extraction and Gas Chromotography with Electron-Capture Detectopm/ EPA/600/4-90/020. Methods for the Determination of Organic Compounds in Drinking Water, Supplement 1 (July 1990).

b Method 5910A. *Standard Methods for the Examination of Water and Waste Water*, 17th edn 1989, Supplement 1

c EPA Method 552 (Same as Reference a).

d Method 409E (Same as Reference b).

e EPA Method 150.1 Methods for the Analysis of Water and Wastewater, EPA/600/4-79-020 (July 1979).

f EPA Method 273.1 (Same as Reference E).

Figure 5 shows the time formation curves for TTHMs and chlorine demand from the two sources.

## MODELING TTHMS

In order to test this mixing hypothesis, the average values for TTHMs at each of the sources were utilized. The average TTHM levels for Stafford Lake water and for the aqueduct were 151.1 μg/L and 20.7 μg/L respectively. These TTHM levels were assumed as conservative and TTHM levels at each sampling station were calculated based on the percentage of water from a given source at that point over time. This assumption was verified with field and bench testing. If, at given point in the system, at a given time 50% of the flow is from Stafford Lake and 50% from the aqueduct, the predicted TTHM values at that point would be 85.9 μg/L. Figure 6 shows the results of this approach at the various sampling points.

As can be seen the NMWD cannot meet existing drinking water standards under current operational conditions when both sources are in use. The remainder of this paper explores various options that might be employed in order to improve water quality in the NMWD water system. The first two options will involve modifications to the network in which new pipes are added in order to improve mixing and blending. The third system modification to be examined will be the construction of a tank to serve as a mixing chamber for the two sources. Finally, the cost of installing advanced treatment of the Stafford Lake water will be considered.

## SYSTEM MODIFICATIONS TO ACHIEVE WATER QUALITY GOALS

Several system modifications were simulated with the goal of reducing TTHM levels in the water delivered to the consumer. The first modification considered was the addition of a transmission pipe from the North Marin Aqueduct to the Lynwood tank, which is the major tank in the system. Figure 7 (modification 1) shows this option. Another approach considered was to isolate the north western quadrant of the network and to install a pipe from the Stafford Lake source to a point that intersects a major transmission line coming from the North Marin Aqueduct (modification 2) as shown in Figure 8. The third option (modification 3) was installation of a tank that would allow for mixing of the two sources, as shown in Figure 9 (modification 3).

In order to demonstrate the water quality effects of these modifications , a time history of water quality at the Lynwood tank and three nodes was selected for examination. The costs associated with each modification were also calculated.

## SYSTEM MODIFICATION ONE

Modification one is a 4,975 meter (16,219 ft) long pipe, 76.2 cm (30 in) in diameter designed to route water from the North Marin Aqueduct to the Lynwood tank. This tank is a major component of the network and routing high quality water directly to the tank will facilitate blending of the two sources. In order to demonstrate the effect of installing this pipe, a time history of water quality at three nodes and the Lynwood tank were examined for a 336 hour (2 weeks) period. Figure 10 shows a time history of water quality at the three nodes and the

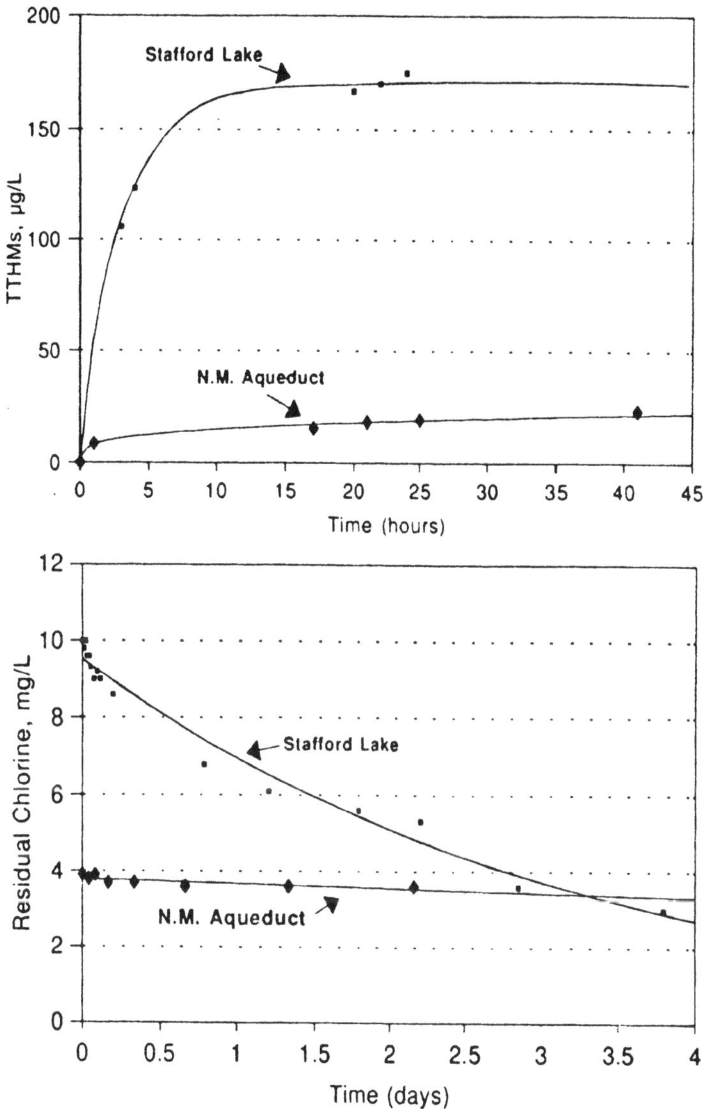


Figure 5. TTHMs formation potential curves and chlorine demand curves for Stafford Lake and North Marin Aqueduct

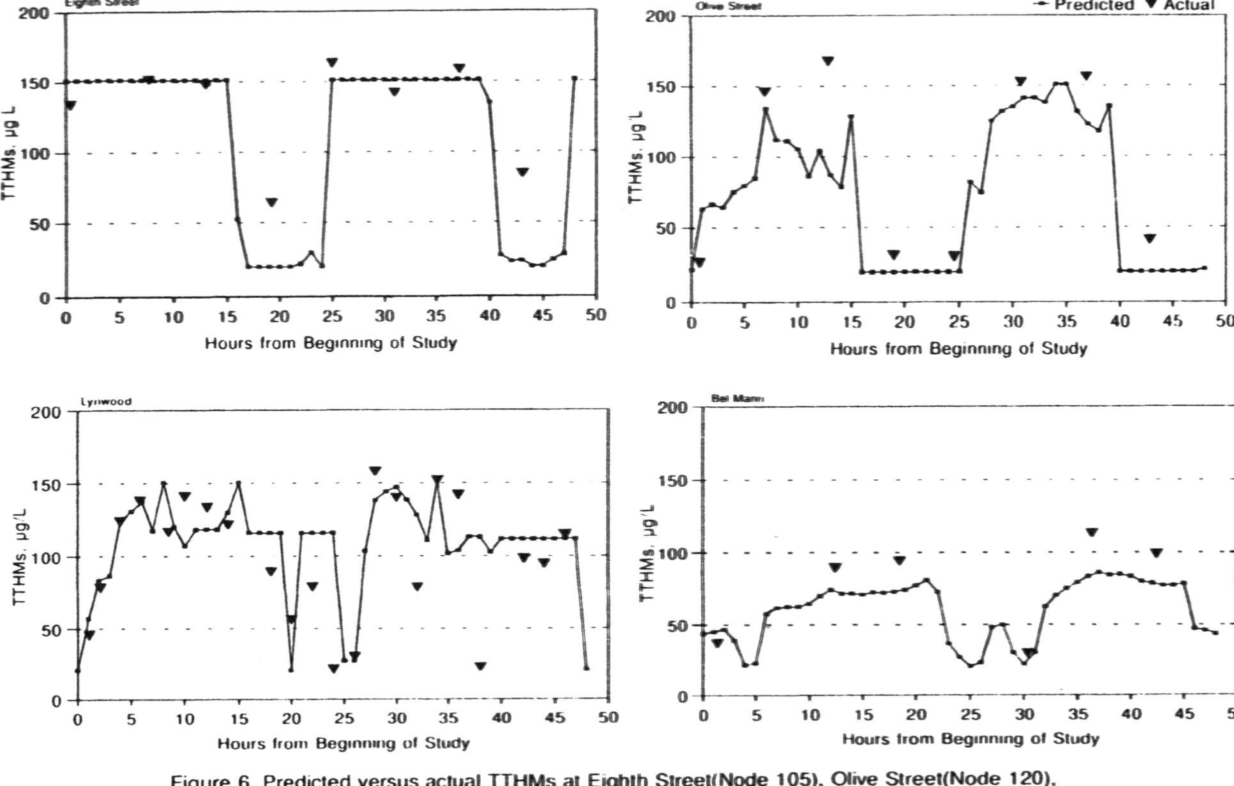


Figure 6. Predicted versus actual TTHMs at Eighth Street(Node 105), Olive Street(Node 120), Lynwood Tank(Node 1), and Bel Marin(Node 215) sampling sites

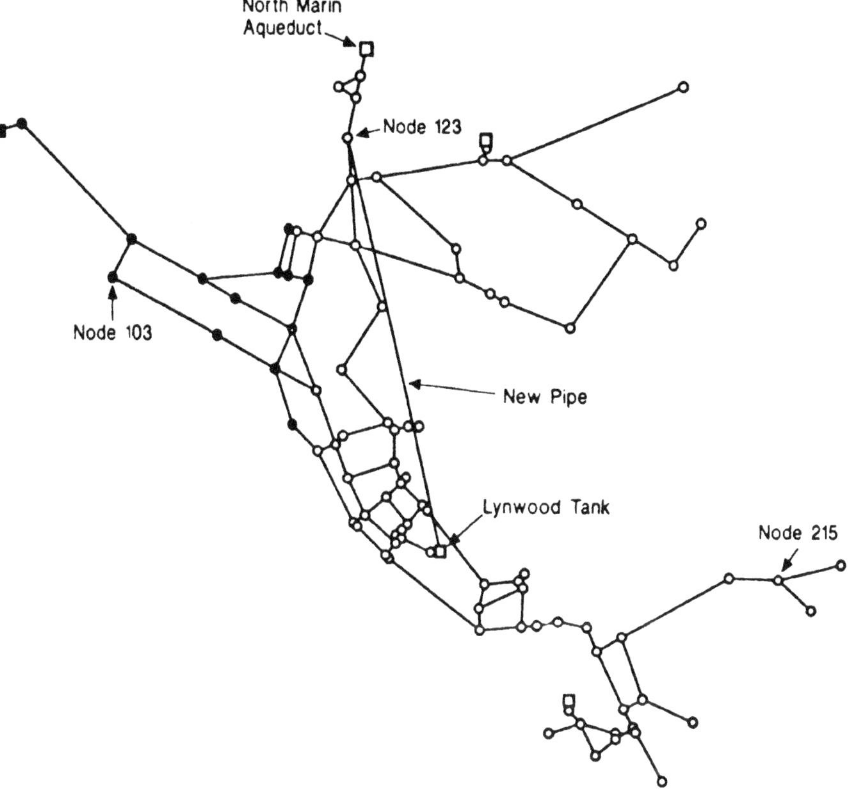


Figure 7. Modification Number 1: Pipe from North Marin Aqueduct (Russian River) Source to Lynwood Tank

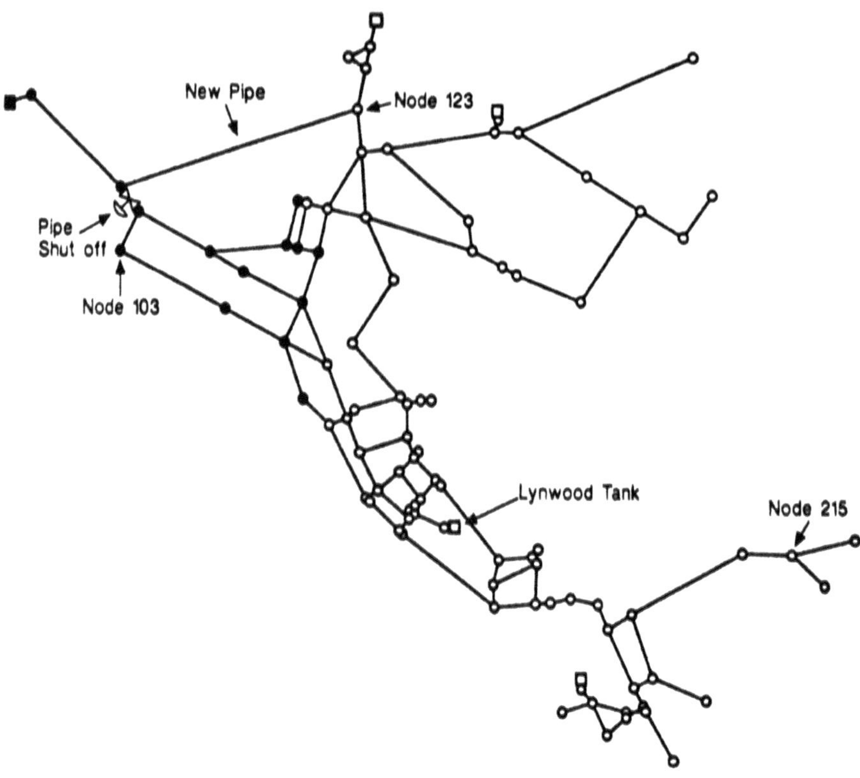


Figure 8. Modification Number 2: Pipe from Lake Stafford Source to Intersect Pipe from North Marin Aqueduct

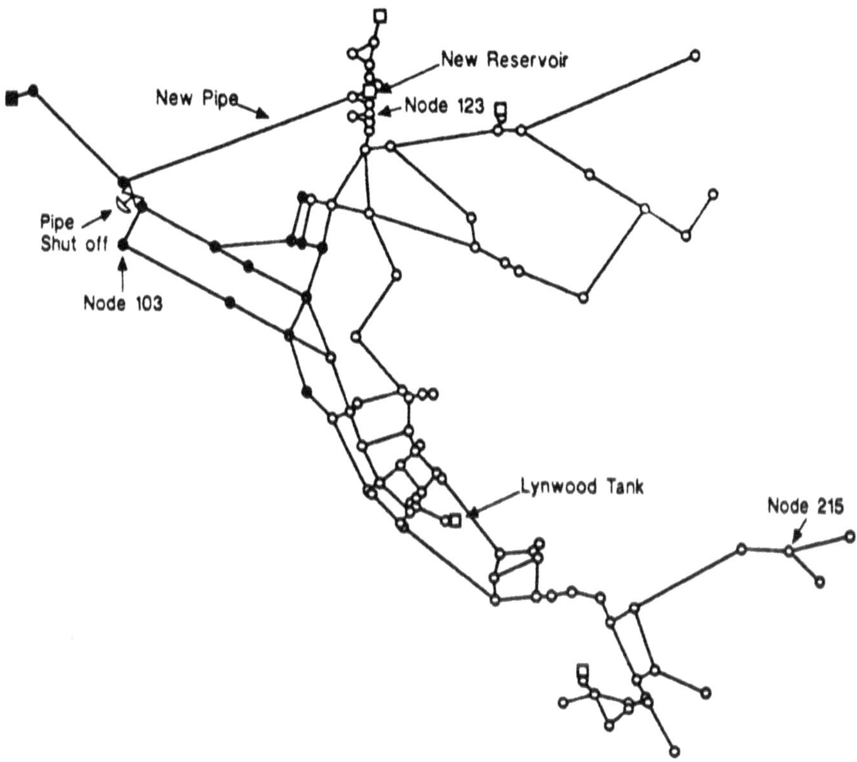


Figure 9. Modification Number 3: Addition of Reservoir to Blend Lake Stafford and North Marin Aqueduct Water

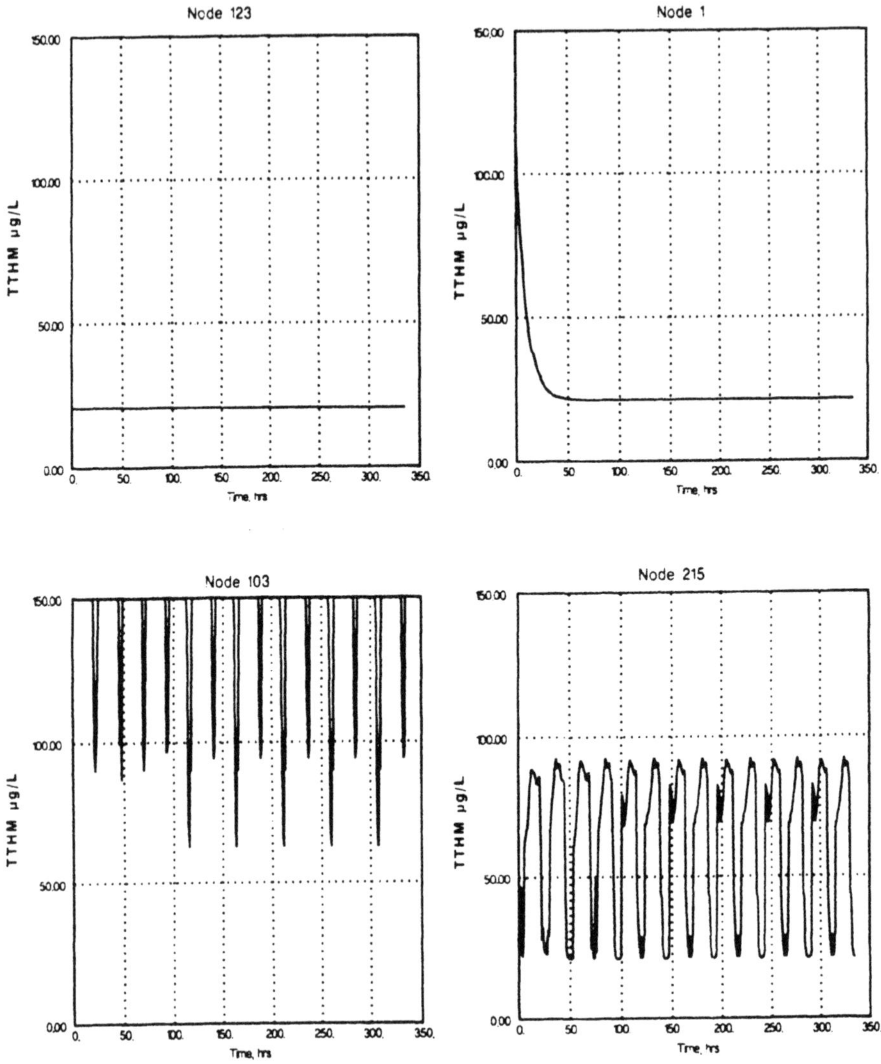


Figure 10. Water Quality Variations Over Time at Nodes 123, 103, 1, and 215 Under Modification 1

Lynwood tank. As can be seen, at times the TTHM levels are below the standard of 100 μg/L especially at nodes 123 and 215. However, the TTHM levels at node 103 in the northwestern quadrant are still significantly above the currently regulated TTHM level.

## SYSTEM MODIFICATION TWO

System modification two consists of the installation of a valve between nodes 501 and 101 and the installation of a 1,829 meter (6,000 ft) long line, 45.72 cm (18 in) in diameter to intersect the transmission line which carries primarily North Marin Aqueduct water. Figure 11, shows a time history of water quality for the three test nodes and the Lynwood tank. As can be seen water quality at all of the test nodes meets a projected standard of 80 μg/L, as proposed under the Stage 1 DBP Rule.

## SYSTEM MODIFICATION THREE

System modification three uses the same network configuration as in modification two with the addition of a storage tank which is 48.78 meters (160 ft) in diameter and 30.49 meters (100 ft) in height. This tank facilitated direct mixing of water from the two sources and is virtually the only option that can meet the proposed Stage 2 DBP regulations by making system modifications alone. Figure 12 shows the results from this configuration. As can be seen this configuration easily meets the Stage 2 targets.

## TREATMENT OPTIONS

In addition to system changes, the option of enhancing treatment at the Lake Stafford Source was explored. Three levels of granular activated carbon treatment were assumed to reach TTHM levels of 100, 80 and 40 μg/L respectively in the effluent from the Stafford Plant. The specific assumptions that underlie the carbon plant design are shown in Table 3.

TABLE 3. GRANULAR ACTIVATED TREATMENT ASSUMPTIONS

| ITEM | VALUE |
|---|---|
| GAC Contactor System Operation | 70% of Capacity |
| Capital Amortization | 10% over 20 years |
| GAC Price | \$0.408/kg (0.90/lb) for 45,360 kg (100,000 lb) |
| Labor Rate | \$15/hr |
| Electric Rate | \$0.08/kwh |
| Natural Gas | \$0.046/SM$^3$ (0.0013/SCF) |

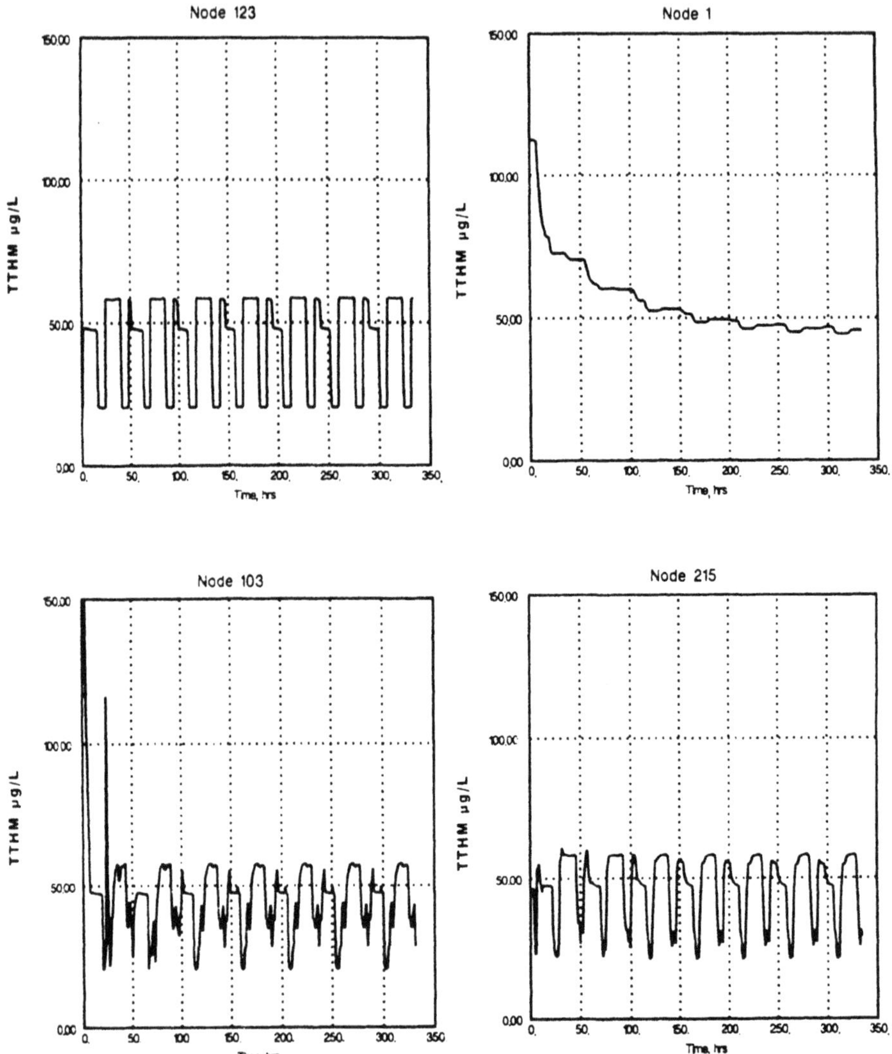


Figure 11. Water Quality Variations Over Time at Nodes 123, 103, 1, and 215 Under Modification 2

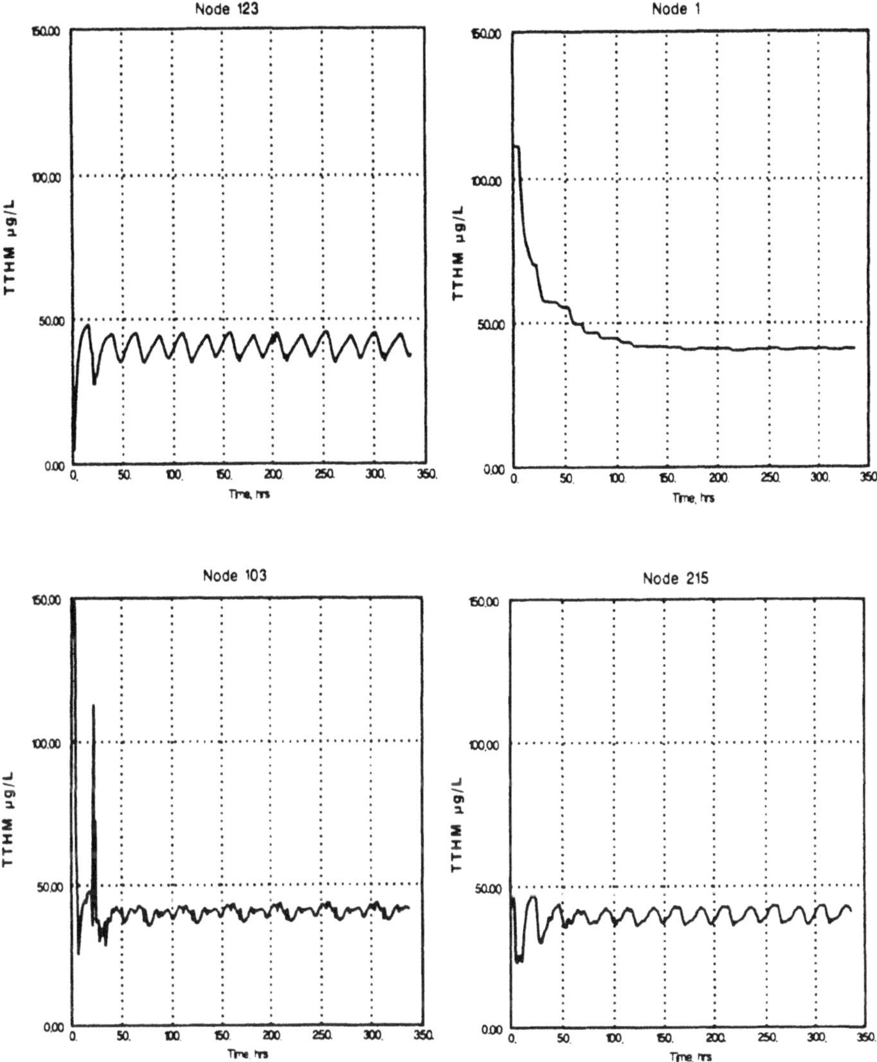


Figure 12 Water Quality Variations Over Time at Nodes 123, 103, 1, and 215 Under Modification 3

The current system configuration was assumed for the treatment analysis. It is possible, however, that some combination of treatment and system modifications could provide a lower cost solution to this problem than the options presented. Three different reaction periods were assumed in order to achieve treated water goals. Figures 13, 14 and 15 show the TTHM levels achieved for each of the three reactivation periods.

## COST COMPARISON

Each of the system modifications options has a cost associated with it. These options are summarized in Table 4. The lowest cost system modification yielded a cost of $0.042/1000 gal and was able to achieve a level of 80 μg/L for TTHMs. Two of the three treatment options are more costly than this system modification. However, a system modification to achieve the projected level of 40 μg/L is more costly than the treatment options to achieve the same level.

## SUMMARY AND CONCLUSIONS

It is increasingly obvious, especially in water short areas, that finding supplies which are adequate in quantity and quality is and will become more difficult. The degradation of existing water supplies due to municipal and industrial discharges, increasing urbanization and population, combined with more stringent drinking water standards is making this process even more difficult. This problem is exemplified by the situation in the North Marin Water District. NMWD is located in an area of low yearly rainfall, especially during the summer. NMWD has one source of very high quality, but must also use water of very low quality to supplement that source. The current blend of water violates U S Drinking Water Standards in certain sections of the distribution system. These standards are going to become more stringent in the future. However there are several options available including system modifications and the installation of treatment. Each of these option has a cost. In the scenarios examined, relatively simple system modifications can achieve the first level of DBP regulations at minimum cost. However, to achieve a more stringent level will require investment in advanced water treatment.

## ACKNOWLEDGEMENT

The authors would like to acknowledge Ms. Jean Lillie and Mr. Steven Waltrip for their assistance in preparing this document.

The authors would also like to acknowledge Mr. Paul Smedshammer of the North Marin Water district for his assistance in preapring this manuscript.

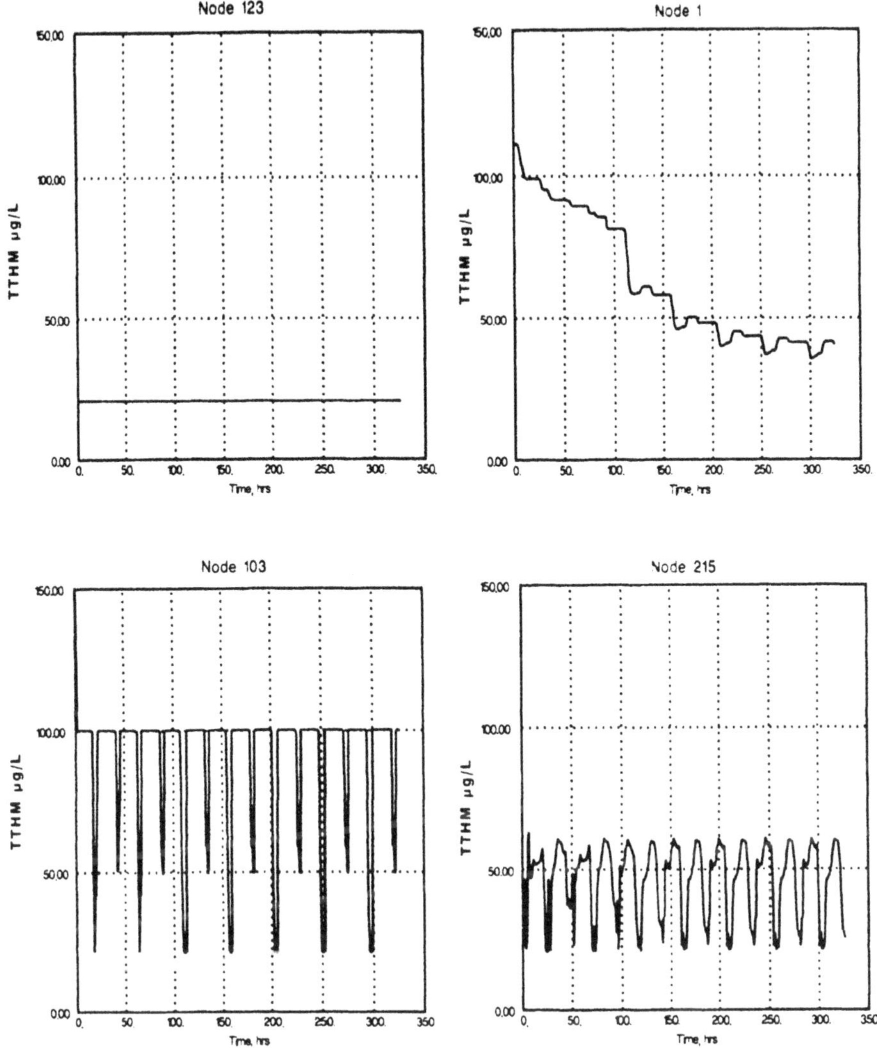


Figure 13. Water Quality Variations Over Time at Nodes 123, 103, 1, and 215
Using GAC with a Once Per Three Month Reactivation Cycle

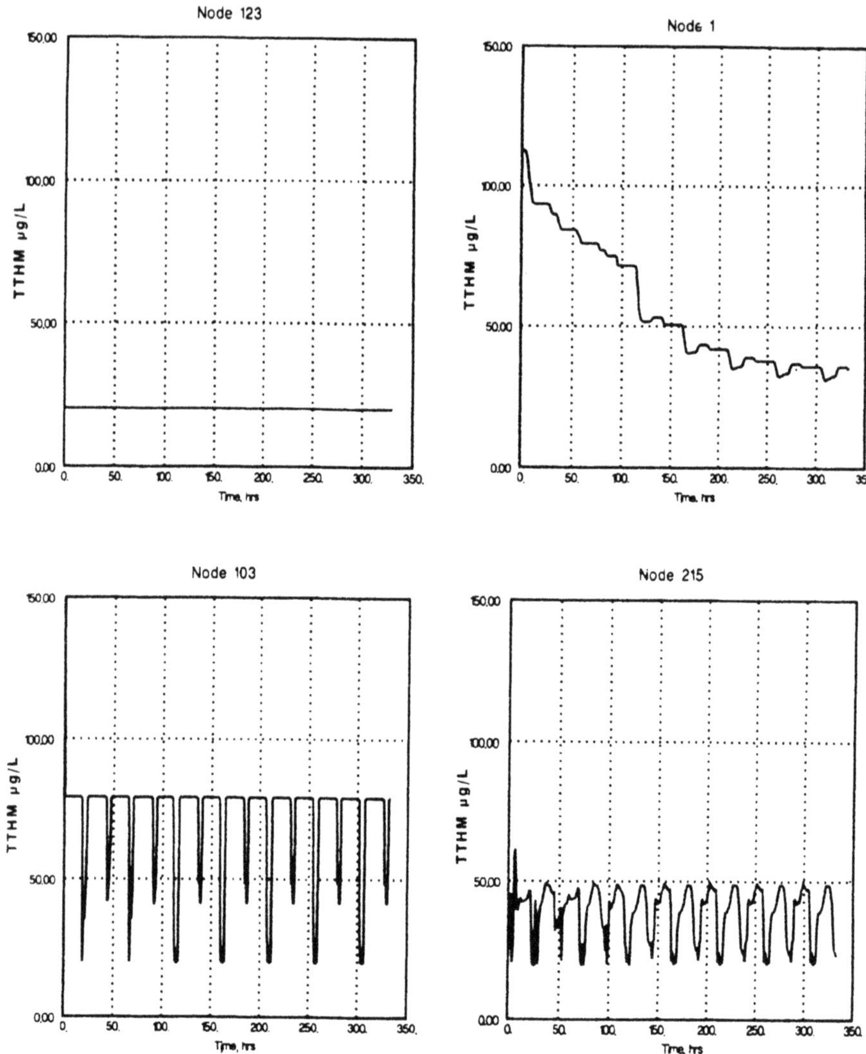


Figure 14. Water Quality Variations Over Time at Nodes 123, 103, 1, and 215 Using GAC with a Once Per Two Month Reactivation Cycle

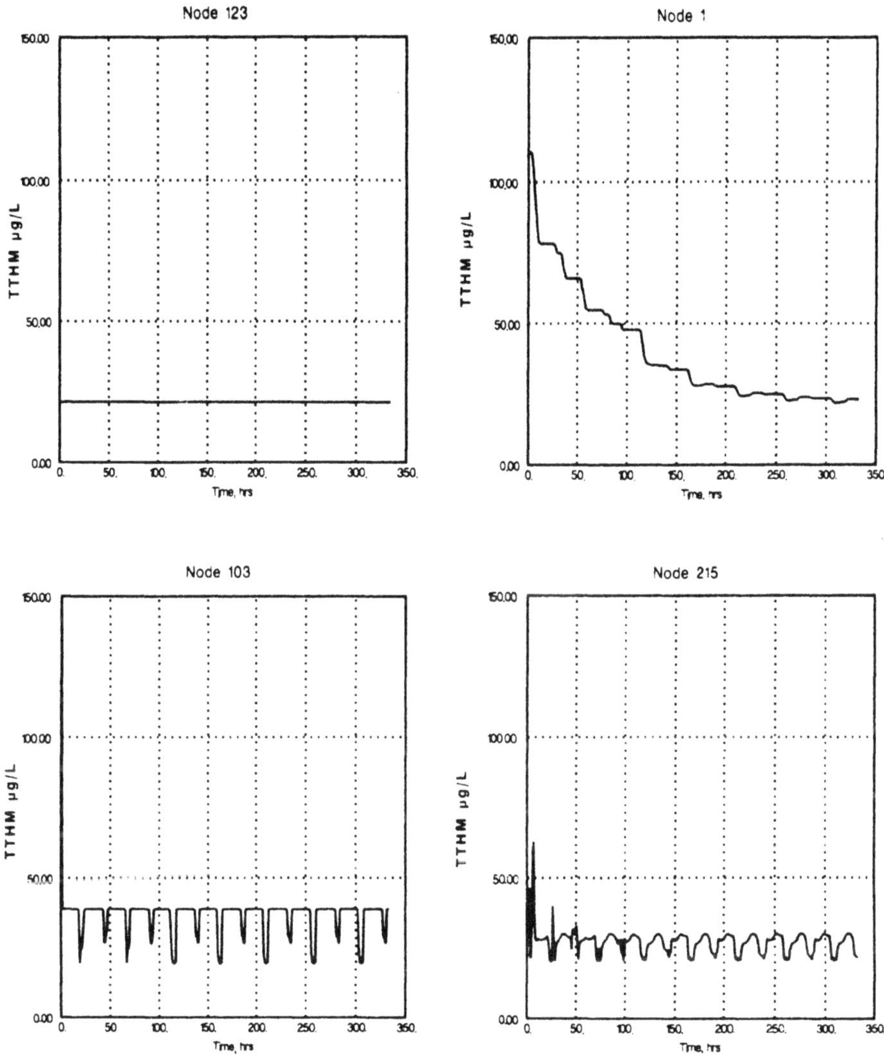


Figure 15. Water Quality Variations Over Time at Nodes 123, 103, 1, and 215 Using GAC with a Once Per Month Reactivation Cycle

TABLE 4. COST OF SYSTEM MODIFICATION AND TREATMENT OPTIONS

| | SYSTEM MODIFICATIONS | | | TREATMENT OPTIONS | | | |
|---|---|---|---|---|---|---|---|
| Modification | Capital Cost in $ | Annual Cost in $/Yr | Unit Cost in $/1000 L ($/1000 Gal) | Reactivation Frequency in Months ($\mu g/L$) | Capital Cost in $ | Annual Cost in $/Yr | Unit Cost in $/1000 L ($/1000 Gal) |
| 1 | 1,040,560 | 122,266 | 0.015 (0.058) | 3 (100) | 1,050,000 | 246,155 | 0.053 (0.20) |
| 2 | 756,770 | 88,920 | 0.011 (0.042) | 2 (80) | 1,224,000 | 300,150 | 0.061 (0.23) |
| 3 | 6,664,770 | 783,110 | 0.098 (0.372 | 1 (40) | 1,589,000 | 429,941 | 0.074 (0.28) |

## REFERENCES

Clark, R.M. and Feige, W.A., "Meeting the Requirements of the Safe Drinking Water Act," in Strategies and Technologies for Meeting the Requirements of the SDWA. Edited by Robert M. Clark and R. Scott Summers, Technomics Publishing Co., Inc., 851 New Holland Ave., Box 3535, Lancaster, PA 17604, U.S.A., 1993.

Clark, R.M., Smalley, G., Goodrich, J.A., Tull, R., Rossman, L.A., Vasconcelos, J.J. and Boulos, P.F. "Managing Water Quality In Distribution Systems: Simulating TTHM and Chlorine Residual Propagation," Journal of Water Supply Reseach and Technology - AQUA, Vol. 43, No. 4, pp 182-191, 1994.

Vasconcelos, J.J., Boulos, P.F., Grayman, W.M., Kiene, L., Wable, O., Biswas, P., Bahri, A., Rossman, L.A., Clark, R.M., and Goodrich, J.A., Characterization and Modeling of Chlorine Decay in Distribution Systems, AWWA Research Foundation, 6666 West Quincy Avenue, Denver CO 80235, 1996.

# Drought Management and Water Transfer Programs: Recent Developments and Research in California

JAY R. LUND,
Department of Civil and Environmental Engineering
University of California, Davis
Davis, CA 95616
USA
jrlund@ucdavis.edu

## Abstract

This paper reviews recent developments in drought management in California. These developments have included increased efforts at agricultural and urban water conservation and shortage management, as well as efforts to reallocate water use via market mechanisms and bi-lateral agreements between water agencies. Many forms of water transfers actively are being examined and pursued by many water agencies throughout the state. In addition, Water management agencies also must assess how to best integrate transferred water with traditional water supply operations and water demand management efforts. This process is ongoing. An economically-based optimization approach to this problem is presented.

## 1. Introduction

Since the 1976-77 drought, water planning in California has been characterized largely by preparations for drought, as opposed to the development of new traditional water supplies, such as reservoirs. Drought preparation has included widespread adoption of long-term water conservation measures and planning for adoption of short-term water conservation measures during droughts. During the most recent drought, 1987-1992, neither California's intensely-developed water infrastructure of reservoirs and conveyance facilities, nor its water conservation efforts were adequate to balance supplies and demands. In the last two years of the drought, water transfers (or water markets) became an important component of local and statewide drought management. The need to resort to water transfer, water marketing, and other forms of water use reallocation arose both from an inherent inability to continue satisfying increasing growth in traditional water demands and the rapid growth of new demands for water for environmental preservation and rehabilitation, stemming largely from Federal and State endangered species requirements. This paper reviews the development of drought management efforts in California, with particular emphasis on recent and emerging use of water transfers, and discusses the use of optimization modeling for drought planning.

## 2. California's Water Geography

Water development and management in California is driven by the natural distribution of water and the growth of human uses of water in the state. California is largely a semi-arid region, with tremendous local and seasonal variation in water supply and water demands. Climatically, the major water sources of California are in the northern parts of the state, in the form of rain and snow in the Northern and Eastern mountains, particularly the Sierra Nevada Mountain range. With a Mediterranean climate, most of California's precipitation falls during five months, from November until March. An almost negligible amount of precipitation is available from May through September. Water availability also can vary substantially between



*E. Cabrera and J. García-Serra (eds.), Drought Management Planning in Water Supply Systems,* 242–260.

years. California's most common drought index is the flow available from the Sacramento River, the state's major river. With a long-term average annual flow of 22.7 $km^3/yr$, the driest year recorded 6.3 $km^3$ of flow and the wettest year about 46 $km^3$.

Water demands for agricultural and urban uses occur mostly in the southern parts of the state and its coastal urban regions. These demands are relatively small during the winter months, and increase dramatically during the summer. Demands tend to increase somewhat during drier years to make up for lost precipitation to crops and urban areas.

Thus, California has a spatial and temporal mis-match in water availability and demands. To reduce the spatial mismatch, major water projects were constructed from the mountains and the north to the major farming and urban areas. To reduce the mismatch of supply and demand between seasons and over wet and dry years, storage reservoirs and groundwater have been developed.

Another geographic feature of California's state-wide water supply system is the importance of the Sacramento-San Joaquin Delta (Figure 1). This confluence of the state's two largest rivers is the site of the largest diversions for the major Federal and State water projects, moving water from the north of California to southern irrigation and urban demands. The Delta itself is a major agricultural area, depending on water withdrawn from the Delta's complex web of channels. The Delta also is the major conduit for fish migration from the sea to inland rivers, particularly for several races of salmon and steelhead trout, requiring habitat and somewhat natural flow regimes, and a major wildfowl migration habitat, guaranteed by international treaties. Most of the state's water-related conflicts, involving agricultural, environmental, and urban water uses, have a major focus on the Delta.

## 3. Water Development in California

California's population and economy have grown rapidly since the mid-1800s, with consequent increases in demands for water. In addition to the normal growth of water demand with population and economic growth, major new water demands have developed with increasing wealth and development of California's society in the form of laws and regulations intended to preserve fish, wildlife, and other aspects of the natural environment. To accommodate these growing and diversifying water demands, water management has gone through several phases.

### WATER INFRASTRUCTURE DEVELOPMENT

The development of water resources in California began with small-scale development of local water sources by local governments and private firms. These systems diverted water directly from nearby streams, primarily by gravity-powered canals, tunnels, and flumes. In the early 1900s, regional water transmission facilities began to be developed, initially with diversions from the Colorado River to nearby irrigation districts, from the Sacramento and San Joaquin Rivers to neighboring agricultural districts, from the Owens Valley to Los Angeles, and from the Sierra Nevada Mountains to San Francisco Bay area cities.

This was followed by a period of larger-scale water infrastructure development, extending from the late 1940s to the late 1970s. By the end of the major water infrastructure development era, water from the northernmost parts of California could be moved for use in the southernmost parts of the state. There are two major inter-regional water storage and distribution systems. The Federal Central Valley Project (CVP) takes northern California water to supply agricultural and some urban uses in the southern and central parts of the state. The State's State Water Project (SWP) is smaller but geographically more extensive, taking water from the north part of the state for urban and agricultural users as far south as the Mexican border.

Over the years, most of the state's large local, regional, and state-wide water storage and distribution systems have become inter-tied, allowing water to be moved with significant, but imperfect, flexibility. The major constraint to water movement in California is limitations on pumping of northern California water south from the Sacramento-San Joaquin Delta. These pumping limitations are placed to reduce water quality and fish migration problems associated with seawater intrusion into the Delta if too little fresh water is allowed to escape to sea (DWR, 1994).

## WATER CONSERVATION

With the completion of water storage and transmission facilities at the most promising locations, additional growth in urban and agricultural water demands had to be fed increasingly by the more efficient management of water use and demand. This began in earnest with the 1976-77 drought, where some of California's largest urban suppliers have shortages on the order of forty per-cent. Subsequent conservation efforts have extended to long and short-term efforts in all urban and most agricultural water systems, reducing water demands per-capita and per unit of crop yield (DWR, 1994).

Today, almost all of the 3,000 local water supply agencies in the state have active programs to reduce water demands. These programs consist of both efforts to increase the efficiency of water use in the long term and efforts to provide additional water savings during drought. Statewide efforts also exist in this area. State legislation on plumbing codes will result in significant savings in water used for sanitation, primarily water closets, with a Statewide standard of 6.8 liters/flush for new toilets. The State also requires that each local agency have drought preparedness plans and provides significant levels of technical assistance for both urban and agricultural water conservation.

Voluntary efforts of local agencies also are significant. The California Urban Water Conservation Council, consisting of about 100 water agencies and 50 advocacy groups, has put forth an aggressive agenda of urban water conservation measures for its signatory agencies, which include the largest urban water districts in the state. A similar effort is underway among many of the state's agricultural water districts.

## EMERGENCE OF WATER TRANSFERS

Since the 1987-92 drought, water transfers have become a significant component to the water plans of most large urban water systems and many agricultural water districts. Recent water plans for most major urban water suppliers in California ascribe an important role for water transfers. Metropolitan Water District of Southern California, the San Diego metropolitan area independently, Westlands Water District (a large irrigation district in the San Joaquin Valley), most other major urban water suppliers, and many other irrigation water suppliers have come to view water marketing mechanisms as major sources of water into the foreseeable future. This increasing reliance on water transfers is compatible with the variability of California's water supplies and demands. If there are to be water shortages, water transfers allow them to be felt with the least overall economic harm and provide those with current transferable water supplies, mostly in the lower-valued agricultural sector, with a direct form of compensation for episodically foregoing their economic uses of water. These transfers are discussed at length later in this paper.

## ENVIRONMENTAL REHABILITATION

In addition to water markets and transfers, within the last decade there has been increasing activity and financial commitment to rehabilitating the habitat of California's native fishes and wildlife. These rehabilitation efforts include increased fish screening at diversions, improved fish ladders at dams and diversions, changes in channel geometries, changes in flow regimes downstream of dams, changes in land use near streams, and other activities. Largely driven by Federal and State endangered species laws and regulations, these habitat rehabilitation programs essentially aim to increase the efficiency and environmental effectiveness of water devoted to environmental purposes.

FIGURE 1: California's Water Geography

**4. Current Water Management in California**

Water management in California is highly de-centralized and pluralistic, managed by several thousand units of government with important involvement from the private sector. These governments include local irrigation, urban water supply, environmental conservation, and flood control districts, State water project and regulatory agencies, and Federal water project and regulatory agencies. Private sector agents include individual farmers or farming corporations, mutual water companies (private irrigation districts), banks, and hydropower producers. Despite the large number of local, State, and Federal agencies involved, the system must function in a coordinated way. This coordination is accomplished through State enabling legislation for local governmental units, Federal regulations governing the operation of Federal agencies, and, perhaps most importantly, an extensive and intricate web of water supply and operations contracts and agreements. These contracts and agreements govern the finance of this highly intertied system, how shortages and surpluses are allocated, how demands are scheduled to be supplied from reservoirs, canals, and pumping plants, and how operations of reservoirs and canals are coordinated within the environmental and water law regulations imposed by State and Federal regulatory agencies.

Among these many diverse parties, there is also significant decentralization of technical information and expertise. Of these several thousand water agencies, about 24 have substantial computer modeling abilities. In addition about a dozen consulting firms make themselves available to serve any party sufficiently interested in technical computer modeling studies. At a more traditional technical level, most water districts have their own engineers, and are served by perhaps a hundred or more engineering consulting firms. Legal expertise is similarly widespread with the many agency and private lawyers. Increasingly, most agencies also have acquired biological and economic expertise, either by adding these professions to their staffs or through hiring private consultants or consulting firms. All major parties have access to substantial technical abilities.

In this largely de-centralized and complex system, water management decisions require considerable time and effort. Much of this work is done bi-laterally, between parties. However, the diversity and number of parties involved and the complexity of the issues often require a longer and more involved process of decision-making. Thus, most major California water management and planning decisions are made over a prolonged period of time through discussions and studies conducted in a variety of forums. The major water management issues in California are addressed in many forums simultaneously, and not as part of a single centralized planning process. Just as there is pluralism in decision-makers, there is also considerable pluralism in decision-making processes. This variety of ongoing decision-making and study forums allows issues to be examined from a wide variety of political, economic, and technical perspectives and allows a wide variety of alternative solutions to be suggested and studied. These simultaneous planning and study processes currently include: CALFED (a joint State-Federal planning effort), State Water Plan update process, CVPIA Programmatic EIS (Federal project environmental re-operation study), Endangered Species Act regulatory actions, State Water Board Water Rights Hearings, local planning or project studies, contract negotiations or re-negotiations, urban water agency consortium studies (CUWA), agriculture-urban water users consortium studies and negotiations ("Ag/Urban"), the Bay-Delta Modeling Forum, and, of course, the law courts.

This pluralistic and complex water management system might seem inefficient, and it is in some ways. However, the system provides an efficient coordination of local, regional, and state-wide management of water supplies and demands, ensuring that each locality's interests have a direct means of becoming involved to their satisfaction. Given the continually changing nature of water demands and supplies in California, this highly de-centralized form of management has been fairly effective overall. A large centralized water agency might be able to move more swiftly under these circumstances, but perhaps with less local effectiveness.

**5. Forms of Water Transfers**

Water transfers are a relatively recent addition to the strategic and drought plans for most California water systems. In the past water transfers or markets have had only a secondary or tertiary role in water system operation and planning, as with the sales of "surplus" water within the federal Central Valley Project (Gray, 1990).

Water transfers can take many forms, as noted in Table 1. The specific needs of the purchasing and selling parties and existing legislation and recent transfer experiences dictate the type of transfer sought and the forum through which transfer arrangements are made. Each transfer form can have different uses in system operation and has different advantages and disadvantages for water buyers, water sellers, and other groups (Lund et al., 1992). The various uses and associated benefits of water transfers are summarized in Table 2. Additionally, water transfers, like many forms of water source diversification, increase the flexibility of a water system's operation, particularly in responding to drought. This flexibility allows new forms of operation not available without transfers and potentially allows operations to vary more significantly and more rapidly than in the past. The following discussion on transfer types focuses on the possible uses and associated benefits of each form of water transfer.

TABLE 1. Major Types of Water Transfers

| TABLE 1. Major Types of Water Transfers |
|---|
| Permanent Transfers |
| Contingent Transfers/Dry-year Options |
| Long-term, Intermediate-term, Short-term |
| Spot Market Transfers |
| Water Banks |
| Transfer of Reclaimed, Conserved, and Surplus Water |
| Water Wheeling or Water Exchanges |
| Operational Wheeling |
| Wheeling to Store Water |
| Trading Waters of Different Qualities |
| Seasonal Wheeling |
| Wheeling to Meet Environmental Constraints |

TABLE 2. Major Benefits and Uses of Transferred Water

| TABLE 2. Major Benefits and Uses of Transferred Water |
|---|
| **Directly Meet Demand and Reduce Costs** |
| Use transferred water to meet demand, either permanently or during drought. |
| Use purchased water to avoid higher cost of developing new sources. |
| Use purchased water to avoid costly demand management measures. |
| Seasonal storage of transferred water to reduce need for peaking capacity. |
| Use drought-contingent transfers to reduce need for overyear storage facilities. |
| Wheeling low-quality water for high-quality water to reduce treatment costs. |
| **Improve System Reliability** |
| Direct use of transferred water to avoid depletion of storage. |
| Overyear storage of transferred water to maintain storage reserves. |
| Drought-contingent contracts to make water available during dry years. |
| Wheeling water to make water available during dry years. |
| **Improve Water Quality** |
| Trade low-quality water for higher quality water. |
| Purchase water to reduce agricultural runoff. |
| **Satisfy Environmental Constraints** |
| Purchasing water to meet environmental constraints. |
| Wheeling water to meet environmental constraints. |
| Using transferred water to avoid impacts of new supply capacity. |

PERMANENT TRANSFERS

A permanent transfer of water involves the acquisition of water rights and a change in ownership of the right. Permanent transfers are a supply augmentation and serve the same needs as capacity expansion projects, including direct use to meet demands and improved system reliability. In some cases, the direct use of permanently transferred water can delay costly water conservation measures or the need for system expansion, which in turn avoids or at least delays potential environmental impacts associated with construction (Table 2).

Most permanent transfers involve the purchase of agricultural water rights by urban interests. These transfers can involve reversion of the farmland to dryland agriculture, immediate or gradual fallowing of farmland, replacement of the farm's water supplies with an alternate supply (possibly of lower quality from an urban use perspective), or lease of the transferred water back to the farmer in wet years when other urban supplies are plentiful. In California, relatively little use has been made of permanent water transfers.

Another form of permanent water transfer requires urban land developers to acquire water rights associated with recently developed, formerly agricultural suburban lands. Some Arizona cities have made provision of such rights to the urban water supplier a pre-requisite for annexation of new suburban developments to urban water systems (MacDonnell, 1990). This ties permanent changes in water use to changes in land use and does not require water rights to be severed from the land, a political and legal difficulty in some cases.

CONTINGENT TRANSFERS/DRY-YEAR OPTIONS

In many cases, potential buyers of water are less interested in acquiring permanent supplies than in increasing the reliability of their water supplies. For these cases temporary transfers contingent on water shortages may be desirable. The appropriate time horizon and conditions for a contingent transfer agreement will depend somewhat on the source of unreliability that the buyer would like to eliminate. Drought-contingent contracts for water are probably best made with holders of more secure water rights, since they are the least likely to be shorted during drought. However, the increased reliability of water from senior rights tends to raise its market value (Lund et al., 1992; *Water*, 1992). An important benefit of contingent transfers is that longer term arrangements allow a more thorough analysis and mitigation of potential third party impacts.

The time horizon of contingent transfers is important. Contingent transfer agreements can be established for a period of several decades. This provides each party long-term assurance of the terms and conditions of water availability. Such long-term agreements can help an urban water utility modify release rules for reservoir storage to maintain less drought storage than would otherwise be desired or reduce the need for new source development. Long-term arrangements also can provide flexibility where future water demands may not meet expectations. However, long-term leasing of water does entail risk for water buyers if water demands meet or exceed current forecasts. Long-term leasing or contingent contracts allow water right owners to retain long-term investment flexibility in anticipation of potentially greater future values for water leasing or sale of a water right.

Intermediate-term (3-10 year) contingent transfer contracts might be used to reduce the susceptibility of the buyer's system to drought during periods prior to the construction or acquisition of new supplies. Short-term (1-2 year) contingent transfer contracts might be used in the midst of a drought by a water agency with depleted storage, preparing for the possibility that the drought might last a year or two longer. This type of short-term contingent transfer contract would enable the buyer to have committed water supplies when their system might be extremely vulnerable.

Advantages of contingent transfers for the seller, typically agricultural interests, are the immediate infusion of cash when the contract is made, the infusion of additional revenues if the contingent transfer option is "called", and an increased ability to predict the conditions and timing of transfers, rather than relying on the vagaries of timing, price, and quantity of a water spot market.

The potential sale of water by farmers during drought affects the need for ground water management (if available as an alternate supply of water) and the special operation of conveyance and storage facilities. The ability of farmers to sell water also might affect the operation rules used by agricultural water suppliers for allocating water from storage to farmers over multi-year droughts. Perhaps additional hedging or overyear storage by agricultural water suppliers will increase farm incomes more than adherence to current reservoir operating rules, by creating a greater scarcity of water and higher water transfer incomes during drought years. Similar issues relate to the overyear use of ground water storage.

In California, contingent transfers appear to attract the greatest interest from urban water agencies. Most urban water systems have adequate water supplies during most years. Contingent transfers allow urban agencies to acquire options for water during non-shortage periods and use the options only during shortages. This lowers the cost of transferred water and improves the atmosphere for negotiating transfers with potential sellers and government regulators, avoiding the crisis atmosphere of droughts.

## SPOT MARKET TRANSFERS

Spot market transfers are short-term transfers or leases, typically agreed to and completed within a single water year. Spot market transfers often are established by some sort of bidding process, often with some conditions for transfer being fixed (e.g., price, quantity). However, spot market transfers can arise from negotiations between individuals or groups of buyers and sellers. A wide variety of bargaining rules for the operation of spot markets have been examined theoretically and through simulation (Saleth et al., 1991). These results illustrate the importance of bargaining rules when there are few buyers and sellers. For large spot markets, the effects of bargaining rules are quickly overshadowed by competition among buyers and sellers.

Spot market purchases can be advantageous in both dry or wet years. During droughts, short-term transfers may be sought to meet demands directly. As with permanent transfers, temporary transfers used to meet demands directly can delay or avoid the costs of developing new supply sources or implementing more stringent demand management measures.

In wet years, water purchased through a spot market can be stored in reservoirs or aquifers as overyear storage. This enhances the yield of the system during drought years by increasing the amount of stored water available upon entering a drought. Overyear storage of transferred water is particularly well suited to acquiring water from junior water rights holders. Junior water rights are typically less expensive than senior (more secure) water rights, although they may only be available during relatively wet years. However, storage of transferred water during wet years may require additional surface or ground water storage capacity, and is subject to evaporative and seepage losses and any costs associated with storage. This approach also may work for seasonal storage.

Spot market transfers are commonly used in California. They improve operational flexibility and provide opportunities to reduce costs, but are not typically seen as a reliable water source. For urban agencies, spot market water is often seen as an inexpensive source of "surplus" water during wet years, allowing more water to be kept in storage for potential future dry years.

## WATER BANKS

Water banks are a relatively constrained form of spot market operated by a central banker. Water is sold to the bank for a fixed price and bought from the bank at a higher fixed price. The difference in prices typically covers the bank's administrative and technical costs. Each user's response to the bank and involvement in the market is largely restricted to the *quantity* of water she is willing to buy or sell at the fixed price. However, sometimes the bank buys and sells "lots" of water at different prices, sometimes negotiated.

The California Drought Emergency Water Banks of 1991 and 1992 are examples of water banks or spot markets where the terms and price of transfer were relatively fixed, with the State acting as a banker (*1991 Drought*, 1992; Howitt et al., 1992). A similar, but smaller water bank was established locally in Solano County, California (Lund et al., 1992). In agricultural regions, it is common for water banks or pools to exist within large irrigation systems. For many existing water pools, sellers avoid only the cost of purchasing unneeded water from the system. Water buyers in these pools pay the system normal wholesale water prices, plus some administrative cost (*Water*, 1992; Gray, 1990). California's State water banking has evolved in recent years to include increasingly longer term "options" (Jercich, 1997). These add to the flexibility of the transfers for buyers and sellers.

Where spot market or water bank transfers have become established, as in California, agencies of all types are likely to plan on these markets being available for either buying or selling water (Lund et al., 1992; Israel and Lund, 1995). The existence of spot markets and water banks during droughts provides incentives for urban water suppliers to rely somewhat less on more expensive forms of conventional water supply capacity expansion and urban water conservation in planning, and also may encourage different designs for new facilities and modified operation of existing facilities. For agricultural water districts, the existence of water banks and spot markets during drought has implications for the wording of water supply contracts and the management of water and cropland during a drought.

WHEELING AND EXCHANGES

In the electric power industry, power is often "wheeled" through the transmission system between power companies and electric generation plants to make power less expensive and more reliable. Water can be similarly "wheeled" or exchanged through water conveyance and storage facilities to improve water system performance. Again, such movements of water involve the institutional transfer of water among water users and agencies. There are a number of forms of wheeling water or water exchanges (Lund et al., 1992).

Sometimes the cost of conveying water or the losses inherent in water conveyance can be reduced by wheeling water through conveyance and storage systems controlled by others. An example would be the use of excess capacity in a parallel lined canal owned by another agency, rather than use an agency's own unlined canal to convey water. Differences in pumping efficiencies might also motivate operational wheeling between conveyance facilities. Similar considerations might apply to decisions on where to store water during a drought, when different reservoirs have different seepage or evaporation rates or if the distribution of hydropower heads is considerable for different storage options.

Seasonal wheeling of water is common in agricultural regions where different sub-areas have complementary demands for water over time. This can provide opportunities for one water user to exchange water with another user during his low-demand season, with repayment coming in the form of additional water during the user's high-demand season.

Also, by paying farmers not to use their rights to water, the consumptive use foregone becomes available for instream demands downstream. This mechanism is particularly applicable to riparian rights which cannot be legally transferred for use away from the riparian lands (Lund et al, 1992). Another application of wheeling to meet environmental constraints could involve the use of storage facilities to release water when desired for instream flows while meeting demands before this time from other reservoirs or ground water.

In many cases, historical happenstance has left agricultural users with rights to high-quality water for irrigation while new urban development is left with remaining water sources of lesser quality. In such cases the additional costs of treating low-quality water for urban use is usually much greater than the costs from slightly lower crop yields from use of the lower quality water. Given reasonable conveyance costs, it therefore becomes desirable for water-quality based trades between agricultural and urban users. Urban users can often afford to make these trades on an uneven basis, trading more low-quality water for less high-quality water or providing a monetary inducement for a volumetrically even trade of water. Lesser quality waters might also be traded for environmental uses of aquifer recharge or habitat maintenance (Lund et al., 1992).

## TRANSFER OF RECLAIMED, CONSERVED, AND SURPLUS WATER

Although not always recognized as such, the purchase of water made available by reclamation or reductions in water demand is a form of water transfer. Numerous urban water utilities have become involved in purchasing water back from their retail customers. Such schemes usually involve rebates to customers for installing low-flow toilets or removing relatively water-intensive forms of landscaping (*Landscape,* 1988). Some cities have developed clever schemes where water transfers are made within their customer base. For instance, Morro Bay, California has a program whereby developers can receive water utility hook-up permits if they cause a more than equivalent reduction in existing water demand through plumbing retrofits, landscaping, or other measures (Laurent, 1992).

Urban areas have taken an interest in financing the conservation of irrigation water to make additional water available for urban supplies. This has primarily been accomplished through the lining of irrigation canals. For example, the transfer of water between the Imperial Irrigation District (IID) and the Metropolitan Water District of Southern California (MWD) involves a 35-year contract for MWD payments for canal lining and other system improvements in IID's irrigation infrastructure in exchange for the water saved by these improvements. The savings are estimated at 123.3 $Mm^3$/year (100,000 ac-ft/year) from IID's Colorado River water supplies (Gray, 1990; Sergent, 1990). This approach can have additional benefits where agricultural seepage and drainage water has led to water quality problems or high water tables, but can create additional problems where canal seepage is used to recharge ground water.

## SOME PROBLEMS WITH WATER TRANSFERS

While water transfers hold out a great deal of promise for water management in California, there are some problems (Howe et al., 1986; Brajer et al., 1989):

- Water rights are often poorly defined. The definition and quantification of water rights is a particularly difficult problem. Technologically, there is probably at least a ±10% range of error in water flow, use, and consumption estimates. When water cannot be sold, this error is not terribly important. When water can be sold, however, this error represents real water and real money to both buyers and sellers.
- Water transfers can have high transaction costs. Particularly given the potential legal entanglements represented by water transfers, there are potentially very high costs and risks associated with pursuing water transfers.
- Water markets will often consist of relatively few buyers and/or sellers. This is a problem mostly for local water transfers within small regions. Statewide, California's extensive plumbing systems provides a wide variety of potential sellers and buyers.
- Water is often costly to convey between willing buyers and sellers. This problem is also fairly local, but can often affect statewide transfers as well. These costs include conveyance, storage, and water treatment costs, as well as interruptions due to flow management for environmental water uses.
- Communication between buyers and sellers may be difficult. This is a problem both for local and statewide water transfers, owing mostly to the early nature of most water transfer activity. This problem has diminished greatly with the entry of the State into water banking.
- Third party effects have been inadequately addressed. This is a major problem which must be resolved at a governmental policy level and such a resolution requires a sound legal basis. Some examples of third parties are listed in Table 3.

To overcome these problems often requires the development of physical, legal, or institutional infrastructure. There will need to be incremental tuning of the entire system of water supply, water demand, water law, and water management to adapt to support and take advantage of water transfers. Such changes may require a decade or so to mature.

TABLE 3. Some Potential Third Parties to Water Transfers

|  |  |
|---|---|
| **Urban** | |
| | Downstream urban users |
| | Landscaping firms and employees |
| | Retailers of lawn and garden supplies |
| **Rural** | |
| | Farm workers |
| | Farm service companies and employees |
| | Rural retailers and service providers |
| | Downstream farmers |
| | Local governments |
| **Environmental** | |
| | Fish and wildlife habitat |
| | Those affected by potential land subsidence, overdraft, and well interference |
| | Those affected by potential ground water quality deterioration |
| **General** | |
| | Taxpayers |

## 6. Optimization for Shortage Management Planning

Computer models have long been used for planning and operation of complex water systems. The addition of water transfers and complex water conservation/demand management measures have increased the utility of computer models for drought management. One problem with conventional simulation models in such cases is the enormous number of combinations of potential decisions that could be made and the difficulty of working through these huge numbers of decisions. Optimization modeling, by using algorithms that suggest promising decisions to maximize an explicit measure of performance, can greatly shorten this analytical process and foster improved integration of a diverse range of design and operating decisions.

In California, optimization models have become common for the estimation of farmer responses to water management decisions and optimization heuristics and explicit optimization are becoming more frequently used for urban drought planning. The State and federal agencies commonly use the Central Valley Production Model (CVPM) for examining the likely response of farmers to changes in water availability and price. This model, based on quadratic programming, assumes that farmers adjust their cropping patterns and use of capital, labor, land, and water resources to maximize profit. In addition to predicting farmer response, the model also provides estimates of the economic impact of alternative water management policies, in terms of changes in farm profits.

For urban drought management, a variety of shortage cost and management models have been proposed. This section examines a relatively recent attempt to reasonably quantify both the costs of unreliability in urban water supplies and suggest economically reasonable long-term and drought-specific shortage management measures, particularly promising demand reduction and water transfer options (Wilchfort and Lund, 1997). This particular modeling approach is not currently widespread, but represents the kind of research-level approach which has increasingly come into practice in California.

### TWO STAGE LINEAR PROGRAMMING

Here a two stage linear programming model is used to represent least-cost shortage management, given hydrologic uncertainty in supply system yield. The model integrates demand management options and supply enhancement measures for long term and short term durations. The first stage decisions in the model represent long term measures such as long term conservation, dry year transfer contracts, additional water treatment, and water reuse. Long-term measures have a long life span, relatively fixed annualized cost, and must usually be implemented before shortages occur. The second stage decisions consist of short term measures available to augment water supplies or reduce demands for particular shortage events. Short term decisions are temporary responses to given shortage levels and their potential may vary with the decisions made in the first stage. The costs of short term measures for each shortage level are weighed by the probability of the shortage.

Inputs to the optimization model include the different long term and short term measures available, their costs and effectiveness in either reducing demands or augmenting supplies. The combined demands may include urban use, withdrawals by senior right holders, and environmental uses. The model also requires a shortage or yield frequency distribution. The shortage exceedence probability distribution is based on a reservoir operation yield model. Usually, a simulation model is used based on seasonal historical inflow data, seasonal demands, a mathematical representation of the system configuration, and operating rules. The yield model provides a time series of shortages that are converted to a probability distribution of shortage events for use in two stage linear programming. The model results provide the least cost combination of long term and short term measures, their expected level of use, and the combined annual cost associated with the shortage probability distribution.

## MODEL LIMITATION

California and much of the North American West experience droughts of long duration (many months to several years). The California climate combined with controlled reservoir operation results in shortages long enough that the reaction time for triggering short-term measures is relatively unimportant. For many droughts in more humid regions, droughts are of short enough duration (weeks or a few months) that establishing the triggering rules for implementing short-term drought management measures can be the most important shortage management decisions. This aspect of shortage management is not addressed by this shortage management modeling approach.

## MODEL FORMULATION

### *Objective Function*

The objective of the shortage management optimization model is to minimize the expected value cost of a combination of long term and short term alternatives required to meet demand for a predefined shortage or yield frequency distribution. The objective function has two components. The first component is the combined costs of all long term measures selected in the first stage. The second component is the sum of all short term measures costs implemented as a response to particular shortages weighed by each shortage probability. Equation 1 is the mathematical representation of the objective function.

$$\text{(1)} \qquad \text{Min } Z = \sum_{i=1}^{m} c_i L_i + \sum_{s=1}^{y} \sum_{e=1}^{r} p_e \sum_{j=1}^{n} c_{j,s} S_{j,e,s}$$

where,

$Z$= Total cost of responding to shortage probability distribution (in \$1000s)

$L_i$= Annual long term measure quantity, $m^3$/year, $\forall$ i,

$S_{j,e,s}$= Seasonal short term measure quantity, $m^3$/season, $\forall$ j,e,s,

$p_e$ = Shortage event probability, $\forall$ i

$c_i$ = Unit cost of long term measure i, \$/$m^3$, $\forall$ i

$c_{j,s}$ = Unit cost of short term measure j for season s, \$/$m^3$, $\forall$ j,s,

s = Season, (of y seasons),

i = Long term measures, (of m measures),
j = Short term measures, (of n measures), and
e = Shortage event, (of r events).

*Decision Variables*

The model decision variables are the long term ($L_i$) and short term ($S_{j,e,s}$) alternatives available to increase supply system reliability. Long term decisions include water reuse, conservation in the form of xeriscaping and water fixture replacement, additional water treatment capacity, and acquiring dry year transfer options. Long term decisions have units of m3 per year. Short term decisions include drought conservation measures, activating dry year transfer options, and purchasing spot market water. Short term decisions are seasonal decisions in response to shortage events and similarly have units appropriate to their implementation. Thus, reductions in landscape watering might have units of $m^3$ per season and event. A more detailed description of decision alternatives appears in the next section.

*The Model Constraints*

The principal model constraints are limits on the long term and short term measures and the requirement of satisfying the demands at each shortage level for each season. The sum of long term measures converted to seasonal water volumes and short term measures must meet or exceed seasonal shortages (equation 2). Long term and short term measures also cannot exceed specified limits (equations 3 and 4, respectively). Limits of conservation measures are based on demand and their effectiveness, i.e. xeriscaping is a function of outdoor use and water fixture retrofit is a function of indoor use. Non-negativity constraints apply to all long term and short term measures (equations 5 and 6).

$$\sum_{i=1}^{m} f_{i,s} L_i + \sum_{j=1}^{n} S_{j,e,s} \geq SH_{s,e}, \quad \forall s,e, \tag{2}$$

$$L_i \leq L_{i,\max}, \quad \forall i, \tag{3}$$

$$S_{j,e,s} \leq S_{j,e,\max}, \quad \forall j,e,s, \tag{4}$$

$$S_{j,e,s} \geq 0, \quad \forall j,e,s, \tag{5}$$

$$L_i \geq 0, \quad \forall i, \tag{6}$$

where,
$SH_{s,e}$ = Shortage: the shortage volume for season s and event e, $m^3$/season
$f_{i,s}$ = Distribution factor for long term measure i in seasons, dimensionless

$$\sum_{s=1}^{y} f_{i,s} = 1, \quad \forall i \tag{7}$$

More specific constraints apply to the relationship between long term and short term measures. Short term conservation efforts often are limited by the long term conservation measures adopted. This constraint type reflects "demand hardening"; as more conservation measures are permanently placed, the effectiveness of short term conservation measures decreases and their relative costs increase (Lund, 1995).

As an example, for our case study, lawn watering reduction in response to a shortage, a short term conservation measure, depends on the level of long term xeriscaping attained (equation 7). Lawn watering reduction can be divided into two segments to reflect the severity of implementing large water reductions. Lawn watering reduction I (measure lw1) is first implemented and lawn water reduction II (measure lw2) is implemented at a much higher cost as needed (equation 8).

Installing water displacement devices to temporarily reduce water demand depends on the reduction due to the long term water fixture retrofitting decision (equation 9). The demand hardening factor ($h_i$) represents the reduction in the effectiveness of short term water conservation as more permanent water fixture retrofitting measures are implemented.

(7) $$S_{lw,e,s} \leq (L_{xe,max} - L_{xe})\, f_{xe,s}, \quad \forall\ s,e$$

(8) $$S_{lw1,e,s} + S_{lw2,e,s} \leq S_{lw,e,s}, \quad \forall\ s,e$$

(9) $$S_{wd,e,s} \leq (L_{rt,max} - L_{rt})\, f_{rt,s}\,, \quad \forall\ s,e$$

where,
$S_{lw,e,s}$ = Lawn watering reduction limit for season s and event e, $m^3$/season
$S_{lw1,e,s}$ = Lawn watering reduction part I for season s and event e, $m^3$/season
$S_{lw2,e,s}$ = Lawn watering reduction part II for season s and event e, $m^3$/season
$S_{wd,e,s}$ = Water displacement device for season s and event e, $m^3$/season
$L_{xe}$ = Xeriscaping annual water savings, $m^3$/year
$L_{rt}$ = Fixture retrofitting annual water savings, $m^3$/year
$h_{rt}$= Demand hardening factor for long term retrofitting, dimensionless
$f_{xe}$ = Xeriscaping seasonal factor, dimensionless
$f_{rt}$ = Fixture retrofitting seasonal factor, dimensionless

Water transfers often are limited by the treatment capacity of the existing water system. Water treatment capacity can be expanded as a long term measure to increase the quantity of water that can be contracted as a dry year transfer option or purchased from spot markets (equation 10). For each shortage level, the amount of dry year option activated depends on the long term decision of the dry year option contract (equation 11). The sum of the spot market purchased and the dry year option activated must not exceed the total transfer limit which might vary with a particular shortage event (equation 12).

(10) $$S_{tt,e,s} \leq (CAP + L_{cap})\, f_{cap,s}, \quad \forall\ s,e$$

(11) $$S_{ta,e,s} \leq L_{tc}\, f_{tc,s}, \quad \forall\ s,e$$

(12) $$S_{ta,e,s} + S_{sm,e,s} \leq S_{tt,e,s}, \quad \forall\ s,e$$

where,
CAP = Available capacity for transferred water treatment, $m^3$/year
$f_{i,s}$ = Distribution factor for long term measure i in seasons, dimensionless
$S_{tt,e,s}$ = Total transfers (dry year option and spot market) for season s and event e, $m^3$/season
$S_{ta,e,s}$ = Activated dry year option for season s and event e, $m^3$/season
$S_{sm,e,s}$ = Spot market purchased for season s and event e, $m^3$/season
$L_{cap}$ = Additional water treatment capacity, $m^3$/year

$L_{tc}$ = Annual dry year option contract, $m^3$/year

## LONG-TERM AND SHORT-TERM MEASURES

Long-term and short-term measures include both demand management and supply enhancement. The following options are included in the model.

### *Modeling Water Conservation*

Water conservation practices are used to reduce water demand, moderate peak consumption to delay or avoid capital expenditures of water system expansion, and reduce the effects of water consumption on the environment. Common water conservation methods include efficient irrigation, xeriscaping, and water fixture retrofits. Water agencies encourage conservation by enacting various forms of rationing such as fixed allotments to customers, percent reduction in supply, adoption of tiered pricing to control consumption, and rotation of service to customers (Lund and Reed, 1995). Education which emphasizes the public benefit of conservation and persistently informs of the consequences of serious water shortages has been shown to have an important effect on the implementation of conservation measures (Cameron and Wright, 1990). Conservation measures can be permanently incorporated into the supply system (water fixture retrofits and xeriscaping) or be adopted as a short term measure in response to a particular shortage event (reduced lawn watering). Short term conservation programs tend to become less effective in mitigating emergency shortages and more expensive as permanent conservation practices are integrated into the water supply system in anticipation of future shortages (Weber, 1993). The total cost of implementing conservation measures includes the cost of implementing the conservation measure as well as the forgone revenue by the water supplier (Mann and Clark, 1993).

### *Modeling Water Reuse*

Reused water can be added to the supply system as either a new source of water supply or for pollution control. Reused water has been used for agricultural and landscaping irrigation, industrial process and cooling water, complying with environmental instream flow requirements, groundwater recharge, and direct consumptive use. The use of reused water has been steadily increasing as a result of severe droughts and stringent Federal Water Pollution Control regulations that generally require a minimum of secondary treatment and in some cases, advanced treatment to meet municipal discharge standards. Reusing water for landscaping application generally requires only secondary treatment and disinfection while reusing water for potable purposes requires much more extensive treatment. In addition to primary and secondary treatment, potable reuse requires treatment processes such as recarbonation, multimedia filtration, selective ion-exchange, carbon adsorption, reverse osmosis, and disinfection. In general, water reuse is more feasible and cost effective for nonpotable purposes than for human consumption (Asano and Madancy, 1984).

In evaluating the cost of reuse as a water supply source, the cost of the required added treatment, the conveyance system, and operation and maintenance should be considered. Generally, the majority of cost associated with wastewater reclamation is attributed to the cost of distribution (approximately \$0.24/$m^3$) to which treatment, operation and maintenance costs must be added. The deferred costs of wastewater effluent discharge permits, an external benefit, should be incorporated to water reuse cost analysis (Asano and Mills, 1990).

### *Modeling Water Transfers*

Water transfers can be used to augment water supply during shortage conditions that are due to droughts, high demands, and interruption of normal supply due to natural disasters. Water transfers can be used to meet demand, increase reliability, improve quality, and satisfy environmental constraints. Various water transfer methods can be integrated into a regional water supply system (Lund and Israel, 1995).

Permanent transfers account for the permanent acquisition of water rights by a water agency to supplement the existing water supply. Contingent transfers or dry year options are long term alternatives in which a contract is made between agricultural senior water rights holder and a water agency to be activated during shortage events. Spot market transfers are short term transfers, usually completed within a year, and can be used either to augment water supply during a shortage event or to increase system reliability in wet years. Water banks are a constrained form of spot market. Water is purchased from agricultural users and sold to urban suppliers at fixed prices. The difference between the buying and selling prices accounts for the bank's technical and administrative costs.

The cost of water transfers varies with market conditions. The total cost of water transfers includes the purchase cost, conveyance modification costs, treatment cost, transaction costs, and costs associated with third party losses such as economic losses to community and increased groundwater pumping. The amount of water actually transferred can vary greatly from the amount contracted due to conveyance losses because of evaporation, seepage, and natural accretion, and due to the uncertainty associated with the amount of water a farmer actually has rights to sell (Lund and Israel, 1995).

## ILLUSTRATIVE MODEL RESULTS

This model was applied to a simplified representation of the East-Bay Municipal Utility District in California (Wilchfort and Lund, 1997). The water supply yield-probability distribution required by the model was found using a reservoir simulation model using 73 years of historical unimpaired streamflows. The resulting shortage probability distribution appears in Table 4.

The least-cost results appear in Tables 5 and 6, with least-cost levels of various long-term (permanent) shortage management measures appearing in Table 5, with a total annualized cost of $6.6 million/year. Least-cost decisions for short-term management vary with season (wet or dry) and drought severity; these appear in Table 6 with a total average annual cost of $2.9 million/year. Thus the total cost imposed on the system of the entire shortage probability distribution in Table 4 is averages $9.5 million/year.

These results can be used not only to suggest economically promising and effective shortage management measures. The economic value (or costs) of changes water yield available to the system can also be assessed. By making separate runs of the model for different reservoir capacities or operating rules, the total shortage costs for each resulting yield or shortage probability distribution can be found. This provides an ability to quantify the shortage management economic benefits of improvements in water yield probabilities.

TABLE 4: Example Shortage Probabilities Used for Illustrative Results

| Shortage Level | Wet Season | Dry Season |
|---|---|---|
| 0% | 0.933 | 0.947 |
| 20% | 0.017 | 0.007 |
| 40% | 0.01 | 0.007 |
| 60% | 0.004 | 0.005 |
| 80% | 0.004 | 0.005 |
| 100% | 0.031 | 0.03 |

TABLE 5: Least-Cost Long-term Shortage Management Decisions from Model

| Option | Annual Implementation ($Mm^3/yr$) |
|---|---|
| Conservation: | |
| Xeriscaping | 0 |
| Plumbing Retrofit | 74 $Mm^3/yr$ |
| New Treatment Capacity | 21 $Mm^3/yr$ |
| Dry Year Option Purchases: | |
| Wet Season: | 0 |
| Dry Season: | 89 $Mm^3/yr$ |
| Wastewater Reuse | 0 |
| Total Annualized Cost: | $6.6 million/yr |

TABLE 6: Least-Cost Short-term Shortage Management Decisions for Different Seasons and Levels of Shortage

| Event | Probability | % Shortage | Short-term measures | Cost ($1000/yr.) |
|---|---|---|---|---|
| 1 Wet season | 0.933 | 0% | none | 0 |
| 2 Wet season | 0.017 | 20% | none | 0 |
| 3 Wet season | 0.01 | 40% | Spot market | 34 |
| 4 Wet season | 0.004 | 60% | Spot market | 43 |
| 5 Wet season | 0.004 | 80% | Spot market, Conservation | 82 |
| 6 Wet season | 0.031 | 100% | Spot market, Conservation | 1,190 |
| 1 Dry season | 0.947 | 0% | none | 0 |
| 2 Dry season | 0.007 | 20% | none | 0 |
| 3 Dry season | 0.007 | 40% | Dry year option | 29 |
| 4 Dry season | 0.005 | 60% | Dry year option, Conservation | 44 |
| 5 Dry season | 0.005 | 80% | Dry year option, Conservation | 107 |
| 6 Dry season | 0.03 | 100% | Dry year option, Conservation | 1,307 |
| Total Average | Short-Term Cost : | | ($million/year) | 2.9 |

## 7. Conclusions

Growth in water demands and the emergence of new environmental water demands have greatly changed water management in California. Management has changed from the development of new reservoirs and canals to management of water demands (water conservation), regulatory establishment and enforcement of environmental water demands, and market-based reallocations of water demands.

Water markets and other forms of water transfers have become increasingly common and sought-after to balance local, regional, and statewide water supplies and demands. Although frequently controversial and often difficult to implement, water transfers show great promise and flexibility for providing an economic balance of supplies and demands.

Computer modeling is an essential aspect of California's water operations, planning, and management. Optimization modeling is increasingly used both for demand estimation as well as system operation models. An optimization approach to probabilistic water shortage management is suggested.

## 8. References


Asano, Takashi and Madancy, Robert S. (1984), "Water Reclamation Efforts in the United States," *Water Reuse* , Ann Arbor Science, pp. 277-291.

Asano, Takashi, and Mills, Richard A. (1990), "Planning and Analysis for Water Reuse Projects," *American Water Works Association Journal*, Vol. 82, pp. 38-47.

Brajer, V., A.L. Church, R. Cummings, and P. Farah (1989), "The Strengths and Weaknesses of Water Markets as They Affect Water Scarcity and Sovereignty Interests in the West," *Natural Resources Journal*, Vol. 29, Spring, pp. 489-509.

Cameron, T.A. and Wright, M.B. (1990), "Determinants of Household Water Conservation Retrofit Activity - A Discrete Choice Model Using Survey Data", *Water Resources Research*, 26(2), 179-188.

Department of Water Resources (DWR) (1994), *California Water Plan Update*, Bulletin 160-93, Department of Water Resources, Sacramento, CA.

Gray, Brian E. (1990), "Water Transfers in California: 1981-1989," in MacDonnell, Lawrence J.(Principal Investigator), *The Water Transfer Process As A Management Option for Meeting Changing Water Demands,* Volume II, USGS Grant Award No. 14-08-0001-G1538, Natural Resources Law Center, University of Colorado, Boulder.

Howe, C.W., D.R. Schurmeier, and W.D. Shaw, Jr. (1986), "Innovative Approaches to Water Allocation: The Potential for Water Markets," *Water Resources Research*, Vol. 22, No. 4, April, pp. 439-445.

Howitt, R., N. Moore, and R.T. Smith (1992), "A Retrospective on California's 1991 Emergency Drought Water Bank", March.

Israel, M. and J.R. Lund (1996), "Recent California Water Transfers: Implications for Water Management, "Natural *Resources Journal*, No. 1.

Jercich, S.A. (1996), "California's 1995 Water Bank Program: Purchasing Water Supply Options," *Journal of Water Resources Planning and Management*, ASCE, Vol. 123, No. 1, January/February, pp. 59-65.

*Landscape Water Conservation Guidebook No. 8*, (1988) California Department of Water Resources, Sacramento, CA.

Laurent, M.L. (1992), "Overview New Development Process/ Water Allocations/ Conservation," City of Morro Bay, CA.

Lund, Jay R. (1995), "Derived Estimation of Willingness to Pay to Avoid Probabilistic Shortage", *Water Resources Research*, 31(5), 1367-1372.

Lund, Jay R. and Israel Morris (1995), "Optimization of Transfers in Urban Water Supply Planning", *Journal of Water Resources Planning and Management*, 121(1), 41-48.

Lund, Jay R. and Reed R.U. (1995), "Drought Water Rationing and Transferable Rations," *Journal of Water Resources Planning and Management*, 121(6), 429-437.

Lund, J.R., M. Israel and R. Kanazawa (1992), *Recent California Water Transfers: Emerging Options in Water Management,* Center for Environmental and Water Resources Eng. Report 92-1, Dept. of Civil and Env. Eng., University of California, Davis.

Maass, A. and R. Anderson (1978), *... And the Desert Shall Rejoice: Conflict, Growth, and Justice in Arid Environments*, MIT Press, Cambridge, MA.

MacDonnell, L.J.(Principal Investigator) (1990), *The Water Transfer Process As A Management Option for Meeting Changing Water Demands,* Volume I, USGS Grant Award No. 14-08-0001-G1538, Natural Resources Law Center, University of Colorado, Boulder.

Mann, Patrick C and Clark, Don M. (1993), "Marginal-Cost Pricing: Its Role in Conservation," *AWWA Journal*, Vol. 85, 71-78.

Saleth, R.M., J.B. Braden, and J.W. Eheart (1991), "Bargaining Rules for a Thin Spot Water Market," *Land Economics*, Vol. 67, No. 3, August, pp. 326-339.

Sergent, M.E. (1990), Water Transfers: The Potential for Managing California's Limited Water Resources, Masters Thesis, Civil Engineering Department, University of California, Davis.

*The 1991 Drought Water Bank*, (1992), California Department of Water Resources, Sacramento, CA, January.

*Water Transfers in the West: Efficiency, Equity, and the Environment*, National Research Council, (1992), National Academy Press, Washington, D.C.

Weber, Jack A. (1993), "Integrating Conservation Targets into Water Demand Projections," *American Water Works Association Journal*, 85, 63-70.

Wilchfort, O. and J.R. Lund (1997), "Shortage Management Modeling for Urban Water Supply Systems," *Journal of Water Resources Planning and Management*, ASCE, Vol. 123, No. 4, July/August.

# URBAN WATER PRICING AND DROUGHT MANAGEMENT: A RISK BASED APPROACH

MESSELE Z. EJETA
*Graduate Student*

LARRY W. MAYS,
*Prof. of Civil and Environmental Engineering*
*Department of Civil and Environmental Engineering,*
*Arizona State University*
*Tempe, Arizona 85287, USA*

**Abstract**

This paper presents a new methodology for urban water pricing during drought condition that expresses urban water pricing as a function of risk. Risk is given as the probability that urban water demand exceeds available water supply. Urban water demand is believed to be elastic to price. However, uncertainty is involved in the estimation of the expected demand and the available supply. Both the demand and the supply, which are also related to the return period of hydrologic conditions, can be represented by probability distribution functions about the expected values. The general trend is that the gap between the available supply and the expected demand diverges as the return period increases. Relationships between urban water price, the return period of hydrologic conditions and the associated risk are developed. Also under sustained drought conditions where the demand exceeds the available supply, a methodology is developed whereby the demand can be adjusted down to the available supply through successive increases in the water price.

## 1. Introduction

Droughts continue to rate as one of the most severe weather induced problems around the world. Global attention to natural hazards reduction includes drought as one of the major hazards. Changnon (1993) gave seven lessons or truths that have emanated out of studying the major droughts from 1932 to 1992 in the U.S. These lessons are summarized below:

1) Major drought is a pervasive condition affecting most portions of the physical environment as well as the socioeconomic structure.



*E. Cabrera and J. García-Serra (eds.), Drought Management Planning in Water Supply Systems*, 261–298.

2) Droughts are a major but unpredictable part of the climate of all parts of the United States. Moreover, they occur infrequently and this results in a decay in the attention to drought preparedness and mitigation.
3) Responses and adjustments to drought problems can be sorted into two classes: a) short-term fixes and b) long-term improvements.
4) Although many long-term adjustments have been made as a result of the major droughts of the last 60 years, many factors make today's society generally more vulnerable to drought than ever before.
5) Agriculture, in general, can not escape from experiencing major drought losses in the future, even with healthier crop strains and increased irrigation.
6) Opportunities for improvement in water management exist and could make the nation's water resources more impervious to drought. However, many water-related problems are localized and at the substate scale and often do not get needed attention.
7) Drought is ubiquitous: everything and everybody is affected, and yet no one (every one) is in charge.

Shortage of water supply during drought periods is such a significant factor for the general welfare that its effect can not be easily undermined. Domestic water supply shortages during these periods in particular have been crucial in some cases and as a result initiated various measures that were taken by different water supply agents. The measures targeted different means of reducing water demand during such periods. These measures, which may be considered as semi-empirical to empirical, include water metering, leak detection and repair, rate structures, regulations on use, educational programs, drought contingency planning, water recycling and reuse, pressure reduction and so on. Such efforts are collectively termed as water conservation, although there has not been a uniform definition among authors.

On the other hand, different researchers and scientists have tried to develop more scientific methods for water conservation during drought periods. These methods have been aimed at water conservation through price increases of the water supply to the customers. The results elucidated the fact that water is more of a commodity than it is a public resource. However, the several models developed so far which relate reduction in demand for water due to the increase in its price, through the price elasticity, used different variables that range from the income of the customers to hydrologic conditions. The relations developed used regression analysis and as a result the differences and the variations of the variables considered are significant that the estimated demand is subject to uncertainty. Thus the demand may be better expressed by an estimated value and a probability distribution.

The basics of the price elasticity approach presumes that the demand can be adjusted to the available supply. This may happen on average basis; however, the demand has a random distribution about the available supply. By similar reasoning, the available supply corresponding to a given return period of weather conditions may have a random distribution about the expected value. All the aforementioned uncertainties call for risk evaluation to determine the probability that the demand exceeds the available supply, for water supply project planning. Conversely, the price of water supply for a given tolerable risk level can be determined.

This paper first discusses various efforts reported in the literature for water conservation and then culminates with the new idea of a risk based approach. The different water conservation practices are briefly discussed, giving a coverage of the price elasticity formulation. The basic reasons which make it necessary for a risk based approach are described. Some risk level indices which have been used for the evaluation and prediction of a drought period are given and their limitations are explained. A method for evaluation of the damage associated with certain levels of drought severity is developed. This new approach relates the risk, the price and the return period. It is found through this relationship that risk is sensitive to the return period and to the price changes.

## 2. Background of Water Conservation

### 2.1. DROUGHT MANAGEMENT OPTIONS

Experiences from past droughts have shown that the action of water managers can greatly influence the magnitude of the monetary and non-monetary losses from drought. There have been a variety of drought management options that have been undertaken in response to anticipated shortages of water, which can be categorized as (Dziegielewski, 1986): 1) demand reduction measures; 2) improvements in efficiency in water supply and distribution system; and 3) emergency water supplies. A topology of drought management options is given in Table 1.

Not only is water conservation necessary during drought periods but the economic merits are also important to consider. In the US, federal mandates urge that opportunities for water conservation be included as a part of the economic evaluation of proposed water supply projects (Griffin and Stoll, 1983). Water conservation during drought periods, however, requires important attention because our demand of water may exceed the available resource in the demand environment. Conservation may be achieved through different activities. According to the US Water Resources Council (1979a), these activities include, but are not limited to:

1. reducing the level and/or altering the time pattern of demand by metering, leak detection and repair, rate structure changes, regulations on use (e.g. plumbing codes), education programs, drought contingency planning;
2. modifying management of existing water development and supplies by recycling, reuse, and pressure reduction; and
3. increasing upstream watershed management and conjunctive use of ground and surface water (Griffin and Stoll, 1983).

Table 1. A Topology of Drought Management Options (Dziegielewski, 1986)

I. Demand Reduction Measures
   1. Public education campaign coupled with appeals for voluntary conservation
   2. Free distribution and/or installation of particular water saving devices:
      2.1 Low-flow showerheads
      2.2 Shower flow restrictors
      2.3 Toilet dams
      2.4 Displacement devices
      2.5 Pressure-reducing valves
   3. Restrictions on non essential uses:
      3.1 Filling of swimming pools
      3.2 Car washing
      3.3 Lawn sprinkling
      3.4 Pavement hosing
      3.5 Water-cooled air conditioning without re-circulation
      3.6 Street flushing
      3.7 Public fountains
      3.8 Park irrigation
      3.9 Irrigation of golf courses
   4. Prohibition of selected commercial and institutional uses:
      4.1 Car washes
      4.2 School showers
   5. Drought emergency pricing:
      5.1 Drought surcharge on total water bills
      5.2 Summer use charge
      5.3 Excess use charge
      5.4 Drought rate (special design)
   6. Rationing programs:
      6.1 Per capita allocation of residential use
      6.2 Per household allocation of residential use
      6.3 Prior use allocation of residential use
      6.4 Percent reduction of commercial and institutional use
      6.5 Percent reduction of industrial use
      6.6 Complete closedown of industries and commercial establishments with heavy uses of water

II. System Improvements
   1. Raw water sources
   2. Water treatment plant
   3. Distribution system:
      3.1 Reduction of system pressure to minimum possible levels
      3.2 Implementation of a leak detection and repair program
      3.3 Discontinuing hydrant and main flushing

Table 1. Cont'd

III. Emergency Water Supplies
- 1. Inter-district transfers:
  - 1.1 Emergency interconnections
  - 1.2 Importation of water by trucks
  - 1.3 Importation of water by railroad cars
- 2. Cross-purpose diversions:
  - 2.1 Reduction of reservoir releases for hydropower production
  - 2.2 Reduction of reservoir releases for flood control
  - 2.3 Diversion of water from recreation water bodies
  - 2.4 Relaxation of minimum streamflow requirements
- 3. Auxiliary emergency sources:
  - 3.1 Utilization of untapped creeks, ponds and quarries
  - 3.2 Utilization of dead reservoir storage
  - 3.3 Construction of a temporary pipeline to an abundant source of water (major river)
  - 3.4 Reactivation of abandoned wells
  - 3.5 Drilling of new wells
  - 3.6 Cloud seeding

The effort to conserve water started out with metering rather than providing a flat rate. Both domestic and sprinkling demands reduced significantly as a result of the introduction of water meters (Hanke, 1970). Grunewald, et al (1976) stated: "Traditionally, water utility managers have adjusted water quantity (rather) than prices as changes in demand occurred."

In general, some of the major measures followed for water conservation efforts with references are enumerated below:

- ◆ Use restrictions (no car washing, or hosing down sidewalks, alternate-day lawn and garden watering and the like (Moncur, 1989)
- ◆ Increasing rate structures, also called "inverted-block rates," "inclining-block rate," "increasing blocks," "inverted pyramid" rates (Jordan, 1994).
- ◆ A lump sum charge and a commodity charge per unit volume imposed in addition to the normal rates (Carver and Boland, 1980).
- ◆ Allowing the market process to operate, that is, adopting marginal cost pricing, even for normal periods, rather than averaging price (Moncur, 1989).
- ◆ Attempt to decrease the amount used by industries by trying to utilize existing technology to design and install production processes using less water per unit of output (Grebenstein and Field, 1979).
- ◆ Reducing withdrawals for production processes by recycling (Grebenstein and Field, 1979).
- ◆ Passing water conservation acts, requiring builders to install ultra-low flow fixtures in all new projects (Jordan, 1994).

- Forcing the public to be bound to treated wastewater for new recreational use. In Phoenix, Arizona, for instance, it has been considered not to allow new recreational lakes unless treated wastewater was used (Maddock and Hines, 1995).

Increasing the price of domestic water supply has been a focus of several studies. These studies were conducted to analyze the effect of urban water pricing and how it contributes to water conservation during a drought period (Agthe and Billings, 1980; Moncur, 1989). However variations have been observed in the approaches followed. According to Jordan (1994), water pricing is an effective way of conserving water, compared to the other measures mentioned above. An increase in the price of water contributes to water conservation because of the fact that customers have limited money. For every percent increase in the price, there is some decrease in the demand, which is explained through the price elasticity. A significant number of studies have been undertaken in different regions to determine price elasticity associated with pricing. The following section explains price elasticity.

## 2.2. PRICE ELASTICITY OF WATER DEMAND

The elasticity of demand is the responsiveness of consumers' purchases to varying price. The most frequently used elasticity concept is price elasticity which is defined as the percentage change in quantity taken if price is changed one percent. Young (1996) states that "the price elasticity of demand for water measures the willingness of consumers to give up water use in the face of rising prices, or conversely, the tendency to use more as price falls". Two different ways have been followed to formulate the price elasticity of demand for water: one based upon average price and the other based upon marginal price. Agthe and Billings (1980) state that the elasticity determined based upon average price overestimates the result. Therefore they recommend (as several others) that the marginal price be used.

Howe and Linaweaver, Jr., (1967) defined the price elasticity of water as

$$\eta_P = \frac{\Delta d}{\bar{d}} \div \frac{\Delta P}{\bar{P}} \tag{1}$$

where $\eta_P$ is the price elasticity, $\bar{d}$ is the average quantity of water demanded, $\bar{P}$ is the average price, $\Delta d$ is the change in the demand and $\Delta P$ is the change in the price. For a continuous demand function, the following more general formula is applicable.

$$\eta_P = \frac{dd}{d} \div \frac{dP}{P} \tag{2}$$

Table 2 is a summary of some of the values of price elasticity of water demand reported in the literature.

Table 2. Summary of some of the price elasticity values from different researches.

| No | Researchers | Research area | Year | Estimated price elasticity | Estimated income elasticity | Remarks |
|---|---|---|---|---|---|---|
| 1 | Howe & Linaweaver | Eastern US | 1967 | -0.860 | | |
| 2 | Howe & Linaweaver | Western US | 1967 | -0.52 | | |
| 3 | Wong | Chicago | 1972 | -0.02 | 0.20 | |
| 4 | Wong | Chicago Suburb | 1972 | -0.28 | 0.26 | |
| 5 | Young | Tucson | 1973 | -0.60 - -0.65 | | Exponential and linear models used |
| 6 | Gibbs | Metropolitan Miami | 1978 | -0.51 | 0.51 | Elasticity measured with the mean marginal price |
| 7 | Gibbs | Metropolitan Miami | 1978 | -0.62 | 0.82 | Elasticity measured with the average price. |
| 8 | Agthe & Billings | Tucson | 1980 | -0.27 - -0.71 | | Long-run model |
| 9 | Agthe & Billings | Tucson | 1980 | -0.18 - -0.36 | | Short-run model |
| 10 | Howe | | 1982 | -0.06 | | |
| 11 | Howe | Eastern US | 1982 | -0.57 | | |
| 12 | Howe | Western US | 1982 | -0.43 | | |
| 13 | Hanke & Maré | Malmö, Sweden | 1982 | -0.15 | | |
| 14 | Jones & Morris | Metropolitan Denver | 1984 | -0.14 - -0.44 | 0.40 - 0.55 | Linear and log-log models used. |
| 15 | Moncur | Honolulu | 1989 | -0.27 | | Short-run model |
| 16 | Moncur | Honolulu | 1989 | -0.35 | | Long-run model |
| 17 | Jordan | Spalding County, Georgia | 1994 | -0.33 | | A price elasticity of -0.07 was also reported for no rate structure, but increased price level |

The use of the price elasticity of water has been applied to some cities with some important achievements having been obtained. The following schematic may depict the general trend of this principle, as derived from the conclusion reached by Jordan (1994).

$$\Uparrow \textit{(Price)} \Rightarrow \Downarrow \textit{(Water demand)} \;\&\; \Uparrow \textit{(Revenue)} \tag{3}$$

An increase by less than 40% of the price resulted in a 10% decrease in the demand in Honolulu, Hawaii - the announced goal of the restrictions imposed in the drought episodes of 1976 to 1978 and in 1984 (Moncur, 1987). This was achieved using a price elasticity of only -0.265. In Tucson, Arizona, an inverted rate structure was claimed to have been credited with reducing public demand from about 200 gallons per capita per day (gpcd) to 140 - 160 gpcd (Maddock and Hines, 1995).

The way in which water utilities are structured is probably the most important factor which complicates the study of price elasticity. For instance, some customers who own homes or who pay for water bill, more or less, react to the price change whereas those who rent apartments or who do not pay for water bill are almost indifferent to it. Furthermore, water necessity for residential, commercial, and industrial purposes are not equally important. Because of this reason, different researchers had to study demand elasticity by categorizing water distribution systems for industrial, commercial and residential uses. The demand patterns under these categories are not uniform. One of the most comprehensive studies on price elasticity of water demand done by Schneider and Whitlatch (1991) for six user categories (residential, commercial, industrial, government, school and total metered) showed different results for these categories. Residential water use is further complicated by different factors: many residents who rent housing do not pay for water and as such are indifferent to demand regulations; the patterns for indoor and outdoor water demand differ quite significantly and hence necessitate different approaches of demand analysis. The climatic conditions of a given area and the time of the year are also worth mentioning. These are probably the reasons why apparently different elasticity values are reported for the eastern and the western U.S. and for winter and summer uses.

From the studies enumerated so far, a general conclusion is reached: that demand is elastic to price increase. Almost all research has reinforced this hypothesis. However, differences exist between the elasticity values calculated for different geographic locations. For instance, Howe (1982) obtained values of -0.57 and -0.43 for the eastern and the western U.S. respectively. On the other hand, no clear consistency exists in the way that elasticity is calculated: some use average price, some use marginal price, and still some include the intramarginal rate structure. Although some of the studies targeted alleviating water shortage problems during drought periods, they did not approach the problem from the perspective of risk analysis.

## 2.3. DEMAND MODELS

It is important to have demand related to the drought severity. Several studies have expressed demand as a function of different variables. Mays and Tung (1992) gave a general form of demand models as

$$d = f(x_1, x_2, \ldots, x_k) + \varepsilon \tag{4}$$

where $f$ is the function of explanatory variables $x_1$, $x_2$, ..., $x_k$ and $\varepsilon$ is a random error (random variable) describing the joint effect on $q$ of all the factors not explicitly considered by the explanatory variables.

Several explicit linear, semi-logarithmic and logarithmic models have been developed through different researches. Billings and Agthe (1980), for example, gave the following water demand function for Tucson, Arizona (notations modified to fit the notations adopted for this study).

$$\ln(d) = -7.36 - 0.267\ln(P) + 1.61\ln(I) - 0.123\ln(DIF) + 0.0897\ln(W) \tag{5}$$

In the above equation, $d$ is the monthly water consumption of the average household in 100 ft$^3$; $P$ is the marginal price facing the average household in cents per 100 ft$^3$; $DIF$ is the difference between the actual water and sewer use bill minus what would have been paid if all water was sold at the marginal rate ($); $I$ is the personal income per household ($/month); and $W$ is the evapotranspiration minus rainfall (inches).

The above equation implicitly relates demand to the hydrologic index, $W$. The positive coefficient of $W$ shows that demand increases exponentially with $W$, which indirectly indicates increases of demand with the dryness of weather conditions. The general trend of the average demand with the return period, therefore, may be shown as given by the demand curve in Fig. 1. Demand increases with the return period of the drought severity because the more severe the drought, the more the customers are prompted to use more water. Different demand curves are illustrated in Fig. 2 for different price levels. As shown in this figure, the higher the price, the lower the demand for a given hydrologic conditions.

Equation (5) may be rearranged as

$$d = 0.00006362 P^{-0.267} I^{1.61} (DIF)^{-0.123} W^{0.0897} \tag{6}$$

or in more general terms,

$$d = a' P^{b'} I^{c'} (DIF)^{d'} W^{e'} \tag{7}$$

where $a'$, $b'$, $c'$, $d'$, and $e'$ are constants. The price elasticity of demand for equation (6) is -0.267. Therefore, changing the price while keeping the other variables constant results in different average demand values, $\bar{d}_{P_i}$. Again, varying $W$ while keeping the other variables constant gives a general relation of the average demand associated with the return period $T$.

As given in equation (7), it can be seen that the demand $d$ is related to the hydrologic index $W$ which is also related to the return period. The available supply (flow) $q$ is also related to the return period (Hudson and Hazen, 1964). Thus the general relationships between demand and return period and supply and return period which are shown in Fig. 1 are based on these trends.

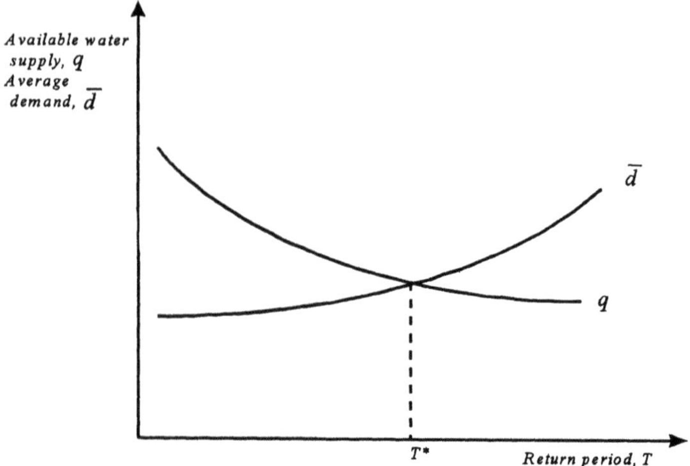


Fig. 1. Water supply availability and average demand as related to the return period, $T$

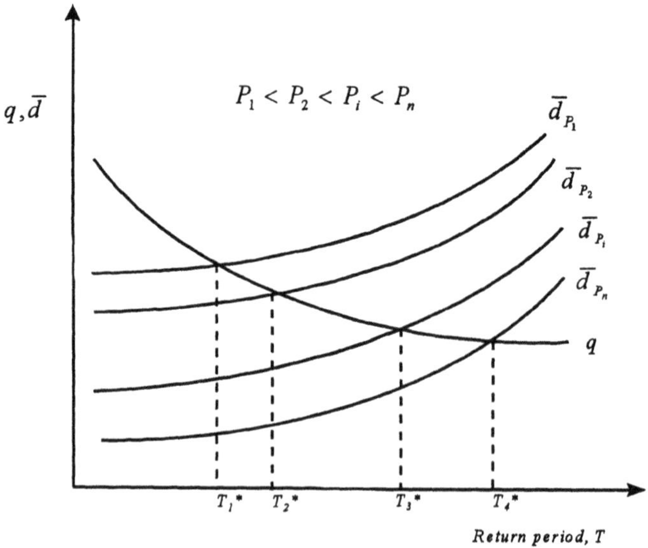


Fig. 2. Water supply and average demands for different price values as related to $T$

## 2.4. THE NEED FOR A RISK BASED APPROACH

A few past studies analyzed risk only by defining it as the monetary (financial) loss. They did not consider the risk as the probability of the supply not meeting the demand. To make these two connotations of risks distinctive, the terms *financial risk* and *probabilistic risk* are introduced and used differently. Many of the previous studies on risk did not explicitly define financial and probabilistic risks.

*Financial risk* can be simply stated as the monetary loss associated with a certain damage. *Probabilistic risk*, which is explicitly used in this paper, may be formulated as the probability *p*( ) that the demand $d_{P_i}$ at price $P_i$ for level *i* exceeds the available supply $q_T$, expressed as

$$Risk = p(d_{P_i} > q_T) \tag{8}$$

Municipal water supply shortage problems have been manifesting themselves in different regions at different times for a long time. A study by Dixon, et al. (1996) for California shows that projections of future water supply and demand (including environmental uses) indicate that the gap between supply and demand will widen to 4.1 million acre-feet in average water years and 7.4 million acre-feet in drought years by 2020. In 1977 in Fairfax County, Virginia, the drought was so severe that drastic measures such as the closing of schools and businesses were actively being considered (Sheer, 1980).

Two major groups of actions are undertaken by water agencies in order to avert some serious consequences of impending water shortages caused by droughts. They are measures that reduce demands and measures that enhance existing supplies. Developing practical methods for determining the necessary prices and devising structures of water rates that would achieve the desired reductions in water use are the most critical needs for establishing effective drought pricing policies (Dziegielewski, et al., 1991).

There are different uncertainties involved with either one of these measures. In trying to reduce demands by increasing the price, uncertainty is involved in that the demand volume may not be equal to the limited available supply. This is simply because the demand depends on so many factors that can not be totally controlled, irrespective of the price increase. On the other hand, enhancing the existing supply may cost more than the risk of not undertaking this task at all. By way of risk analysis, it is possible to optimize between the economic loss and the cost of enhancing the existing supply – such as emergency supply construction. Many scientists in different professions agree that the level of risk as a decision support system is a good indicator for sound decisions. Decisions in which the effects are portrayed relatively in the long run may finally result in adverse effects. Such effects are incurred at the expense of non conservative risk level designs.

Suter (1993) gave the followings as the reasons for risk assessment approach for decision making:

1. the cost of estimating all environmental effects of human activities is impossibly high; and
2. regulatory decisions must be made on the basis of incomplete scientific information.

He concluded that a risk based approach balances the degree of risk to be permitted against the cost of risk reduction and against competing risks. Lansey, et al. (1989) also suggested that reliability analysis (a complement of risk analysis) be viewed as an alternative to making a decision without an analytical structure.

It has not been a common practice by responsible bodies to systematically incorporate in the decision process the risk of water supply shortages during sustained drought periods. Bruins (1993) indicated that "... governments often respond to drought through crisis management rather than pre-planned programs (i.e., risk management)." Wilhite (1993) also criticized that until recently, nations had devoted little effort toward drought planning, preferring instead the crisis management approach.

A consensus among water managers and researchers regarding water supply during drought is that the key to adequate management in urban areas lies in pre-drought preparation, especially as it relates to conservation and planning for future water needs (Dziegielewski, et al., 1991). All the above accounts prompt us to focus on the necessity of risk based design, especially when we deal with such phenomenon as drought that is very difficult to predict accurately enough its timing and magnitude.

## 2.5. DROUGHT SEVERITY AS RISK INDICES

Every natural phenomenon with which detrimental effects to human beings and their environment are associated need our keen attention of how and when it occurs. Unfortunately, the degree of some such phenomena including drought is difficult to determine, as accurately as desirable, before they occur. A study by the National Research Council (1986) indicated that there is not a firm rationale or explanation of the drought mechanism. It adds that though empirical relations have been documented so far, why and when these relations trigger the occurrence of significant drought is not understood.

In the absence of such rationale, it is worth studying the degree (level) of risk, such as in the case of droughts, based on the available indices. The level of risk is apparently reflected by the severity of the drought. Severe drought implies a relative shortage of the required water supply which in turn can be expressed by a certain level of the risk that the demand is not met. Thus calibrating drought severity may be used to indirectly determine the risk level.

No single definite method has been in use as a drought severity indicator. Nonetheless there are some which are being used in different fields. According to Wilhite (1993), the simplest drought index in widespread use is the percent of normal precipitation. This, indeed, is a good approach to infer the status of the available supply. However, it does not render an obvious forecast to enable a risk management body to be prepared to a forthcoming drought period.

Sheer (1980) tried to calculate the risk that the reservoir of a water supply system becomes empty by blending together the severest hydrologic and hydraulic conditions of different time periods. Specifically, he considered a condition in which the demands were the highest, the reservoir storage the lowest and the date when these conditions occurred the beginning of one of the worst drought years. This simulation resulted in four years out of 26 in which the reservoir was empty and it is concluded that the risk is $4/26$.

Although the approach is reasonable enough to indicate what would have happened had the conditions been met, the authors hardly believe that it fully reflects the realistic situation. One simple reason is that if the actual conditions were as worse as the ones selected, the demand could be higher and might result in more years of an empty reservoir, since the demands under such conditions would be much higher over the considered time span. Another reason may be that the risk in that study is not fully analogous with the usual convention. This is to say that the risk is based on the demand exceeding the supply, which is reached long before the reservoir becomes empty.

As the best alternative, risk analysis may be viewed in relation to the uncertainty associated with the different variables. Tung (1996) points out that the most complete and ideal description of uncertainty is the probability density function of the quantity subject to uncertainty. It is, therefore, very feasible to consider the probability density function of the demands about a fixed available supply during drought and thus derive the risk as the cumulative probability function of the supply being exceeded.

To be able to calculate the risk, the level of the drought severity must be determined (forecast). There are several drought severity indices which have been used so far. Some of them are used to assess an already happened drought event's severity while a few others are used for forecasting. The Palmer Drought Severity Index (*PDSI*) and the Sheer Steila Drought Index (*DI*) are examples of the former category while the Surface Water Supply Index (*SWSI*) and the Southern Oscillation Index (*SOI*) are examples of the indices that are used for drought forecasting.

Palmer (1965) expressed the severity of a drought event by developing the following equation (Steila, 1972; Puckett, 1981).

$$PDSI_i = 0.897 PDSI_{i-1} + \tfrac{1}{3} Z_i \tag{9}$$

where $PDSI$ is the Palmer Drought Severity Index and $Z$ is an adjustment to soil moisture for carryover from one month to the next, expressed as

$$Z_i = k_j \left[ PPT_i - \left( \alpha_j PE_i + \beta_j G_i + \gamma_j R_i - \delta_j L_i \right) \right] \tag{10}$$

in which the subscript $j$ represents one of the calendar months and $i$ is a particular month in a series of months. $PPT_i$ is the precipitation, $PE_i$ is the potential evapotranspiration, $G_i$ is the soil moisture recharge, $R_i$ is the surface runoff (excess precipitation), and $L_i$ is the soil moisture loss for month $i$. The coefficients $\alpha_j$, $\beta_j$, $\gamma_j$, and $\delta_j$ are the ratios for long-term averages of actual to potential magnitudes for $E$, $G$, $R$, and $L$ based on a standard 30-year climatic period.

The Surface Water Supply Index (*SWSI*) gives a forecast of a drought event. It is a weighted index that generally expresses the potential availability of the forthcoming season's water supply (U.S. Soil Conservation Service, 1988). It is formulated as a re-scaled weighted of nonexcedence probabilities of four hydrologic components: snowpack, precipitation, streamflow and reservoir storage (Garen, 1993).

$$SWSI = \frac{\alpha p_{snow} + \beta p_{prec} + \gamma p_{strm} + \omega p_{resv} - 50}{12} \tag{11}$$

where $\alpha$, $\beta$, $\gamma$ and $\omega$ are weights for each hydrologic component and add up to unity; $p_i$ is the probability of nonexcedence (in percent) for component $i$; and the subscripts *snow*, *prec*, *strm*, and *resv* stand for the snowpack, precipitation, streamflow and reservoir storage hydrologic components, respectively. This index has a numerical value for a given basin which varies between -4.17 to +4.17. The following are the ranges for the index for practical purposes: +2 or above, -2 - +2, -3 - -2, -4 - -3 and -4 or below. These ranges are associated with the qualitative expressions of abundant water supply, near normal, moderate drought, severe drought and extreme drought conditions, respectively.

The *SWSI* has been in use to forecast different basins' monthly surface water supply forcasts (see, for example, the Colorado Water Supply Outlook, U.S. Soil Conservation Service, Jan. 1988). In fact, it gives a forecast of both wet and dry (drought) months. On the other hand, Wilhite (1993) reports that several scientists agree that it has been possible to forecast drought for up to six months in Australia by using the *SOI*, which is based on forecast meteorological conditions.

## 3. **Risk-Price Relationship**

Risk can be defined as the probability that the loading exceeds the resistance (Chow, et al., 1988; Mays and Tung, 1992). Analogously, the risk in water distribution systems is defined as the probability that the demand exceeds the available supply where the demand is considered as the loading and the supply as the resistance. For future planning purposes, it is not certain when a drought event of a certain severity level will occur.

In planning for urban water supply projects, therefore, it is important to determine the probability distribution parameters of the demand and the supply. Both demand and supply are related to hydrologic indices. Also, operation/management of an existing water distribution system can be handled better through a risk analysis approach when the forthcoming period's (say month) conditions of weather or water supply availability can be predicted ahead.

One of the common ways to represent uncertain events such as demand and supply under drought conditions is using an appropriate probability distribution of these variables. On the other hand, both variables are related to the return period $T$ of the drought. The available supply data of many years can be arranged in descending order of magnitude for drought indication. These arranged flow data can be plotted versus the return period $T$, which is a measure of hydrologic conditions, as shown in Fig. 1 (Section 2.3). Two basic ways can be considered for selecting the representative flow data in relation to the return period. The first one is selecting one extreme value for each unit of time, e.g., the lowest monthly flows in a period of years. The second is selecting the lowest monthly flows in a period of years (Hudson and Hazen, 1964). Both of these procedures give a general relationship between available supply (flow) and its corresponding return period, as given in Fig. 1 (Section 2.3).

## 3.1. DEVELOPING RISK-PRICE RELATIONSHIPS

Demand depends on many uncertain factors and consequently is uncertain for a given return period drought event. The uncertainty can be represented through a probability distribution function as illustrated in Fig. 3 which indicates the risk at two different return periods.

For decision purposes, the design may be fixed at the condition where the demand equals the available supply for a given price level. Beyond this point the demand exceeds the supply and there will be some associated risk. As shown in Fig. 2 (Section 2.3), the intersection points and the region beyond represent different values of water price and the associated risk. The illustrations in Figs. 1 and 2 (Section 2.3) show that for return periods larger than the critical return period $T^*$ at the intersection point of supply and demand, the demand at the given price is greater than the supply. As the price decreases, the shortage volume increases thereby increasing the risk. Thus a graph of risk versus price may be plotted as shown in Fig. 4.

The regions beyond each of the intersection points in Fig. 2 have some corresponding risk levels, that is, the probability $p(\ )$ that the available supply falls below the demand corresponding to the price adopted at level $i$, $i = 1, 2, 3, \ldots$ n. Such a relation can help water supply planners to determine a municipal water supply price based on a predetermined tolerable risk or to assess the risk associated with a certain price level. Although not yet demonstrated by data analysis, the price-risk relationship indicates that price is infinite at no risk and risk is close to 1.00 at zero price (Fig. 4).

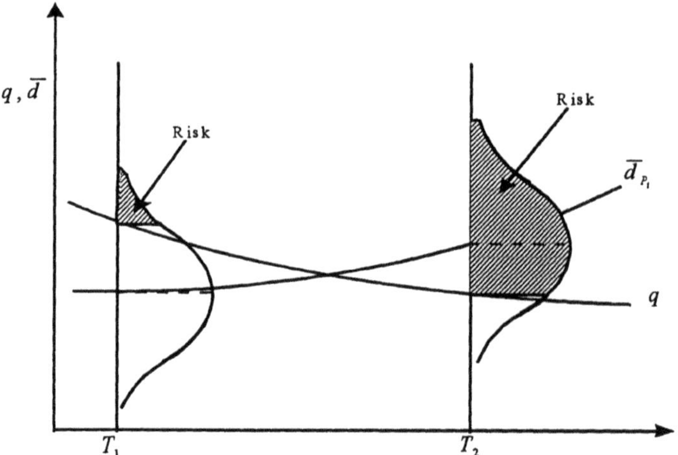


Fig. 3. Probability distribution of demand at different return periods.

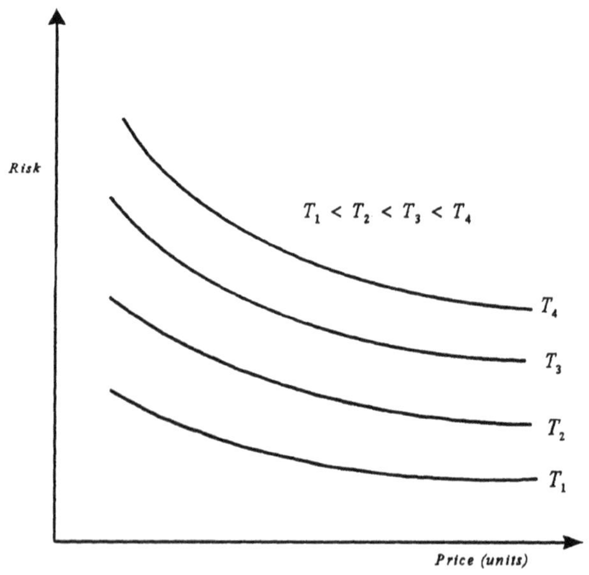


Fig. 4. Risk-price relationships for different return periods

3.1.1. *Risk Evaluation Procedures*

The general procedures for evaluating the risk of a system's loading exceeding a system's capacity are considered under two different scenarios. For water supply systems, the demand may be considered as the loading and the supply as the capacity. The two scenarios include: 1) when the loading is uncertain and the capacity is certain, and 2) when both the loading and the capacity are uncertain. Risk evaluation in the first case involves consideration of the probability density function of one variable (the demand) which is computationally more simpler. The second one involves composite risk evaluation.

Suppose that the probability density function of loading $L$ is $f(L)$. The probability $p$ that the loading will exceed a fixed and known capacity $C^*$ is given as (Chow, et al., 1988; Mays and Tung, 1992)

$$p(L > C^*) = \int_{C^*}^{\infty} f(L)dL \tag{12}$$

This relationship holds true when the capacity $C$ is a deterministic quantity, which corresponds to the first scenario. Analogously, if the probability density function of demand $d_{P_i}$ at price level $P_i$ is $f(d_{P_i})$, the risk of demand $d_{P_i}$ at price level $P_i$ exceeding the supply $q_T$ for return period $T$ is expressed as

$$Risk_{(P_i,T)} = \int_{q_T}^{\infty} f(d_{P_i})dd \tag{13}$$

Using this definition for risk, the risk-price relationship may be developed for each $T$. The higher the price the lower the demand is, and consequently the lower the risk.

When the capacity is also uncertain but may be represented by a probability density function $g(C)$, i.e., the second scenario, the composite risk is used. The general formula for risk in this case is (Fig. 5)

$$Risk = \int_{-\infty}^{\infty}\left[\int_{C^*}^{\infty} f(L)dL\right]g(C)dC \tag{14}$$

Again in similar analogy, the corresponding composite risk where both demand and supply are considered to be uncertain (for a given price and return period) is expressed as (Fig. 6)

$$\begin{aligned} Risk_{(P_i,T)} &= p(\bar{d}_{P_i} > \bar{q}_T) \\ &= \int_{-\infty}^{\infty}\left[\int_{\bar{q}_T}^{\infty} f(q_T)dq\right]f(d_{P_i})dd \end{aligned} \tag{15}$$

A similar relationship as the one shown in Fig. 4 can also be developed for the composite risk from these relationships. In both cases (equations (13) and (15)), the risk at given price and return period is computed as the probability of the demand exceeding the supply. The difference is in the certainty of the supply in the former equation and its uncertainty in the latter one.

### 3.1.2. *Methodology of Risk Evaluation*

Numerical evaluations of risk using the above equations call for the approach to determine the quantitative values of different statistical parameters of the loading and/or the capacity. The risk equations consist of complex probability distribution functions which become difficult to integrate. Due to this reason, alternative ways of

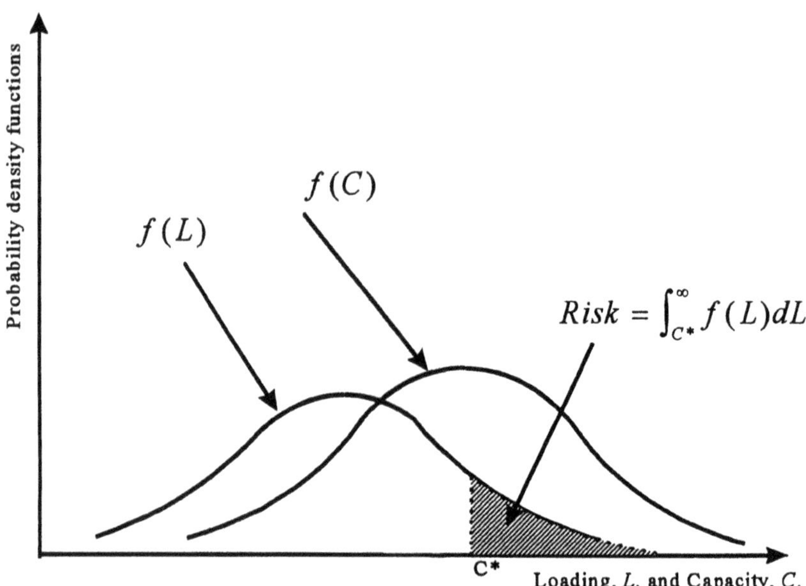


Fig. 5. Probability distribution functions of loading and capacity.

evaluating the value of risk are often utilized. The safety margin and safety factor approaches (see Chow, et al., 1988, Mays and Tung, 1992) are generally used for the computation of the risk from the probability distributions of the loading and/or the capacity. The safety margin approach is illustrated below with numerical data and the safety factor approach will be introduced in Section 4.5.

The safety margin $SM$ is generally given as the difference between the loading and the capacity or $SM = C - L$. Thus the risk in terms of the safety margin is given as

$$Risk = p(C - L < 0) = p(SM < 0) \tag{16}$$

If $C$ and $L$ are independent random variables, the mean value and the standard deviation of $SM$ are given respectively as

$$\mu_{SM} = \mu_C - \mu_L \tag{17}$$

$$\sigma_{SM}{}^2 = \sigma_C{}^2 + \sigma_L{}^2 \tag{18}$$

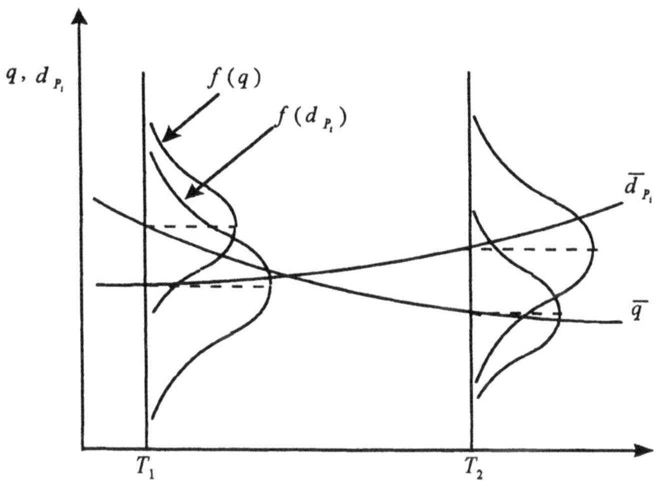


Fig. 6. Probability distributions of both demand and supply at different return periods.

By taking water demand and the available supply as the loading and the capacity respectively, the risks at different price levels for different return periods can be easily computed. Using the safety margin approach,

$$Risk = p(d_{P_1} - q_T) < 0) = p(SM < 0) \tag{19}$$

$$\mu_{SM} = \mu_q - \mu_{d_p} \tag{20}$$

$$\sigma_{SM}^2 = \sigma_q^2 + \sigma_{d_p}^2 \tag{21}$$

where $\mu_q$ is the mean supply and $\mu_{d_P}$ the mean demand at price level $P$, respectively. Assuming that the safety margin is normally distributed, the risk is expressed as

$$Risk = p\left(\frac{SM - \mu_{SM}}{\sigma_{SM}} < \frac{0 - \mu_{SM}}{\sigma_{SM}}\right) = p\left(z < \frac{-\mu_{SM}}{\sigma_{SM}}\right) = p\left(z < \frac{-(\mu_q - \mu_d)}{\left(\sigma_q^2 + \sigma_d^2\right)^{1/2}}\right)$$

$$= \Phi_z\left(-\frac{\mu_{SM}}{\sigma_{SM}}\right) = \Phi_z\left(-\frac{(\mu_q - \mu_d)}{\left(\sigma_q^2 + \sigma_d^2\right)^{1/2}}\right) \qquad (22)$$

where $z$ is the standard normal variable with mean 0 and standard deviation 1.

However, before using these equations it is further required that the mean and the standard deviation estimates of demand and/or supply must be estimated. The expected value of demand at different price levels can be estimated using the price elasticity formula. Its standard deviation, on the other hand, can be estimated from the first order analysis of uncertainty of the demand model (equation). If a dependent variable $Y$ is a function of independent variables $\mathbf{X}$ $(\mathbf{X} = X_1, X_2, \ldots, X_k)$ such that $Y = g(\mathbf{X})$, the first order approximation of $Y$ is given as

$$Y \approx g(\bar{\mathbf{x}}) + \sum_{i=1}^{k}\left[\frac{\partial g}{\partial X_i}\right]_{\bar{\mathbf{x}}}(X_i - \bar{x}_i) \qquad (23)$$

in which $\bar{\mathbf{x}} = (\bar{x}_1, \bar{x}_2, \ldots, \bar{x}_k)$, a vector containing the means of $k$ random variables (Mays and Tung, 1992). The variance of $Y$, $Var[Y]$ or $\sigma_Y^2$, is estimated by equation (24), which can be derived from the first order analysis of uncertainty of equation (23).

$$\sigma_Y^2 = Var[Y] \approx \sum_{i=1}^{k} a_i^2\sigma_i^2 + 2\sum_{1<}^{k}\sum_{j}^{k} a_i a_j Cov[X_i, X_j] \qquad (24)$$

where $a_i = \left[\frac{\partial g}{\partial X_i}\right]_{\bar{x}}$ and $\sigma_i^2$ is the variance corresponding to random variable $X_i$.

When the $X_i$'s are independent random variables, $Cov[X_i, X_j] = 0$.

The foregoing discussion in general indicated that for a given return period for design, water supply planners can decide the price of the water supply for an affordable risk level or can determine the risk at a given affordable water price. The flow chart given in Fig. 7 summarizes the basic steps used to develop the risk-price-return period relationships.

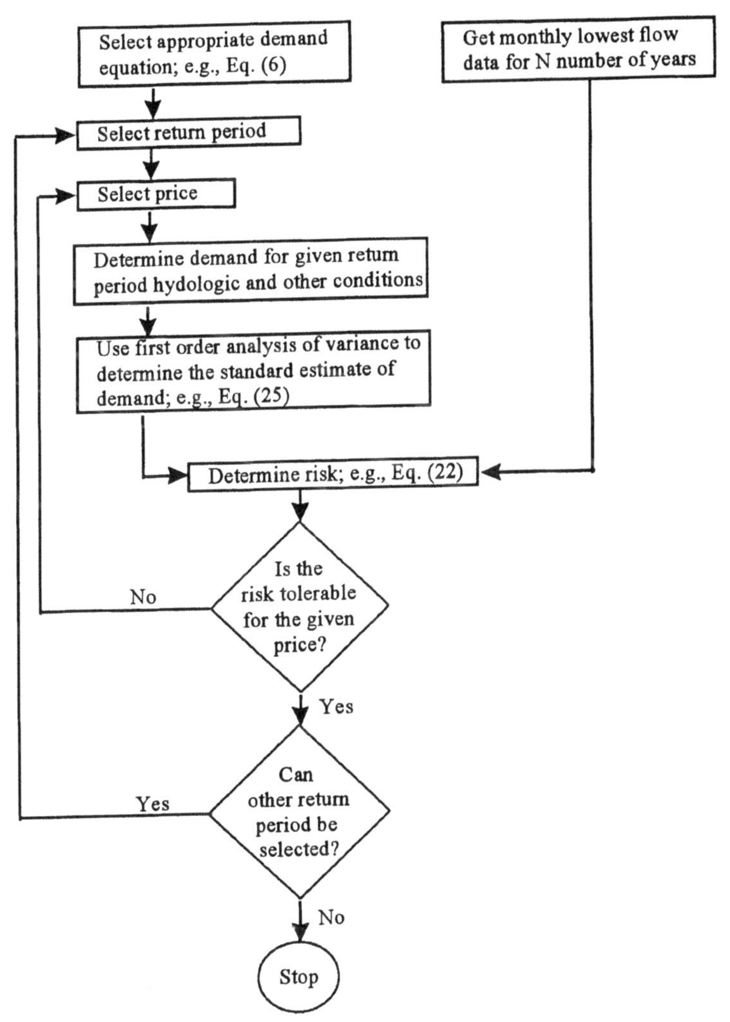


Fig. 7. Flow chart for the proposed planning procedure

3.1.3. *Risk Evaluation Example*

Based on the safety margin analysis given by the above equations and the price elasticity of demand definition (equation (1)), it is possible to determine the risk values for given return period and different price levels. For a given return period of drought, the expected demand when the price is increased by a certain amount can be determined by equation (26). Table 3 lists demand for an initial price level $P_1$ and also the available supply for different return periods. Equation (1) for the price elasticity of demand is rearranged to solve for $d_i$ as,

$$d_i = \frac{d_{i-1}\left[1+\eta_P\left(\frac{P_i - P_{i-1}}{P_i + P_{i-1}}\right)\right]}{1-\eta_P\left(\frac{P_i - P_{i-1}}{P_i + P_{i-1}}\right)} \tag{26}$$

Table 3. (Hypothetical) data of demand and supply for different return periods

| Return period, $T$ (years) | Demand at price level, $P_1$, (units) | Available flow, $q$ (units) |
|---|---|---|
| 1 | 8.0 | 12.0 |
| 5 | 8.5 | 11.0 |
| 10 | 9.5 | 9.5 |
| 25 | 11.0 | 8.0 |
| 50 | 13.0 | 7.0 |

Equation (26) is used to determine the demand at a given price level and a given return period. Price increases of up to 200% and a price elasticity of -0.5 are used to compute the demand reduction due to the increases in the price for each of the return periods. The risks associated with different price levels and different return periods are determined based on approximate estimates of the standard error for supply as 2.0 units and for demand as 4.0 units, for which $\sigma_{SM}$ equals 4.47. The results thus obtained are given in Table 4 and also plotted as shown in Fig. 8. The plots show that the risk is not significantly sensitive to the price change for small price increases. The plot in Fig. 9 shows how risk is so sensitive to the return period $T$. It is inferred from these two plots that planning and/or overcoming shortage of water supply during drought periods requires strong commitment to increase the price sufficiently.

Table 4. Risk values for different return periods and price increases of up to 200%.

| Price (unit) | Return Period, T, (years) | | | | |
|---|---|---|---|---|---|
| | 1 | 5 | 10 | 25 | 50 |
| 1.0 | 0.183 | 0.288 | 0.500 | 0.749 | 0.910 |
| 1.2 | 0.147 | 0.236 | 0.425 | 0.674 | 0.864 |
| 1.4 | 0.123 | 0.198 | 0.367 | 0.614 | 0.813 |
| 1.6 | 0.102 | 0.169 | 0.326 | 0.564 | 0.770 |
| 1.8 | 0.090 | 0.147 | 0.295 | 0.516 | 0.726 |
| 2.0 | 0.079 | 0.131 | 0.264 | 0.480 | 0.674 |
| 2.2 | 0.069 | 0.119 | 0.245 | 0.448 | 0.655 |
| 2.4 | 0.064 | 0.109 | 0.224 | 0.421 | 0.622 |
| 2.6 | 0.058 | 0.102 | 0.212 | 0.394 | 0.599 |
| 2.8 | 0.054 | 0.093 | 0.198 | 0.378 | 0.568 |
| 3.0 | 0.050 | 0.087 | 0.187 | 0.359 | 0.544 |

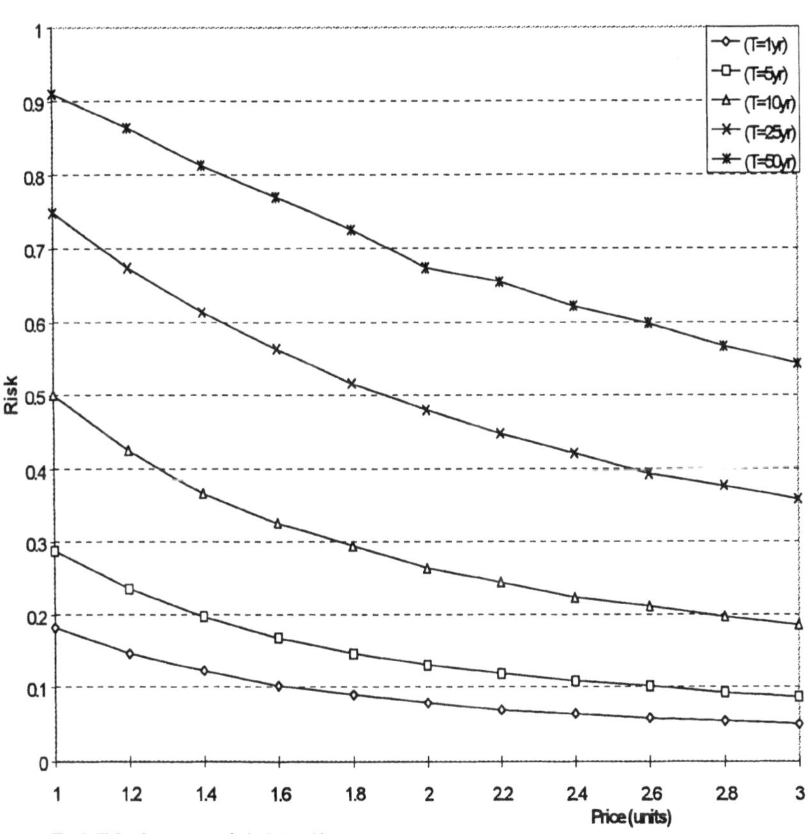


Fig. 8. Risk-price-return period relationships

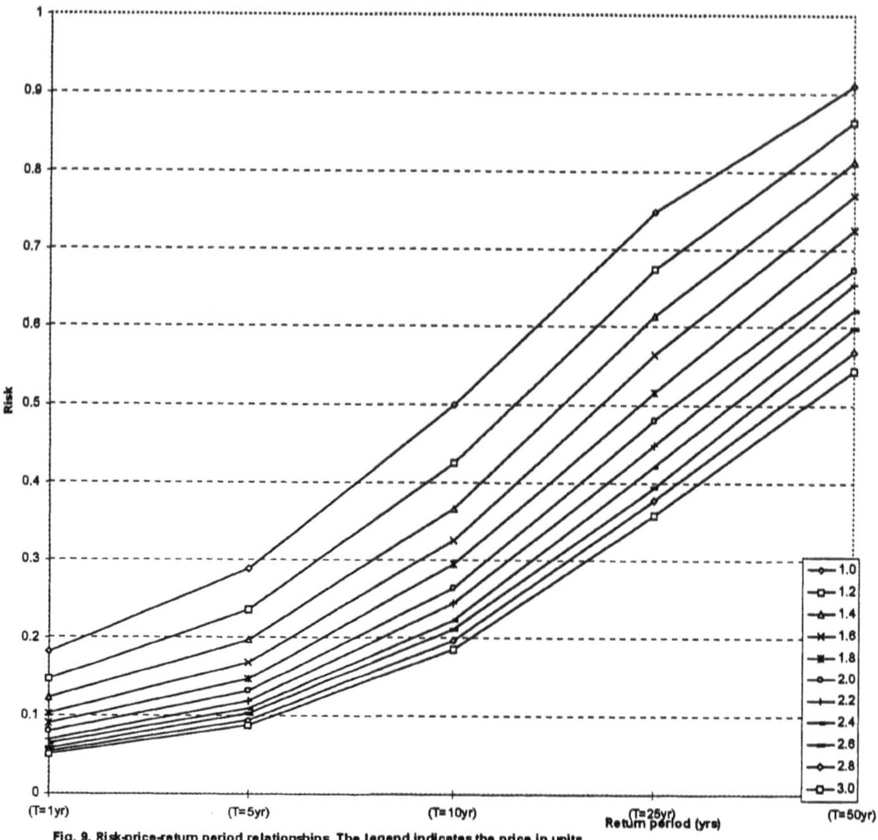


Fig. 9. Risk-price-return period relationships. The legend indicates the price in units

## 4. Operation/Management Planning Under Sustained Drought Conditions

The price elasticity formulation indicates that when the available supply is less than the demand, the latter can be adjusted to the former by increasing the price. In other words, for a drought event of severity index greater than the one at which the demand equals the available supply (Fig.10), it is possible to force the demand curve down to the supply curve by increasing the price. However, the fact that it has not been easy to forecast drought conditions well ahead of time and the uncertainty in its magnitude and length requires operation/management of water supply systems that will attempt to smooth out the effect of the drought. Such operation/management efforts will be based on data of short time interval. The efforts in effect are a supplement to the planning procedure already mentioned above. The planning basically turns out to be a one time decision while operation/management especially under sustained drought conditions involve routine decisions. The severity of the drought could be so high that emergency water supply construction projects may be considered. It is required to estimate the damage that would result from a sustained drought period the result of which, most of all, may be used to determine if an emergency water supply should be implemented.

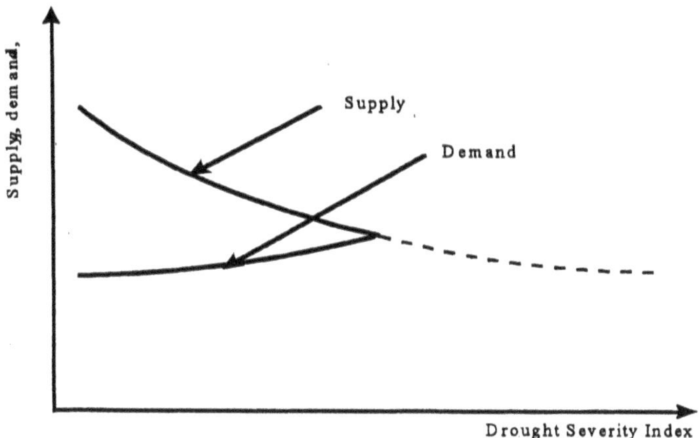


Fig. 10. Water demand during a sustained drought period as adjusted to the available supply (the broken line shows the adjustment).

## 4.1. ECONOMIC ASPECTS OF WATER SHORTAGE

Shortage of water supply during drought periods results in different types of losses in the economy including, but not limited to, agricultural, commercial and industrial. In agriculture, lack of water supply results in crop failures, in the commerce it may result in a recession of the business and in the industry it may result in under production of commodities. The loss in each production or service sector depends on the purpose of the sector. For instance, the economic impact of drought on agriculture depends on the crop type, etc. (Easterling, 1993). There is no single common way of assessing the economic impact of drought on any one of the sectors. Evaluating and comparing what actually happens during a drought period with what would have happened had there been no drought may be one way of assessing the effects of drought (Dixon, et al., 1996). Dixon, et al. (1996) adopted the concept of willingness-to-pay to value changes in well-being. They define willingness-to-pay as the maximum individuals would have been willing to pay to avoid the drought management strategies imposed by water agencies.

On the other hand, since water is supplied during a drought period at a greater price, it can be viewed as a revenue generator. Therefore, when the demand exceeds the available supply, the revenue collected by the water supply agency will be less than what could have been collected had there been more supply than that actually available. In other words, if the demand exceeds the supply, the problem is not only limited to lack of water but there will also be economic loss since the customers would pay for more supply if there were enough. Depending on the risk level, it is possible to reach a decision of whether supply augmentation is necessary or the pressure for more demand could be tolerated with the available supply.

Some water shortage relief efforts can be undertaken so that emergency water supplies may be made available to the users. This can be implemented by well drilling, trucking in potable supplies, or transporting water through small diameter emergency water lines. In such cases, it may be required that the emergency supply construction costs be paid by the users (Dziegielewski, et al., 1991). The estimation of the expected financial loss can be used to determine and inform the users of its extent and advise them of the necessity, if any, of paying for the emergency supply construction costs.

If the option for emergency supply construction is justified, then the design needs to take into consideration the possibilities of optimization. The construction can be designed such that the financial risk and the cost of construction are at optimum. Fig. 11 illustrates this optimization process.

The economic loss (damage) can be calculated with the help of equations (30) and (31) (given in the next section) and the cost of emergency construction must be determined from the physical conditions at the disposal of the water supply agency.

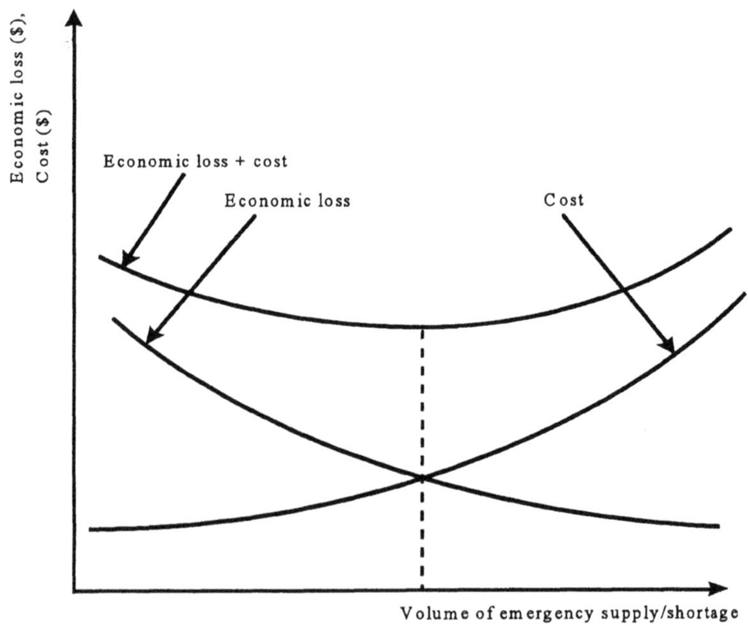


Fig. 11. Optimization for emergency water supply construction

## 4.2. DAMAGE ASSESSMENT

The damage that would result if a certain drought event occurred helps as one of the main decision factors. Since the time of occurrence of the drought event that causes the damage is difficult to determine, only the expected value is assessed by associating its magnitude with its probability of occurrence.

Chow, et al. (1988) define the expected annual damage cost $D_T$ for the event $x > x_T$ as

$$D_T = \int_{x_T}^{\infty} D(x) f(x) dx \qquad (27)$$

where $f(x)dx$ is the probability that an event of magnitude $x$ will occur in any given year and $D(x)$ is the damage cost that would result from that event. The event $x$ in this case can be assumed as the demand and $x_T$ can be the available supply during a drought event of return period $T$.

Breaking down the expected damage cost into intervals,

$$\Delta D_i = \int_{x_{i-1}}^{x_i} D(x) f(x) dx \qquad (28)$$

from which the finite difference approximation is obtained as

$$\Delta D_i = \left[\frac{D(x_{i-1}) + D(x_i)}{2}\right] \int_{x_{i-1}}^{x_i} f(x) dx$$

$$= \left[\frac{D(x_{i-1}) + D(x_i)}{2}\right] \left[p(x \geq x_{i-1}) - p(x \geq x_i)\right] \qquad (29)$$

Thus the annual damage cost for a structure designed for a return period $T$ is given as

$$D_T = \sum_{i=1}^{\infty} \left[\frac{D(x_{i-1}) + D(x_i)}{2}\right] \left[p(x \geq x_{i-1}) - p(x \geq x_i)\right] \qquad (30)$$

To determine the annual expected damage in the above equation, the damage that results from drought events of different severity levels must be quantified.

The magnitude of the drought (in monetary units) may be obtained by estimating the volume of water shortage that would result from that drought. In other words, not having the water results in some financial loss to the water supply customer.

The resulting financial loss to the customer from a certain drought event is thus considered as the damage from that drought event.

As shown in Fig. 1 in Section 2.3, after the critical return period $T^*$ the divergence between the demand and the supply increases with the return period. Expressing the demand and the supply as a function of return period $T$ of drought events enables one to estimate the annual expected water supply shortage volume as given by equation (31):

$$S_V = \int_{T^*}^{T} \left[ d(T) - q(T) \right] dt \tag{31}$$

The shortage volume $S_V$ is illustrated by the shaded area in Fig. 12. The shortage volume for a drought event of a higher return period above the critical one results in higher shortage volume and consequently a higher associated damage. The relationship between the shortage volume and the associated damage generally depends on several factors including water use category – residential, industrial, commercial, agricultural and so on. To use the procedure presented herein for assessing the damage that results from certain water shortage volume, the damage given by equation (30) must be developed for a specific user category.

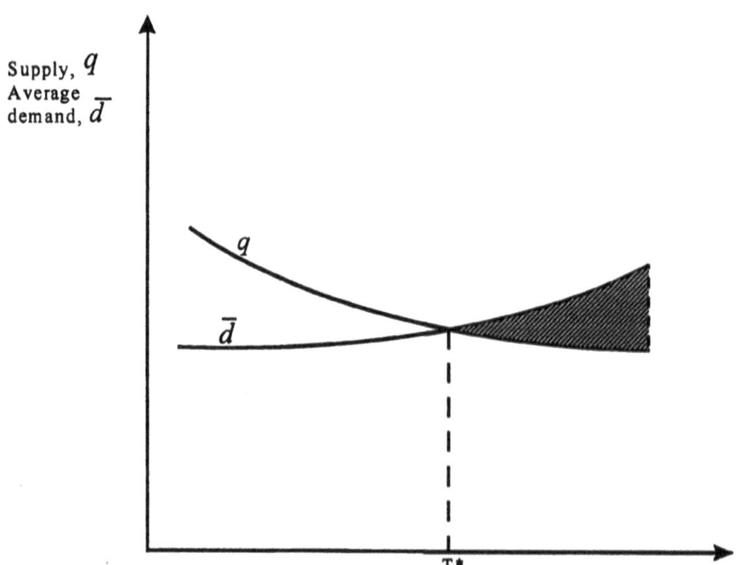


Fig. 12. Demand and supply showing water shortage volume when demand exceeds supply

### 4.3. OPERATION/MANAGEMENT

For operation/management of an existing municipal water supply system during a sustained drought period, administrative decisions may be based on short time forecasting of the hydrologic conditions. A forecast of, say one month ahead of the available supply, helps the supply managers to pre-adjust the expected demand to the forecast available supply by increasing the price. In other words, the expected demand can be, in principle, suppressed to forecast available supply by increasing the price. Howe (1993) points out that since price presumably affects the quantities users demand, price can be used to adjust demand to the available supply. The basic factor in the decision will be the damage that would occur if the adjustment were not undertaken. This is the reason why we need to focus on the assessment of such damages.

It may be easily conceived from the above reasoning that it is possible to express price increase as some function of damage. If $dP$ is an elementary increase in price due to a certain level of drought, the following general relationship may be formulated:

$$dP = \phi(\xi) \tag{32}$$

where $\xi$ is an implicit variable for drought severity level.

The amount of decrease in the demand attained as a result of the increase in the price may be determined from the concept of price elasticity of demand for water, which is rewritten in finite difference form as

$$\Delta d - \frac{\eta_P \left[ \dfrac{d_{x_{i-1}} + d_{x_i}}{2} \right]}{\left[ \dfrac{P_{x_{i-1}} + P_{x_i}}{2} \right]} \Delta P \tag{33}$$

The equation for the increased price $P_{x_i}$ is obtained from equation (33) as

$$P_{x_i} = \left[ \frac{P_{x_{i-1}} + P_{x_{i-1}} \left( \dfrac{d_{x_i} - d_{x_{i-1}}}{\eta_P \left( d_{x_{i-1}} + d_{x_i} \right)} \right)}{\left[ 1 - \dfrac{\left( d_{x_i} - d_{x_{i-1}} \right)}{\eta_P \left( d_{x_{i-1}} + d_{x_i} \right)} \right]} \right] \tag{34}$$

Also equation (32) can be written in finite difference form as

$$\Delta P = P_{x_i} - P_{x_{i-1}} = \phi(\xi) \tag{35}$$

$$P_{x_i} = P_{x_{i-1}} + \phi(\xi) \tag{36}$$

A close-up look at equation (34) indicates that the price $P_{x_i}$ at drought event level $x_i$ is greater than the price $P_{x_{i-1}}$ at drought event level $x_{i-1}$, as expected. To achieve this, the price must increase from $P_{x_{i-1}}$ to $P_{x_i}$ by the amount $\phi(\xi)$, as shown by equation (36). Thus by increasing the price, the supply deficiency of water during sustained drought periods can be overcome or minimized. In fact, the price can be forced to rise to the level that limits the demand of water to that amount which is available. Doing so will theoretically enable us to adjust the portion of the demand curve beyond the critical drought severity index (Fig. 11) down to the supply curve. However, this may not be readily accepted by the customers and thus arises the uncertainty. In essence, there will result a positively skewed distribution tendency of the customers for the demand, and hence the analysis of the associated uncertainty comes into picture.

### 4.4. THE $\phi(\xi)$ FUNCTION

To fully make use of equation (34) or equation (36) for water demand abatement through price increase, an explicit form of the drought function, $\phi(\xi)$, in which $\xi$ is the drought severity index must be determined. Different approaches have been followed to develop indices for a drought event. Presuming that the $SWSI$ is one of the alternatives available to forecast a drought severity level then $\phi(SWSI)$ will be used herein. The subscripts of $d$ and $P$ may be substituted by the numerical values of $SWSI$. For instance, if a drought month of $SWSI$ = -2.00 is forecast to follow a normal month of $SWSI$ = 0.00, the price $P_{-2.00}$ can be determined based on equation (34), with the price during the normal month $P_{0.00}$ known. Since the $SWSI$ depends on $p_{snow}$, $p_{prec}$, $p_{strm}$ and $p_{resv}$, the following general relation between $SWSI$ and the variables may be conceived.

$$SWSI = \psi(p_{snow}, p_{prec}, p_{strm}, p_{resv}) \tag{37}$$

Apparently, then,

$$\phi(\xi) \equiv \phi(SWSI) = \phi'(p_{snow}, p_{prec}, p_{strm}, p_{resv}) \tag{38}$$

Once a fully explicit model is developed for equation (38), it becomes possible to re-compute the expected demand using equation (26). As an alternative for this, the following equation may also be derived from equation (33):

$$P_{x_i} = P_{x_{i-1}} + \frac{\left(d_{x_i} - d_{x_{i-1}}\right)\left[\dfrac{P_{x_i} + P_{x_{i-1}}}{2}\right]}{\eta_P\left[\dfrac{d_{x_i} + d_{x_{i-1}}}{2}\right]} \tag{39}$$

The second term on the right hand side in equation (39) above is equivalent to the $\phi(\xi)$ function mentioned earlier. It is to be noted that equation (34) gives an explicit equation to determine the price at drought event level $x_i$ while equation (39) gives a term equivalent to the $\phi(\xi)$. The $SWSI$ can be used to indicate if a drought may occur and to determine its severity level if it occurs. It is to be recalled that, as indicated by equation (34), the demand at drought event level $x_i$ can be adjusted to the estimated available supply $q_{x_i}$ by increasing the price from its value at drought event level $x_{i-1}$ to a new value at drought event level $x_i$.

## 4.5. UNCERTAINTY AND RISK IN DEMAND

Although it is presumed that demands can be adjusted to the available supply there is uncertainty. Demand is a variable and may not meet the available supply irrespective of the increase in the price. Hobbs (1989) points out that future demands are random because they depend upon weather, consumer tastes and preferences, household income, water rates and level of economic development. These reasons naturally cause the demand to have some positively skewed probabilistic distribution.

Some organizations and researchers have used different probability distributions for demand. Charles Howard and Associates (in 1984) and Norrie (in 1983) used a gamma distribution for demand for Seattle, Washington (Hobbs, 1989). Also, it may be possible that the statistics of the distribution of the demand about the available supply is not uniform at different drought severity levels.

A general trend of the supply with the drought severity index and the distribution of the demand about the supply may be represented as shown in Fig. 13. A general gamma probability density function which is given by equation (40) (Montgomery and Runger, 1994) is assumed.

$$f_X(x;\lambda,r) = \frac{\lambda^r x^{r-1} e^{-\lambda x}}{\Gamma(r)} \quad x > 0,\ \lambda > 0 \text{ and } r > 0. \tag{40}$$

where

$$\Gamma(r) = \int_0^\infty x^{r-1} e^{-x} dx \quad r > 0 \tag{41}$$

Taking the $SWSI$ as the drought severity level indicator and the demand as the variable $x$ in the gamma function given above, a general relationship between the supply $q$, the $SWSI$ and the density function of the demand $f(d)$ can be given as illustrated in Fig. 13. Negative values of the $SWSI$ values normally adopted are used in this Figure to indicate the increase of severity with the index in absolute terms.

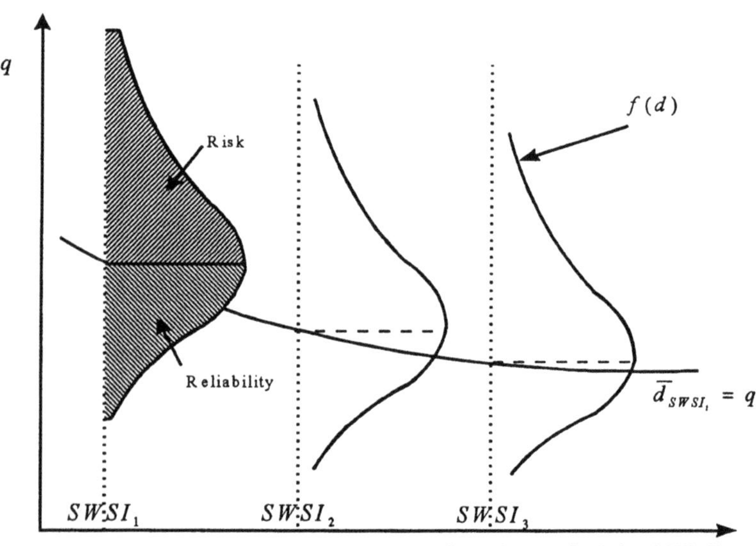


Fig. 13. Expected water demand as adjusted to the available supply under sustained drought conditions and its probability distribution.

Let $d_{x_i}$ be the random demand at drought severity level $x_i$ and by implication $q_{x_i}$ be the corresponding known available supply at drought severity level $x_i$. Tung (1996) defines reliability as the probability that the resistance is greater than the loading. In a similar analogy, the reliability of a water supply system may be defined as the probability that the available supply is greater than the expected demand. Thus, the reliability of supply is the probability that $q_{x_i}$ is greater than $d_{x_i}$, expressed as

$$R = p(q_{x_i} > d_{x_i}) \tag{42}$$

and the risk is

$$Risk = 1 - p(q_{x_i} > d_{x_i}) = p(d_{x_i} > q_{x_i}) \tag{43}$$

where $R$ is the reliability that the available supply is greater than the estimated demand and $p$ is the probability. As illustrated in Fig. 13, a higher demand above the available supply implies a higher risk and a lower reliability. The risk defined by equation (43) can also be expressed in terms of the *safety factor*, *SF*, which may be defined as

$$SF = \frac{q_{x_i}}{d_{x_i}} \tag{44}$$

where the corresponding risk formula is

$$Risk = p(SF < 1) \tag{45}$$

For different drought severity indices, different risk-safety factor relationships can be developed. This is illustrated in Figures 14 and 15 below. Once such relationships are developed, it is easier for water supply managers to decide the tolerable risk for a given drought severity index. It is to be noted here that the reliability analysis is just complementary to the risk analysis whereas the safety factor approach is simply an alternative to the safety margin approach discussed in Section 3.1.2.

## 4.6. OPERATION/MANAGEMENT STRATEGY

It is indicated in the foregoing sections that urban water supply operation/management during sustained drought periods requires preparation at least by the water supply agents. Sound preparation procedures entail good strategy to be used. Most of all, collection of enough data affecting the water supply during a forthcoming period of time enables the supply agents to be prepared better for smoothing out the effect from a forecast drought event. Such efforts must be undertaken continuously during a sustained drought period. The flow chart shown in Fig. 16 will help water supply operation/management during drought periods.

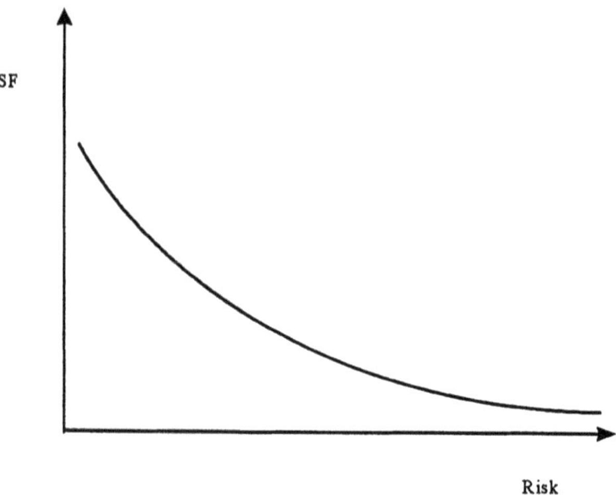


Fig. 14. General illustration of risk-safety factor relationship

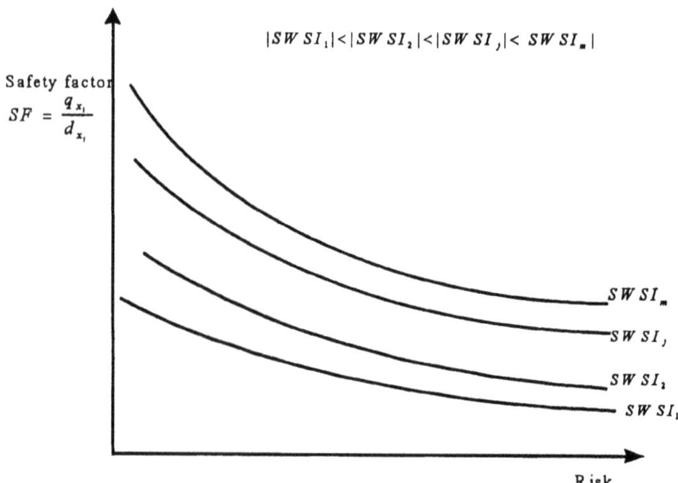


Fig. 15. Risk-safety factor relationship for different *SWSI*'s

As indicated in the flow chart, the important data to forecast are the forthcoming period's (say month) weather conditions or available flow and the expected demand. A comparison between the expected demand (after some necessary price increase) and the available flow clearly indicates if there will be a drought event.

If the expected demand is found to be greater than the available flow, it implies that a drought event exists and hence necessary measure(s) should be sought.

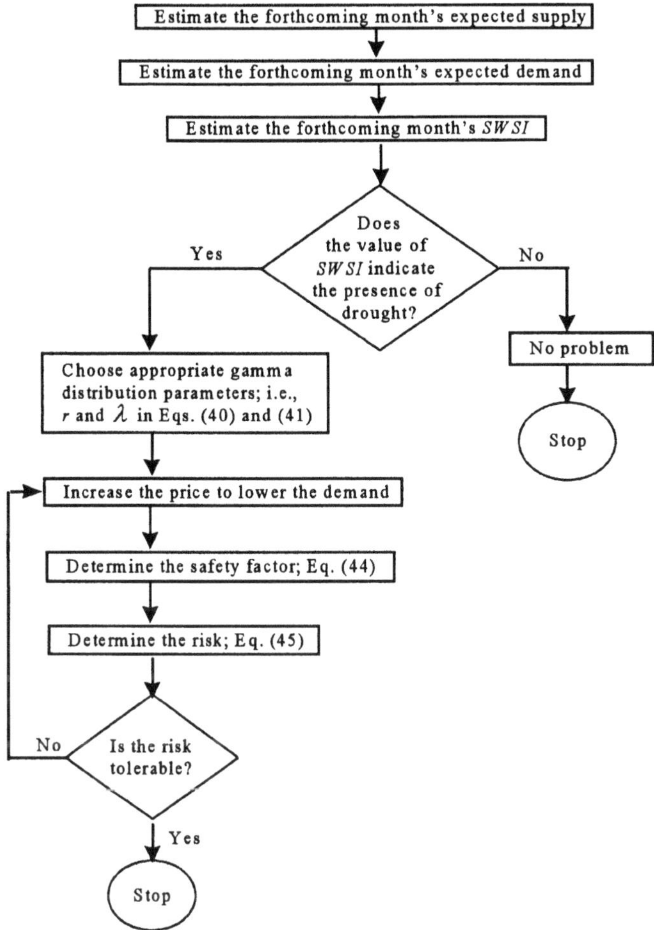


Fig. 16. Flow chart for adjusting demand to available supply under sustained drought conditions

## 5. Summary and Conclusions

The damages caused due to lack of urban water supply during drought conditions may result in adverse and undesirable effects to the general welfare. The efforts which may be taken basically include water conservation practices and/or new water development projects. In the case when the availability of this resource is limited due to the drought, water conservation is a viable target. Marginal pricing of water demand is found to be one of the best alternatives in urban water conservation efforts. This is achieved due to the fact that urban water demand is elastic to price changes. In other words, an increase in urban water price decreases its demand.

For urban water planning purposes during a drought condition, the inclusion of a third dimension termed as the risk, the probability that the demand exceeds the supply, plays important role in deciding, for a given drought condition, the tolerable risk for a given affordable price or vice versa. Thus the risk-price-return period relationship developed will help as one of the decision support systems for urban water planning under drought conditions.

Although water demand is believed to be elastic to price, it is apparently not significant for small price increases. The fact that the risk is highly sensitive to the return period and less sensitive to the price demands strong commitment to conserve water under adverse drought conditions. The price elasticity study has been geographically limited, mainly to the United States. It must attract more research in the regions with different socioeconomic status and hydrometeorological setup, that is, in the regions where the income of the customers is relatively low and the weather conditions are more uncertain and/or the drought occurrences are more frequent. It is likely that the use of the concept of the price elasticity is more effective in such areas.

## References

Agthe, Donald E. and Billings, R. Bruce (1980) Dynamic Models of Residential Water Demand, *Water Resources Research* **3**, 476-480.

Anderson, Jock R. and Dillon, John L. (1992) Risk Analysis in Dryland Farming Systems, *FAO*, Rome: Italy.

Billings, R. Bruce and Agthe, Donald E. (1980) Price Elasticities for Water: A Case of Increasing Block Rates, *Land Economics, Vol.* 56, **1**, 73-84.

Bruins, Hendrick (1993) Drought Risk and Water Management in Israel: Planning for the Future, in Donald A. Wilhite (ed). *Drought Assessment, Management and Planning: Theory and Case Studies*, Kluwer Academic Publishers, Boston.

Changnon, Stanley (1993) Are We Doomed to Fail in Drought Management?, in Martin Ruess (ed). *Water Resources Administraiton in the United States: Policy, Practice, and Emerging Issues*, Michigan State University Press, East Lansing: Michigan, 194-202.

Chow, Ven Te, Maidment, David R., and Mays, Larry W. (1988) *Applied Hydrology*, McGraw Hill, Inc., New York.

Carver, Philip H, and Boland, John J. (1980) Short- and Long-Run Effects of Price on Municipal Water Use, *Water Resources Research, Vol. 16*, **4**; 609-616.

Dixon, Lloyd S., et al. (1996) *Drought Management Policies and Economic Effects in Urban Areas of California, 1987 - 1992*, RAND.

Dziegielewski, Benedykt, et al. (1986) Drought Management Option, *Drought Management and Its Impact on Public Water System*, Report on a Colloquium Sponsored by the Water Science and Technology Board, Sept. 5, 1985, National Academy Press, Washington, D. C.;

Dziegielewski, Benedykt, et al. (1991) *National Study of Water Management During Drought: A Research Assessment for the US Army Corps of Engineers Water Resources Support Center*, Institute for Water Resources, IWR Report 91-NDS-3.

Easterling, William E. (1993), Assessing the Regional Consequences of Drought: Putting the MINK Methodology to Work on Today's Problems, Donald A. Wilhite (ed). in *Drought Assessment, Management and Planning: Theory and Case Studies*, Kluwer Academic Publishers, Boston.

Garen, David C. (1993), Revised Surface-Water Supply Index for Western United States, *Journal of Water Resources Planning and Management, ASCE, Vol. 119*, **4**, 437-454

Gibbs, Kenneth (1978) Price Variables in Residential Water Demand Models, *Water Resources Research, Vol. 14*, **1**, 15-18.

Grebenstein, Charles R. and Field, Barry C. (1979) Substituting for Water Inputs in U.S. Manufacturing, *Water Resources Research, Vol. 15*, **2**, 228-232.

Griffin, Ronald C., and Stoll, John (1983) The Enhanced Role of Water Conservation in the Cost-Benefit Analysis of Water Projects, *Water Resources Bulletin, American Water Resources Association, Vol. 19*, **3**, 447-457

Hanke, Steve H.; 1970; Demand for Water under Dynamic Conditions, *Water Resources Research, Vol. 6*, **5**, 1253-1261.

Hanke, Steve H. and de Maré, Lennart (1982) Residential Water Demand: A Pooled, Time Series, Cross Section Study of Malmö, Sweden, *Water Resources Bulletin, American Water Resources Association, Vol. 18*, **4**, 621-625.

Hobbs, Benjamin F., (1989) An Overview of Integrated Water Supply System Availability, in Larry W. Mays (ed)., *Reliability Analysis of Water Distribution Systems*, ASCE, New York.

Howe, Charles W. (1993) Water Pricing: An Overview, *Water Resources Update, The Universities Council in Water Resources*, **92**, 3-6.

Howe, Charles W. (1982) The Impact of Price on Residential Water Demand: Some New Insights, *Water Resources Research, Vol. 18*, **4**, 713-716.

Howe, Charles W., and Linaweaver, F. P.(Jr) (1967) The Impact of Price on Residential Water Demand and Its Relation to System Design and Price Structure, *Water Resources Research, Vol. 3*, **1**, 13-32.

Hudson, H.E. (Jr). and Hazen, Richard (1964), Drought and Low Stream Flows, in Ven Te Chow (ed)., *Handbook of Applied Hydrology*, McGraw Hill, Inc., New York.

Jones, C. Vaughan and Morris, John R. (1984) Instrumental Price Estimates and Residential Water Demand, *Water Resources Research, Vol. 20*, **2**, 197-202.

Jordan, Jeffrey L. (1994) The Effectiveness of Pricing as a Stand-Alone Water Conservation Program, *Journal of the American Water Resources Association, Vol. 30*, **5**, 871-877.

Lansey, Kevin E., et al. (1989) Methods to Analyze Replacement-Rehabilitation of Water Distribution System Components, in Larry W. Mays (ed)., *Reliability Analysis of Water Distribution Systems*, ASCE, New York.

Maddock, Thomas S. and Hines, Walter G. (1995), Meeting Future Public Water Supply Needs: A Southwest Perspective, *Journal of the American Water Resources Association Vol. 31*, **2**, 317-329.

Mays, Larry W. (ed). (1996) *Water Resources Handbook*, McGraw Hill, Inc., New York.

Mays, Larry W., and Tung, Yeou-Koung (1992) *Hydrosystems Engineering and Management*, McGraw Hill, Inc., New York.

Moncur, James E. T., (1987) Urban Water Pricing and Drought Management, *Water Resources Research, Vol. 23*, **3**, 393-398.

Moncur, James E. T., (1989) Drought Episodes Management: The Role of Price, *Water Resources Bulletin, The American Water Resources Association Vol. 25*, **3**, 499-505.

Montgomery, Douglas C. and Runger, George C. (1994) *Applied Statistics and Probability for Engineers*, John Wiley & Sons, Inc., New York.

National Research Council (1986) *Drought Management and Its Impact on Public Water Systems, Colloquium 1*, National Academy Press, Washington, D.C.

Puckett, Larry J., (1981) *Dendroclimatic Estimates of a Drought Index for Northern Virginia: Geological Survey Water Supply Paper 2080*, United States Government Printing Office, Washington.

Schneider, Michael L. and Whitlatch, E. Earl (1991) User-Specific Water Demand Elasticities, *Water Resources Planning and Management, Vol. 117*, **1**, 52-73

Sheer, D. P. (1980) Analyzing the Risk of Drought: The Occoquan Experience, *Journal of American Water Works Association Vol. 72*, **12**, 246-253.

Steila, Donald (1972) *Drought in Arizona: A Drought Identification Methodology and Analysis*, University of Arizona, Tucson, Arizona.

Suter II, Glenn W. (1993) *Ecological Risk Assessment*, Lewis Publishers, Chelsea, Michigan.

Tung, Yeou-Koung (1996) Uncertainty and Reliability Analysis, Larry W. Mays (ed)., *Water Resources Handbook*, McGraw Hill, Inc., New York.

U.S. Soil Conservation Service Monthly Report (Jan. 1988), Colorado Water Supply Outlook, United States Department of Agriculture, US Government Documents.

Wilhite, Donald A. (1993) Planning for Drought: A Methodology, in Wilhite, Donald A. (ed). *Drought Assessment, Management and Planning: Theory and Case Studies, Kluwer Academic Publishers,* Norwell, Massachusetts, 87-108.

Wong, S.T. (1972) A Model on Municipal Water Demand: A Case Study of Northern Illinois, *Land Economics,* 34-44.

Young, Robert A. (1973) Price Elasticity of Demand for Municipal Water: A Case Study of Tucson, Arizona, *Water Resources Research, Vol. 9,* **4**, 1068-1072

Young, Robert A. (1996) Water Economics, in Larry W. Mays (ed)., *Water Resources Handbook*, McGraw-Hill, Inc., N. Y.

# DROUGHT MANAGEMENT AND WATER SUPPLY SYSTEMS IN ISRAEL

HENDRIK J. BRUINS
*Ben-Gurion University of the Negev, Jacob Blaustein Institute for Desert Research, Social Studies Center,*
*Department of Geography and Environmental Development*
*Negev Center for Regional Development*
*Sede Boker Campus, 84990, Israel*

## 1. Introduction

The Mediterranean region, from the Iberian peninsula in the west to the Levant in the east, is situated at the northern margin of the largest desert belt on earth. Such a transitional geo-climatic position, between the desert in the south and the humid zone of the westerlies in the north, causes significant inter-annual variations in the amount of precipitation. Drought is, therefore, a common and recurring part of climatic variability, that must be taken into account in the management of water supply systems. The availability of fresh-water resources is usually a function of precipitation. Climatic patterns in different parts of the Mediterranean region show of course dissimilar attributes. Winter precipitation in central-eastern Spain represents only 1-15% of the annual total, while it makes up more than 40% in Israel and other areas in the south-eastern Mediterranean region (Trewartha 1961:240).

More than 60% of Israel has a desert climate. The transition from hyper arid to semi-arid occurs over a distance of merely 40-150 km. Both aridity and drought are, therefore, more pronounced in Israel than in the Iberian peninsula. The average number of rainy days per year varies in Spain from about 60 in the south-east to 180 in the Pyrenees, and in Israel from less than 20 in the south to about 70 in the north (Trewartha 1961:239). Spain has in summer on average about 6 rainy days in the south to 30 in the north. However, in Israel the late spring, summer and early autumn seasons are usually completely dry from May to October.

It became clear in Israel during the 1950s that the various regional water projects would not be adequate to meet the national water needs in the years ahead, because most of the water resources are concentrated in the wetter northern part of the country. The National Water Carrier was planned to integrate most of the regional water resources, in order to ensure a reliable water supply to most of the population during the rainless summer season and during drought years. Construction of this large engineering project began in 1953 and the system was completed in 1964. Water is pumped from Lake Kinneret and transported southward, as the lake supplies about 50% of potable



*E. Cabrera and J. García-Serra (eds.), Drought Management Planning in Water Supply Systems,* 299–321.

water in the country. The National Water Carrier system integrates surface and groundwater resources. Surplus water is transported to underground reservoirs for storage in the aquifers as buffer stocks for use in the dry summer season and during drought years (Ministry of Agriculture, 1973).

The great importance of the National Water Carrier system does not change, however, the basic reality of limited fresh-water water resources in Israel (Table 1), which are used to the maximum and beyond. Overexploitation of the groundwater aquifers occurred during the 1970's and 1980's, which has resulted in a decrease in groundwater quantity and quality (State Comptroller 1990; Schwarz 1990; Bruins, 1993). Although the National Water Carrier system does indeed protect the country against meteorological drought of light to medium severity, the available fresh-water reserves appear not sufficient as a buffer against severe multi-annual drought. Enhanced water resources management and proactive drought planning need to be developed, including large-scale seawater desalination for the domestic sector. Such measurements can ensure water quantity and quality in the aquifers, and protect the country against the impact of severe drought, which will surely come in the years ahead.

Table 1. Amount of fresh water per capita in the Near East.

| | Population (million) | Fresh Water (mcm/yr) | Fresh water per capita (cm/p/yr) |
|---|---|---|---|
| Israel | 5 | 1,600 | 320 |
| West Bank, Gaza | 2 | 250 | 125 |
| Jordan | 4 | 750 | 188 |
| Syria | 13 | 10,500 | 800 |
| Lebanon | 3 | 3,700 | 1,230 |
| Turkey | 59 | 105,000 | 1,800 |
| Egypt | 55 | 60,000 | 1,100 |

Source: Fishelson, G. (1995) International Conference on the Peace Process and the Environment, Tel-Aviv University, Book of Abstracts, pp. 45-48.

## Drought and Aridity: Concepts and Definitions [1]

The phenomenon of drought has been described as the most complex and least understood of all natural disasters (Wilhite, 1993). Drought does not strike suddenly with violent force like an earthquake or hurricane. It is a "creeping phenomenon" (Gillette, 1950) as drought conditions develop passively in a non-dramatic, gradual way, due to lack of precipitation. The impact of drought can, nevertheless, be devastating. Drought in the Sahel in the mid-1980s was the main environmental factor that contributed to desertification and the displacement of some 10 million people. "No other environmental problem has caused so much migration." (Warren 1997:117).

---

[1] This section is largely based on Bruins (in press).

Drought is a recurring, natural part of climatic variability, but scientific prediction of both its beginning and termination are virtually impossible (Wilhite and Glantz, 1985; Wilhite and Easterling, 1987). The frequency and severity of drought can, however, be linked to the various types of climate on earth. Drought can occur in each climatic zone, including in very wet regions. The frequency of its occurrence is usually related to an important empirically observed rule: the drier the climate the larger the inter-annual variability of precipitation. Drought, therefore, tends to occur more often in subhumid and dry climates.

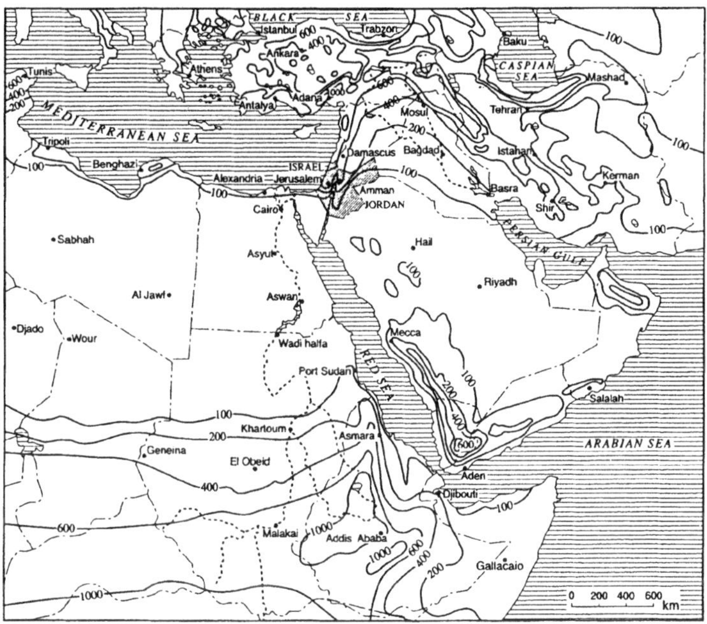


Figure 1. Israel is situated at the northern fringe of the largest desert belt on Earth.

How dry is dry? It is important to classify bioclimatic aridity in a globally comparable way as a basis for land-use planning, proactive drought planning and interactive management (Bruins and Lithwick, 1998). A numerical representation is able to express dryness in much greater detail than language. Average annual precipitation (P) divided by average annual potential evapotranspiration (ETP) is the preferred mathematical way to classify bioclimatic aridity (or humidity), according to the classification system developed by UNESCO (1979). The sophisticated Penman (1948) approach was adopted by UNESCO (1979) as an accurate estimate for potential evapotranspiration. However, certain geophysical data required to calculate Penman's formula are often not available at every site. UNEP (1992) opted, therefore, in its World Atlas of Desertification for the simplified Thorthwaite (1948) approach, which is a step backward.

Table 2. Comparison between the UNESCO (1979) and UNEP (1992) definitions for bioclimatic aridity and the boundaries between the various arid and sub-humid zones

| Climatic Zone | P/ETP Ratio (Penman method) UNESCO (1979) | P/ETP Ratio (Thornthwaite method) UNEP (1992) | Interannual Rainfall Variability |
|---|---|---|---|
| Hyper-arid | < 0.03 | < 0.05 | ~ 100% |
| Arid | 0.03 - 0.20 | 0.05 – 0.20 | 50-100% |
| Semi-arid | 0.20 - 0.50 | 0.20 – 0.50 | 25-50% |
| Sub-humid | 0.50 - 0.75 | 0.50 – 0.65 | < 25% |

Source: Bruins and Berliner (1997)

Bruins and Berliner (1997) prefer to retain the numerical boundaries between the above arid zones as established by UNESCO (1979), basing P/ETP ratio's on Penman related methods only. Discussing the above problems, they suggested a more convenient way to establish ETP data in line with Penman's approach. Berliner (manuscript in preparation) collected measurements which clearly show that Penman's ETP is very well represented by data from Class-A evaporation pans, which are widely available throughout the world. "We, therefore, suggest to base globally comparable P/ETP ratio calculations in the future on Class-A evaporation pan data as a standardized system to define bioclimatic aridity throughout the world" (Bruins and Berliner 1998:102).

Israel comprises large areas with a hyperarid (P/ETP < 0.03) and arid climate (F/ETP 0.03-0.20). These areas, unsuited for rainfed agriculture, have been used since time immemorial for extensive livestock raising by (nomadic) pastoralists, such as the Bedouin. Central Israel has a semi-ar0id climate (P/ETP 0.20-0.50), in which rainfed farming is feasible but quite sensitive to drought. The transition from hyper-arid to semiarid occurs in Israel over a rather short distance of 40-150 km. The sub-humid zone (P/ETP 0.50-0.75) occurs in northern Israel. Drought frequency is somewhat less than in the semi-arid zone, rendering the sub-humid zone more secure for rainfed agriculture.

Wilhite and Glantz (1985) distinguish four main categories of drought according to disciplinary or causative perspectives: meteorological, agricultural, hydrological, and socio-economic drought. Two additional categories are added here: *"pastural drought"* and *"human-made calamity drought"*. Meteorological drought can be regarded as the most basic form of drought, resulting from dynamic and complex geophysical processes in the Earth atmosphere. Drought develops passively as the rains fail to come or if the amount of precipitation is significantly less during a certain time period than the average amount in a specific area. More precise drought definitons are necessarily site specific. Precipitation records enable time-series analysis of drought occurrence and severity (Lee et al., 1986; Sharma, 1994). The Palmer Drought Severity Index (PDSI) is widely used in the United States (Palmer 1965; Wilhite and Glantz 1985).

A certain year cannot be classified as a meteorological drought year, if total annual precipitation reaches the average amount or above. Yet drought damage may have occurred to crops as a result of bad rainfall distribution within that year. The term agricultural drought (Wilhite and Glantz 1985) becomes useful to characterize such dry periods in areas where farming is practiced. Livestock raising is a rational land-use in many parts of the world, often taking place in areas which are too dry for rainfed agriculture. It would be inappropriate to speak of agricultural drought in such regions, where natural pasture rather than agricultural crops are affected by dry spells. *Pastural drought* would seem the logical term in such cases (Bruins, 1996).

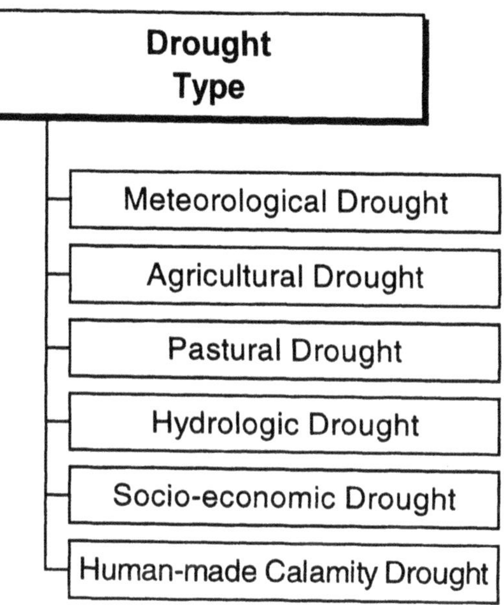


*Figure 2. Overview of different categories of drought (Wihite and Glantz 1985; Bruins 1996)*

Hydrologic drought relates to shortages in the flow of water at the surface or in the subsurface, which causes problems for societies depending on such water (Wilhite and Glantz 1985). Egypt is a classic example of a civilization receiving virtually all its water supply from a so-called allochthonous or exotic river, the Nile. Egyptian agriculture, therefore, is not affected by local meteorological drought in the Eastern Mediterranean or North African region. However, an extended meteorological drought in the catchment of the Nile in Central and East Africa can cause a hydrologic drought in Egypt.

Socio-economic drought, as defined by Wilhite and Glantz (1985), is in fact a human-made drought, although not unrelated to the basic structure of meteorological water supply and geo-hyrological water reserves. It occurs when the government and/or the private sector, in the economic development of a region, create a demand for more water than is normally available (Hoyt, 1942; Wilhite and Glantz, 1985). The relationship between supply and demand of water is not static, but may vary with time in response to changes in government policy, such as legislation to prevent overpumping of aquifers, development of hitherto untapped water resources, including desalination of brackish water and seawater, as well as waterprice and other socio-economic criteria.

If constructive, albeit unwise human activities in the socio-economic realm can lead to drought in developed water management and delivery systems, it is but one step to move in conceptual classification and definition to destructive human activities as the cause for drought. Modern chemical industrial production and the problem of waste disposal may lead to pollution of groundwater, which can render such water unfit for human consumption or agricultural use. This may lead to a shortage of potable water in a certain area. Moreover, modern warfare and weapons of mass-destruction pose a veritable threat not only in terms of direct lethal capability and destructiveness, but also in terms of environmental pollution. The concept of *human-made calamity drought* is introduced (Bruins, 1996) to identify shortages of potable water resulting from destructive human activities. Human-made calamity drought is hereby defined as pollution of water resources and/or destruction of water-supply networks in a certain area, either accidentally or purposefully, resulting in a shortage of potable water for domestic and agricultural use (Bruins, in press).

## Meteorological Drought in Israel

Precipitation in the region is usually related to depression systems coming from the west in autumn, winter and spring. Precipitation in Israel decreases from north to south and from west to east. Average annual precipitation ranges from only 30 mm near the Red Sea resort of Eilat in the south to 800 mm in the elevated parts of Galilee in the north. Rainfall at the Mediterranean coastal plain ranges from 587 mm in Haifa (elevation 10 m), 556 mm in Tel-Aviv (elevation 20 m) to 371 mm in Gaza (elevation 45 m) in the south (average annual amounts for the period 1921-1951, Katsnelson 1964). The Old City of Jerusalem (elevation 760 m) in the central hills, has an average annual rainfall of 561 mm (for the period 1846-1953, Amiran 1994), whereas Jericho, just 25 km to the east, has only 143 mm (1921-1951, Katsnelson 1964), because it lies in the Rift Valley at an elevation of -260 m below ocean level.

*Figure 3. The geographic position of Israel at the south-eastern corner of the Mediterranean Sea.*

Israel lies near the southern limit of cyclonic rains. The number and trajectories of these depressions often determine the amount of rainfall. Large variations occur over Israel in the amount of interannual precipitation (Katsnelson, 1964), which result in the regular occurrence of meteorological drought. It is possible to distinguish three main types of meteorological drought or dryness in Israel: (1) annually recurring seasonal summer dryness; (2) regularly occurring annual to multi-annual drought of light to medium severity; and (3) occasional multi-annual drought of great severity (Bruins, 1993). Time spectrum analysis of Jerusalem rainfall data by Zangvil (1979) revealed a prominent peak at 3.0 to 3.3 years, similar to the frequency of drought years in the Negev based on rainfall-runoff data (Bruins *et al.,* 1986, 1987). A study about rainfall variability in Israel by Amiran (1994) showed the average rainfall amount in drought years to be 30-40% less than the long-term average.

The overall rainfall average in Jerusalem for the period 1846-1993 amounts to 556 mm (Amiran 1994). The three wettest years during this period occurred in 1873/74 (1004 mm), 1877/78 (1091 mm) and 1991/92 (1134 mm). The three driest years occurred in 1950/51 (247 mm), 1959/60 (206 mm) and 1962/63 (227 mm). The Jerusalem rainfall record was studied by Zangvil (1979) for the period 1846-1954. Applying a 10-year running mean, he distinguished a relativey wet period with above-average rainfall during 1868-1911 (690 mm) and a relatively dry period during 1912-1937 (412 mm). Analyzing the precipitation years of Jerusalem for the period 1846-1993 at face value, without running means or filters, Amiran (1994) distinguished six dry periods of three consecutive years or more (Table 3) in which at least one year shows a precipitation decline of more than 33%.

Table 3. Drought periods of three or more consecutive years, having at least one year with a drop in precipitation of 33% or more, based on Jerusalem rainfall records for the period 1846-1993.

| Drought period of three years or more | Number of Consecutive Years | Average rainfall decline (%) | Dryest year in this period | Largest rainfall decline (%) |
|---|---|---|---|---|
| 1869/70-1872/73 | 4 | -21 | 1869/70 | -43 |
| 1898/99-1901/02 | 4 | -16 | 1900/01 | -39 |
| 1922/23-1935/36 | 14 | -27 | 1932/33 | -53 |
| 1945/46-1954/55 | 10 | -14 | 1950/51 | -55 |
| 1957/58-1962/63 | 6 | -37 | 1959/60 | -63 |
| 1983/84-1985/86 | 3 | -22 | 1985/86 | -33 |

Source: Amiran 1994

Ben-Zvi (1987) studied hydrologic drought in Israel over the period 1937 to 1984, analyzing 14 different rivers in the central and northern part of the country. Severity in terms of decline in the flow of water, duration of the hydrologic drought and geographic extent are the principle variables. Severe and extensive hydrologic drought in Israel, affecting the majority of the 14 studied streams, occurred in 1950/51, 1958/59, 1972/73 and 1978/79. Multi-annual continuous shortages occurred approximately in the periods

1955-1961 and 1971-1979, with variations at the beginning and end of those periods with respect to the various streams studied. A third period around 1967-1972 can be recognized for a few streams in central Israel.

Instrumental meteorological and hydrological records are obviously limited to the recent past. Drought risk analysis and impact assessment require a larger time scale than provided by detailed meteorological records. Climatic variations in the past can be studied through proxy paleo-climatic indicators. The Middle Bronze Age, for example, which lasted approximately from 2000-1580 BC (Bruins and Van der Plicht, 1996) seems to have coincided with one of the driest climatic periods within historical times (Bruins, 1994), as indicated by the low level of the Dead Sea (Frumkin *et al.*, 1991).

However, one has to distinguish between specific drought years and an overall dry climatic period of several hundred years duration, in which common agricultural land use and pastoralism was still possible. Evidence for drought years, which caused famine in the past, cannot usually be reconstructed from proxy paleoclimatic indicators, but only from literary data and sometimes from tree rings. Israel has unique literary data about severe multi-annual droughts or famines in the past, as recorded in the Bible: Genesis 12:10; 26:1; 45:6; 47:13; Ruth 1:1; II Samuel 21:1; I Kings 17:1; I Kings 18:1-2; James 5:17-18. The occurrence of droughts is also mentioned in the writings of Rabbinic sources (Sperber, 1978) during the period AD 200-400.

## Water Resources and their Management

The main water resources of Israel are concentrated in three principal catchment basins (Figure 4): Lake Kinneret, the Coastal Aquifer, and the Western Mountain Aquifer (or Limestone Aquifer). About 80% of all the water used in the country is derived from these three catchment basins. Other important water resources are formed by the aquifers of Western Galilee, the Carmel region, the Eastern Galilee, the Golan, the Yizreel Valley and Beit Shean Valley, and the aquifers in the Negev and the Arava Valley (Gilead and Bachmat, 1973; Grinwald and Bibas, 1989). Interception of storm runoff and floodwater in local reservoirs has expanded rapidly since 1987 (Grinwald and Bibas, 1989).

### LAKE KINNERET

Situated deep in the Rift Valley lies Lake Kinneret, also known as Lake Galilee, which is the principal natural storage reservoir of fresh surface water in Israel. The lake covers an area of 165 $km^2$ and holds at maximum capacity almost 4,000 mcm (million cubic meter) of water. Water is supplied to the lake by the Jordan River and its tributaries, and to a lesser extent by wadis and springs. Lake Kinneret remains the main water source of the National Water Carrier. An average safe yield of 450 mcm is pumped from the lake annually. Each meter of water in the lake represents *ca.* 170 mcm. The maximum permissible water level, above which flooding would occur in the town of Tiberias and other villages around the lake, is -208.9 m (below ocean level). The overflow into the Jordan River at the southern end of the lake is controlled by a sluice gate, in order to manage the outflow of water. The minimum acceptable lake level, the so-called red line, has been set at -213 m, below which water quality in the lake, particularly salinity, is expected to deteriorate to unacceptable levels (Grinwald and Bibas, 1989; Schwarz, 1990).

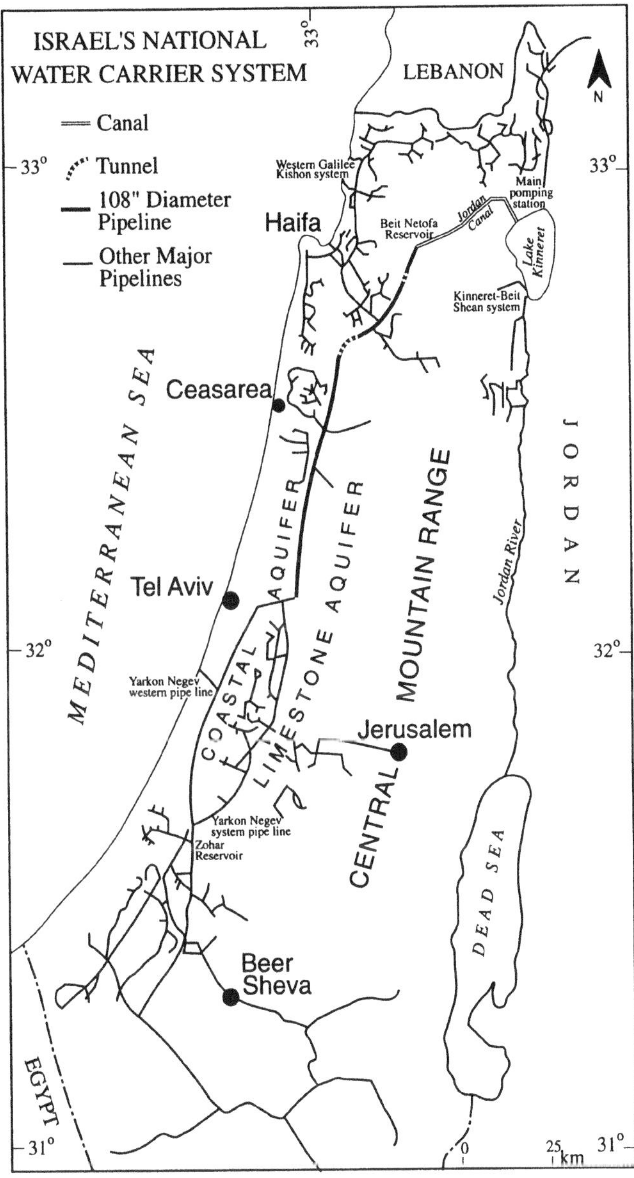


*Figure 4. The National Water Carrier and its distribution system (after Doron, 1993 and Shanan, 1997)*

The original salt content of Lake Kinneret was about 400 mg/l (or ppm), which is above the accepted national standard and World Health Organization (1988) guideline of 250 mg/l. The salinity is caused by brackish waters flowing naturally into the lake from nearby springs. A special canal was built to divert 20 mcm of saline spring water annually into the lower Jordan River Thus the salinity of Lake Kinneret was brought down to acceptable levels of 205-230 mg/l, which is a considerable achievement (Ministry of Agriculture, 1973; Grinwald and Bibas, 1989; Bruins, 1993). The drawback of this solution is the resulting low water quality in the lower Jordan River, which has become unsuited for agriculture.

The 1994 Peace Treaty between Israel and Jordan contains articles to improve water quality in the lower Jordan River. Israel is to desalinate, within four years from the entry into force of the Peace Treaty, the 20 mcm of saline spring water currently diverted into the Jordan River, thereby removing the major cause of pollution. Half the amount of this desalinated water (10 mcm) is to be supplied to Jordan. Until the desalination facilities are operational, Israel will supply Jordan during the winter period with 10 mcm of good-quality Jordan River water, derived upstream from the Deganya gates (Ministries of Foreign Affairs & Finance, 1996; Bruins, in press).

## COASTAL AQUIFER

An important aquifer is found in the subsurface of the sandy coastal plain along the Mediterranean Sea (Figure 4). The Coastal Aquifer has a north-south length of *ca.* 140 km and an average width of 15 km, comprising a total area of some 1,800 $km^2$. The aquifer is composed of Late Tertiary to Quaternary sand and sandstone with intervening layers of clay and fossil soils, which cause divisions in a number of sub-aquifers (Issar, 1968). The sandy deposits attain a thickness of about 120 m near the coast, gradually becoming thinner toward the east. The basis of the aquifer is formed by a thick deposit of Tertiary clays.

Use of the Coastal Aquifer started at the beginning of the 20th century, supplying most of the water in the country until the 1950s. The amount of wells gradually increased and so did the amount of water pumped from the aquifer, rising from 250 mcm in 1948 to a record 493 mcm in 1958. The opening of the National Water Carrier in 1964 allowed a reduction in water withdrawals from the Coastal Aquifer. However, pumping later increased again, reaching 470 mcm in the year 1984-85 (Grinwald and Bibas, 1989). The continuous overpumping of the coastal aquifer has resulted in a drop of the water table, increased penetration of sea water, and a general reduction in water quality (Melloul and Bibas, 1990; Schwarz, 1990).

The hydrological deficit of the Coastal Aquifer at the end of the 1980's was estimated at 1100 mcm (Schwarz 1990). The extraordinary wet year 1991/92 replenished some of this deficit. The average safe yield is about 283 mcm per year (Melloul and Bibas, 1990; State Comptroller, 1990). However, even less water should be withdrawn, *ca.* 210 mcm/year, in order to restore the aquifer (Schwarz, 1990). This will be further enhanced by artificial recharge with water from Lake Kinneret supplied through the National Water Carrier. Other measures required, according to Schwarz

(1990), are additional installations for artificial recharge and to supply existing consumers with alternative water sources.

## WESTERN MOUNTAIN AQUIFER

The western part of the Mountain Aquifer (Limestone Aquifer) is situated east of the Coastal Plain (Figure 4). This aquifer is composed of hard calcareous rocks of Cenomanian-Turonian age, which crop out in the central mountain ridge of the country. Its main reserve is situated in the Yarkon-Tanninim basin east of the Coastal Plain, extending south to Beer-Sheva. The aquifer is of a karstic nature with high conductivity and swift flows (Grinwald and Bibas, 1989). Use of the Yarkon-Tanninim aquifer increased rapidly since the 1950s, reaching annual withdrawal levels of more than 400 mcm. These figures are higher than the long-term safe yield estimated at 310 mcm per year (Schwarz 1990). Water levels in the aquifer dropped by more than 8 m in the period 1970-1990, passing the so-called red line in 1986 and 1990, while the water quality deteriorated due to pollution. Every meter in the level of the aquifer represents an estimated amount of 100 mcm water. Lowering of the water level in the aquifer beyond the red line may introduce salinity problems in this largest groundwater reservoir of the country. The extraordinary rainy season 1991/92, the wettest year since recordings began in Jerusalem in 1846, resulted in considerable recharge of the aquifer.

The Westeren Mountain Aquifer constitutes an important source of fresh water in Israel. It functions as a seasonal and interannual buffer, in conjunction with Lake Kinneret and the National Water Carrier system, to ensure water supply in the country. The amount of storage in the aquifer between the uppermost level, in which the water escapes through natural springs, and the lowest level or red line amounts to about 900 mcm. Preservation of its water quality is of great importance.

Table 4. Short-term water potential in Israel in million cubic meters (mcm) per year, as compared to actual supply in 1984/85 and planned supply for the year 2000.

| Water Source | Fresh Water | Saline Water | Total | 1984/85 | 2000 |
|---|---|---|---|---|---|
| Boreholes | 768 | 132 | 900 | | |
| Springs | 82 | 100 | 182 | | |
| TOTAL GROUNDWATER | 850 | 232 | 1,082 | 1,340 | 1,115 |
| Hula Valley Use | 122 | | 122 | | |
| Lake Kinneret | 490 | | 490 | | |
| Saline water | | 20 | 20 | | |
| Outflow Kinneret | -20 | | 20 | | |
| TOTAL KINNERET BASIN | 592 | 20 | 612 | 620 | 660 |
| Floodwater | 160 | | 160 | 40 | 80 |
| Recycled Wastewater | 241 | | 241 | 110 | 275 |
| Losses | | | | -60 | -40 |
| TOTAL WATER SUPPLY | 1,843 | 252 | 2,095 | 2,050 | 2,090 |

Source: State Comptroller (1990) Report on the management of water in Israel; Jerusalem.

Schwarz, J. (1990) Israel Journal of Earth Sciences 39:57-65.

The demand for fresh water by the domestic sector is expected to rise to 640 mcm by the year 2000, as compared to 420 mcm in 1984/85. These regional aquifers in Israel produce 510 mcm/year, an amount that is expected to increase to 560 mcm/year. Schwarz (1990:59) notes: "Most of these aquifers suffer from overexploitation and part of them, especially in the inner valleys, suffer from salinity increase. pumping will be reduced in these aquifers. However, in other aquifers that are not yet fully utilized due to excessive costs, pumping will increase in the future." The supply of fresh water to agriculture is supposed to decline to 740 mcm in 2000 as compared to 1200 mcm in 1984/85. This loss is to be only partly compensated by an increased amount of reclaimed wastewater of 320 mcm for agriculture in the year 2000. The potential of treated sewage and runoff water is expected to reach ca. 500 mcm in 2010 (Schwarz 1990).

## Water Laws

The State of Israel has regarded water as a scarce national resource since independence in 1948. Therefore, water resources in Israel are centrally governed by the state, based on the Water Law enacted in August 1959. The first clause of the law states: "The water resources in the State are public property, under the control of the State, and intended for the needs of its residents and the development of the country." Quantitative information about water supply and consumption is of crucial importance in a situation of scarcity. The Water Metering Law, enacted in 1955, underlines this vital aspect, as it preceded the Water Law. The former law ensures rigorous information about all water production, water supply and water consumption in the country (Arlosoroff, 1974).

The Minister of Agriculture is responsible for and in charge of implementation of the water legislation. A Water Commissioner is appointed under the Minister of Agriculture in order to implement policies and manage the water affairs of the state through a Water Commission. The latter body consists of five main departments: (1) Allocation and licensing, (2) Drainage, (3) Hydrological Service, (4) Efficient use of water, and (5) Head office (Ministry of Agriculture, 1973; Grinwald and Bibas, 1989).

Many countries have complicated water rights, which gradually evolved in the past. Legislating water as public property under control of the State, enabled Israel to become a leader in the development of an integrated national plan involving both surface water and groundwater. Those using water from private wells, surface or groundwater prior to the enactment of the Water Law received rights to continue using these waters up to a certain allocated amount. Such State interference was considered necessary to prevent over-exploitation, as private costs do not reflect social costs, such as replenishment and aquifer pollution (Shanan and Berkowicz, 1995).

Public participation in matters related to water is organized through a water board, which has a maximum number of 39 representatives. The Water Board has no power of decision. Its main function is to give advice and recommendations to the minister of agriculture (Ministry of Agriculture, 1973).

## Water Supply, Water Allocation and Water Price

The government established in 1952 a National Water Planning Company, a public corporate body, called *Tahal*. Its main tasks as the government planning agency include the design of all major national and regional water projects. The *Mekorot* Water Company Limited, founded in 1937, operates the water supply of the country and also has constructed its infrastructure. The government used to be one of the shareholders in the company, which is now being privatized. The building of the National Water Carrier has been one of its major achievements.

Most of the water resources of Israel are integrated through the the National Water Carrier system. There is important operational flexibility in the system, as surface water and groundwater can be transferred interregionally. Artificial recharge of both the Coastal Aquifer and the Mountain Aquifer can be done in winter with water pumped from Lake Kinneret, about 35 mcm/month (Schwarz 1990). Another important task is the conveyance of water from the wetter north to drier regions in the south of the country.

The National Water Carrier pumps an average annual amount of 450 mcm water from Lake Kinneret, lifting the precious liquid by about 362 m, from the lake level at *ca.* 210 m below ocean level to an elevation of 152 m. From this point the water flows by gravity to the Coastal Plain. The National Water Carrier system consists of pumping stations, tunnels, canals, but the main network is built of pre-pressed concrete pipes, which are 70 to 108 inches in diameter. Transport to the central hills and to the Negev in the south is carried out with additional pump lifts. The National Water Carrier supplies water to local systems rather than delivering individual consumers directly from the main system. Its peak delivery capacity is 20 $m^3$ per second (Shanan and Berkowicz, 1995).

The National Water Carrier was designed as a pressure pipe system, so that water supply could be recorded with standard water meters, which must be maintained by law in good working condition. Water distribution networks are usually made of concrete-lined steel pipes with an external asphalt covering. The pipes are placed about one meter underground and are designed to withstand at least 6 atmospheres of working pressure. Gate valves and air release valves are installed above the ground surface in order to simplify maintenance and operation of the networks. The maintenance of water networks inside villages and towns is for the latter responsibility.

The national water supply network is managed and operated by the governmnent through the *Mekorot* Company. Water is delivered to the consumers on the basis of seasonal and monthly allocations. Municipalities and rural villages have to plan their secondary distribution network from the point of the *Mekorot* outlet. They are responsible for supply to the individual farmers and homes, based on bimonthly quotas which must be monitored. Also private wells are monitored bimonthly by the authorities to ensure that they are operated within the range of their respective water production allocations. The management of water supply to the agricultural sector is rather unique in Israel, as the majority of farmers are living in cooperative rural villages: a moshav or kibbutz. Water is supplied by Mekorot to each of these villages as a single gross allocation, which has to be divided to the individual farmers or units. The moshav council or the kibbutz management is responsible for local deliveries and monitoring. A moshav is a cooperative village of about 300-350 ha, in which families form independent financial units, having their own farms. However, purchasing and marketing are cooperatively organized. A kibbutz is a collective community of about 150-400 families, living on 300-500 ha of land. These families own the kibbutz together without having private property. They have a communally organized system of production and consumption (Shanan and Berkowicz, 1995).

It is important to realize, also with regard to socio-economic drought in Israel discussed below, that agricultural development has largely been determined by the central government, including location and size of agricultural land, water quantity, water price, credit terms and prices of agricultural products. Water price, therefore, is not simply based on marginal cost considerations, but in relation to the whole system. Three "types" of water have been differentiated in Israel, according to prices & costs in 1992. *Low cost* water from shallow wells or surface water, requiring low conveyance and distribution investments, costs US$ 0.10 - 0.15 $m^3$. *Moderate cost* water from deep wells or surface water, requiring high distribution and pumping investments, costs US$ 0.30 - 0.80 per $m^3$. *High cost* water, due to pumping unto high elevations or desalination, costs more than US$ 0.80 $m^3$.

The rather uniform price of water in most regions of the country, whether close to a water source or far away, reflects the egalitarian approach to supply water, as public property, to the population of the State. The Water Law provided for the establishment of an equalization fund to reduce differences in water price in various parts of the country. This fund is partly financed by levies on water allocations to users. The water quota to a user is non-transferable in order to ensure equitable water allocation (Shanan and Berkowicz, 1995).

Water use per quota is economized and controlled through a progressive pricing system. A basic price (rate A) is determined for the first part (50-80%) of the water quota, while the price of the remaining part (50-20%) of the quota is more expensive (rate B). Those who consume water in excess of the allocated quota have to pay an even higher price (rate C), which is a substitute for mere severe penalties, also permissible in accordance with the Water Law (Arlosoroff, 1974). The bimonthly quotas for agriculture are/were related to the evapotranspiration rates and crop water requirements for the respective period of two months. These water delivery policies encouraged efficient use of water, particularly in the season of peak demand. An excellent review of water costs and prices in Israel is quoted at length from Shanan and Berkowicz (1995:8):

> "In Israel, the cost of water is related to the cost of electricity. The National Carrier starts from Lake Kinneret and requires the pumping of about 300-400 mcm/year with about 360 meters of initial static lift at a cost of about 1.2 kWh of power for every cubic meter delivered. Furthermore, since all irrigation in Israel uses either sprinkler or drip technologies, water supplied to a consumer outlet must be at a pressure of at least 2.5 atmospheres. Consequently, by the time water reaches the furthest and highest delivery points, the electric power expended on each marginal cubic meter of water amounts to about 4 kWh. In 1992, the water system used 1,955 million kWh out of a total production of

24,019 million kWh, *i.e.* 8 percent of the electricity generated in Israel (Statistical Abstracts of Israel, 1993).

The pricing of water to the different sectors is indirectly related to the three levels of development costs described above. The price of domestic water (after the municipalities have added the approved levies for operation, maintenance and waste disposal) is about US$ 0.70 - 1.00 per cubic meter. Since domestic consumption of water in Israel is based on an allocation of 100-180 $m^3$ of water per family per year, the average family spends up to US$ 150 annually on water (excluding the watering of lawns and gardens). This represents about 1 percent of the annual expenditure of an average family. A family that uses more than its allocation pays about US$ 1.60 per cubic meter of extra water.

In the industrial sector, while the food processing and paper industries with high water use requirements are sensitive to the cost of water, the "dry" industries such as diamond, furniture manifacturing, tourism, and pharmaceutical and chemical production can easily bear the cost of "medium cost" water. Industrial water has been priced at about US$ 0.20 per cubic meter for the allocated volume of water, at US$ 0.40 per cubic meter for the use of extra water up to 10 percent more of the allocation, and at US$ 0.60 per cubic meter for further water supplies.

Where agriculture is concerned, the water production value of profitable agriculture is generally more than US$ 0.12 per cubic meter. Hence agricultural water has been priced at about US$ 0.10 per cubic meter for the first 50 percent of the allocation and at US$ 0.14 per cubic meter for the remainder. For the first 10 percent addition above the allocation, water is priced at about US$ 0.26 per cubic meter and at about US$ 0.50 per cubic meter for further excessive use. Considering that average allocated water use of crops varies from 3,000 to 7,000 $m^3$/ha, annual water costs are about US$ 360 to US$ 480 per hectare. Only high production can justify these water costs. Cotton farmers, for example, must produce at least 4.5 tons/ha, and citrus farmers 5.0 tons/ha in order to maintain their long-term economic stability. On the other hand, a farm family, without using hired labour, can cultivate 0.3 to 0.4 ha under intensive greenhouse agriculture producing mainly export vegetables and flowers. A greenhouse of this size would use about 5,000 $m^3$ of water per year and gross annual returns would be about US$45,000. Since the cost of water would not exceed US$ 600 per year, water costs are not a production constraint." (Shanan and Berkowicz, 1995:8).

## Water Supply Management, Socio-economic Drought and Meteorological Drought

An interesting interaction can be observed in Israel between meteorological drought and socio-economic drought in relation to available water reserves in the main aquifers. Figure 5 shows the water consumption in Israel for the period 1958-1994, *i.e.* the amount of water sold by the state to the domestic, agricultural and industrial sector.

Assessing the effect of past drought periods of three years or more (Table 6) on water deliveries by the state (Figure 5), it seems apparent that the drought of 1957/58-1962/63 caused only minor change in water supply to the agricultural and domestic sector. Following the severe drought year of 1959/60,characterized by a 63% drop in precipitation (Amiran 1994), only a slight drop can be observed the next year in water supply to the agricultural sector, as well as during 1962/63, the last year of the extended drought period. At that time there were still considerable water reserves in the aquifers, particularly in the western Mountain Aquifer. However, socio-economic drought gradually developed in Israel during the 1970s and 1980s, as the state sold more water from the groundwater reserves than the so-called safe yield, *i.e.* the amount replenished annually by rainfall. A sustained decline resulted in the levels of groundwater in the aquifers. When the meteorological drought of 1983/84-1985/86 came to pass, groundwater levels in the two main aquifers had already approached the red line. As the meteorological drought exacerbated the creeping socio-economic drought, there was no alternative for the state in 1985/86 but to make a deep cut in the supply of water to the agricultural sector. However, water supply to the domestic sector dropped only slightly (see Figure 5).

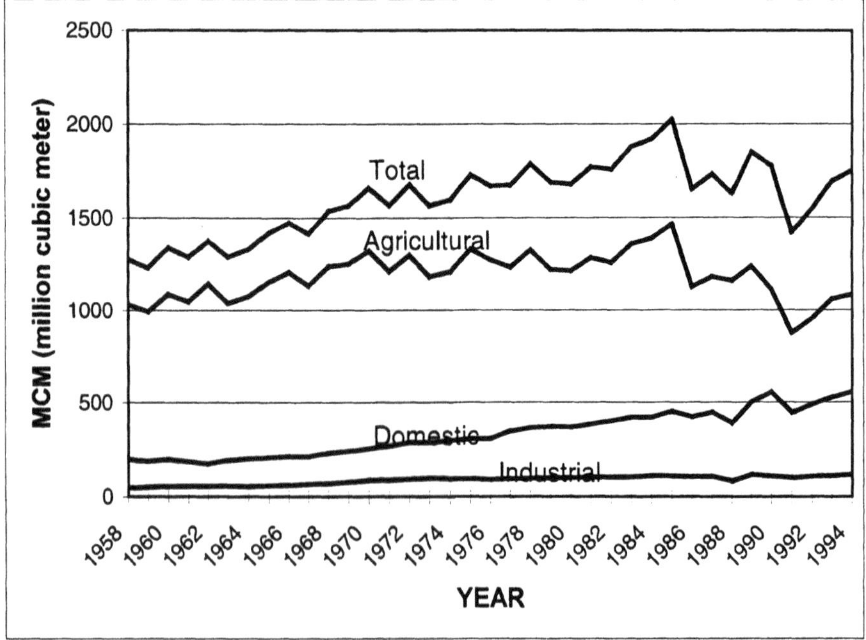


*Figure 5. Water consumption in Israel from 1958-1994 in million cubic meter (mcm), based on data provided by Michael Bibas (Water Commission, Water Allocation Department; personal communication, 1996).*

The unique development in Israel of a government controlled centralized water supply system, headed by the Minister of Agriculture, did not prevent a certain rate of over-utilization of the water resources. The size of irrigated land increased from 17,000 ha in 1948 to about 205,700 ha in 1990. Gradually too much water was sold during the 1970s and 1980s, particularly to the agricultural sector, which led to overpumping accompanied by a depletion of water quantity and quality in the two main aquifers. The integrated water supply system was planned to buffer the annually recurring dry summer season of about five to six months without any rainfall and also to buffer the country against meteorological drought. However, the water reserves in the aquifers were down to the red line in the late 1980s. The socio-economic drought was exposed by a meteorological drought in 1989-1990 and by a report of the state comptroller (1990) in the first days of January 1991.

In her 62-page report, the state comptroller Miriam Ben-Porat concluded that the country is on the verge of a catastrophic water shortage after 25 years of irresponsible management, which has resulted in a very serious overdraft from the aquifers of 1600 mcm of water. This deficit is equivalent to the water requirement of the country for one year. The vital water reserves in Israel have been depleted to dangerously low levels, and the water quality has been seriously damaged. The report warns of the very real danger that Israel may not be able to supply enough minimal quality water to satisfy demand in the near future. Some of the groundwater reservoirs in the country may have been damaged irreparably.

The general public was suddenly alerted, through the considerable degree of media attention, about the grave situation of Israel's water resources. The *Jerusalem Post* (January 3, 1991) dedicated its entire front page to the subject under the banner headline: "Ben-Porat: Water catastrophe looming". The findings of the report have been summarized succinctly by Hellman and Rudge (1991) in the *Jerusalem Post*, quoted hereby at length:

> Ben-Porat blames the catastrophe on a system that permits interested parties, such as the agricultural lobby, through its control of the powerful Knesset Water Committee, to make major decisions regarding the pricing, subsidizing, and allocation of water. The comptroller therefore recommends that water management be removed from the Agriculture Ministry and put under the control of an objective state authority. Such a body would be concerned with the country's overall needs, and would not be influenced by vested interest groups... The report stresses that previous agriculture ministers and water commissoners had deliberately ignored dire warnings of an impending water catastrophe, and that their decision to ignore recommendations voiced repeatedly over the past 25 years was 'erroneous and highly dangerous to the state's water resources'.

Mekorot, the national water carrier, and the State Water Commission's own hydrological service are not blamed by the report, which notes that these bodies were among those that warned of impending shortages. A particular concern expressed in the report refers to the low price charged to agricultural concerns for water consumption. The report blames the policy by the Ministry of Agriculture of subsidizing agricultural water consumption for much of the current crisis. Water subsidies were passed on to foreign consumers through cheap agricultural export prices, leading to what the report terms "essentially, the export of water at a loss." (Hellman and Rudge, 1991).

Kally (1973) had warned about the danger of mining the one-time reserve in the Yarqon-Tanninim aquifer. He already advocated in the early 1970s to reduce agricultural water consumption, lest the national water reserve will be depleted by an amount of the order of magnitude of 1000 mcm. The forecast and warning of Kally proved correct. By 1991 the deficit had risen to 1,600 mcm, as the reserves in the aquifer were depleted below the red line, while Lake Kinneret reached its lowest level in living memory. The increased public awareness and media attention finally caused a shift in government policy, which was long overdue. Water allocations were cut, while the price of water was increased. These measures caused the most severe cut in water consumption by the agricultural sector, declining to levels prior to 1958, and the first significant drop in water consumption by the domestic sector. Total water consumption in 1991 was down to 1420 million cubic meter (mcm) in 1991, as compared to 2024 mcm in 1985. However, just when the water crisis had reached its most serious condition, extraordinary mitigation arrived in 1991/92 with the highest recorded precipitation amounts of the last 150 years since recordings began in Jerusalem in 1846. These rains replenished the aquifers to a certain extent. Data since 1992 (Figure 5) show again an increase in water consumption, but the amount of fresh water made available to agriculture is not allowed to rise to the same levels as before, because the increasing population requires more and more water (Bruins, in press).

## Water Supply System in the Arava Rift Valley and Southern Negev

The National Water Carrier system does not reach the southern Negev and the Arava Rift Valley, which have hyper-arid climates. The relatively thin population and large distance from Lake Kinneret do not justify supply to those regions. Local groundwater and seawater are used in combination with desalination plants to supply potable water to the population.

The year 1962 marks the decision-making start for seawater desalination in Israel, as it was decided to construct a Multi-Stage Flash distillation plant with a capacity of 3,800 $m^3$/day at the town of Eilat, situated along the red Sea in the southernmost part of Israel. The actual supply of desalinated seawater to the town began in 1965. The first Reverse Osmosis plant to desalinate brackish water was constructed in 1968 at Yotvata, 40 km north of Eilat, having a capacity of 200 $m^3$/day. Another Multi-Stage Flash distillation plant for seawater desalination in Eilat was constructed in 1969 to supply more water to the growing population (Levite, 1973; Arad, Glueckstern and Kantor, 1973).

The cost difference between desalination of relatively low salinity brackish water and seawater is in the range of 1 to 4. In 1996 there were 44 Reverse Osmosis units operational in the entire Arava Rift Valley and a few other places in the country. Resources of brackish water are limited, however, so that seawater desalination plants are needed for water supply to urban centers. Large advanced Reverse Osmosis (SWRO) and hybrid Multi-Effect Distillation / Reverse Osmosis (MED/SWRO) coupled to Diesel power stations are probably preferred system for sea shore plants. All the brackish groundwater resources in the Eilat region have now been exhausted and there was a need in this hyper-arid region to return again to the sea. A new SWRO seawater desalination plant has begun operating in Eilat in 1997, producing 8000 $m^3$/day at a cost of $ 0.76 per cubic meter of water. (Glueckstern, 1996).

**Recommendations for Drought and Water Supply Management in the Future**

The possible occurrence of severe meteorological drought and human-made calamity drought must be taken into consideration in the management of Israel's water resources. An important element of any proactive drought mitigation planning is the formation of significant groundwater reserves. The amount of water required to sustain the population at current consumption levels through a severe drought of about 3 to 4 years duration is about 7,000 mcm. This amount exceeds the potential for ground-water storage, which is about 1,600 to 2,000 mcm, depending on the level of red lines and sustainable management concepts.

Additional water resources have to be added to the water supply systems. There is no realistic alternative source but the desalination of sea water. The quantity of sea water is obviously unaffected by meteorological drought and constitutes as such a very important mitigation factor in proactive drought planning. Development plans to desalinate seawater on a rather large scale, about 100 mcm per year, were not approved in the early 1970s, because agriculture would not be able to pay for such water (Kally, 1979). However, the inevitability of the matter is gradually dawning on the Water Commission, which has commenced planning the future integration of seawater desalination plants (Hoffman and Zfati, 1996).

Israel already has some major sewage treatment facilities. A further expansion in this sector is also necessary for health reasons, to combat pollution and to preserve water quality in the aquifers. The combination of seawater desalination for the domestic sector with subsequent treatment of the municipal wastewater allows for a dual contribution of a completely new water supply source from desalinated sea water: (a) first for the domestic sector; (b) then as reclaimed sewage for the agricultural sector. Such a policy would allow the make real and significant cutbacks in the quantitiy of fresh-water pumped from the groundwater aquifers, so that groundwater reserves can be built up for drought years.

The Nubian sandstone aquifer in the south of the country contains water, which is mainly fossil and derived from a wetter climate in the past (Issar, 1985; Issar and Bruins, 1983). Its present use amounts to about 30 mcm per year (Issar et al., 1972; Issar, 1990). The utilization of this aquifer should be enhanced for mitigation (Issar, 1990; Issar et al., 1995) of severe meteorological drought and human-made calamity drought (Bruins, in press). About 200 mcm can be drawn from this aquifer on a long-term basis, according to calculations by Issar (1990). Some of the water is only slightly brackish and is already used successfully in sophisticated irrigation agriculture. However, the more brackish part would require desalination, which, due to the comparatively lower salt content, will be cheaper than the desalination of seawater. The

Nubian sandstone water constitutes a very important resource to mitigate human-made calamity drought. This fossil water aquifer is generally well protected from recharge with modern water, which is important in case of widespread chemical contamination of the environment.

Local, decentralized water supply systems need to be developed in parallel with the centralized system. Regional and local water planning should provide self-reliance, independent of the centralized system (Gradus, 1984), in order to decrease vulnerability. Such dual systems can make meaningful contributions in regular years, while their existance becomes absolutely crucial in times of severe meteorological and human-made calamity drought (Bruins 1996, Bruins, in press).

## References

Amiran, D. (1994) Rainfall and Water Policy in Israel. The Jerusalem Institute for Israel Studies, No. 55 (in Hebrew).

Arad, N., Glueckstern, P. and Kantor, Y. (1973) Desalting plants in operation and under construction in Israel. In J. Bonné, S. Grossman-Pines & Z. Grinwald (eds.) Water in Israel - Part A, Selected Articles, Tel-Aviv: Ministry of Agriculture, Water Commission, Water Allocation Department, pp. 143-151.

Arlosoroff, S. (1974) Legal, administrative and economical means for the preservation and efficient use of water in Israel. In Water in Israel (Reprints), Ministry of Agrculture, Water Commission.

Ben-Zvi, A. (1987) Indices of hydrological drought in Israel. *Journal of Hydrology* 92:179-191.

Bruins, H.J (1993) Drought risk and water management in Israel: planning for the future. In: D.A. Wilhite (ed.) *Drought Assessment, Management and Planning: Theory and Case Studies,* Boston: Kluwer Academic Publishers, Chapter 8, pp. 133-155.

Bruins, H.J. (1994) Comparative chronology of climatic and human history in the southern Levant from the late Chalcolithic to the Early Arab Period. In: O. Bar-Yosef & R. Kra (eds.) *Late Quaternary Chronology and Paleoclimates of the Eastern Mediterranean,* Tucson: Radiocarbon, Dept. of Geosciences, University of Arizona & Cambridge (MA): Peabody Museum, Harvard University, pp. 301-314.

Bruins, H.J. (1996) A Rationale for Drought Contingency Planning in Israel. In Y. Gradus and G. Lipshitz (eds.) *The Mosaic of Israeli Geography,* Beersheva: Ben-Gurion University of the Negev Press, The Negev Center for Regional Development, pp. 345-353.

Bruins, H.J. (in press) Drought hazards in Israel and Jordan: policy recommendations for disaster mitigation. In D.A. Wilhite (ed.) *Drought. Hazards and Disasters: A Series of Definitive Major Works.* Routledge, London.

Bruins, H.J., Evenari, M. and Nessler, U. (1986) Rainwater-harvesting agriculture for food production in arid zones: the challenge of the African famine. *Applied Geography* 6:13-32.

Bruins, H.J., Evenari, M. and Rogel, A. (1987) Run-off Farming Management and Climate. In L. Berkofsky and M.G. Wurtele (eds.) *Progress in Desert Research* (Totowa NJ: Rowman & Littlefield, Chapter 1, pp. 3-14.

Bruins, H.J. and J. Van der Plicht (1996) The Exodus enigma. *Nature* 382:213-214.

Bruins, H.J. and Lithwick, H. (1998) Proactive Planning and Interactive Management in Arid Frontier Development. In H.J. Bruins and H. Lithwick (eds.) *The Arid Frontier: Interactive Management of Environment and Development,* Dordrecht: Kluwer Academic Publishers, Chapter 1, pp. 3-29.

Bruins, H.J. and Berliner, P.R. (1998) Bioclimatic Aridity, Climatic Variability, Drought and Desertification. In H.J. Bruins and H. Lithwick (eds.) *The Arid Frontier: Interactive Management of Environment and Development,* Dordrecht: Kluwer Academic Publishers, Chapter 5, pp. 97-116.

Doron, P. (1993) *Development: The eventful life and travels of an engineer.* Gefen, Jerusalem.

Fishelson, G. (1995) International Conference on the Peace Process and the Environment, Tel-Aviv University, Book of Abstracts, pp. 45-48.

Frumkin, A., M. Margaritz, I. Carmi and I. Zak (1991) The Holocene climatic record of the salt caves of Mount Sedom, Israel. *The Holocene* 1:190-200.

Gilead, D. and Y. Bachmat (1973) Israel's groundwater basins. In J. Bonné, S. Grossman-Pines & Z. Grinwald, eds.) Water in Israel - Part A, Selected Articles, Tel-Aviv: Ministry of Agriculture, Water Commission, Water Allocation Department, pp. 37-51.

Gillette, H.P. (1950) A creeping drought under way. Water and Sewage Works, March 1950: 104-195.

Glueckstern, P. (1996) Short and long term desalination options for enhancement and quality improvement of water supply. International Conference of Water resource Management Strategies in the Middle East. Association of Engineers and Architects in Israel, Society of Water Engineers, Herzlia, Book of Abstracts.

Gradus, Y. (1984) The emergence of regionalism in a centralized system: the case of Israel. *Environment and Planning D: Society and Space,* Vol 1, pp. 87-100.

Grinwald, Z. and M. Bibas (1989) Water in Israel. Tel Aviv: Ministry of Agriculture, Water Commission, Water Allocation Department.

Hellman, Z. and D. Rudge (1991) State Comptroller's report says 25 years of irresponsible mismanagement to blame. The Jerusalem Post, Thursday, January 3, 1991, page 1.

Hoffman, D. and Zfati, A. (1996) Considerations governing the selection and design of optimal seawater desalination plants for integration within conventional regional water supply systems. International Conference of Water resource Management Strategies in the Middle East. Association of Engineers and Architects in Israel, Society of Water Engineers, Herzlia, Book of Abstracts.

Hoyt, W.G. (1942) Droughts. In O.E. Meinzer (ed.) *Hydrology,* New York: Dover Publications, p. 579.

Issar, A.S. (1968) The Pleistocene geology of the Central Coastal Plain of Israel. *Israel Journal of Earth Sciences* 17:16-29.

Issar, A.S. (1985) Fossil Water under the Sinai-Negev Peninsula. *Scientific American* 253 (1):82-88.

Issar, A.S. (1990) Climatic changes in the Levant and the possibility of their mitigation. In R. Paepe, R.W. Fairbridge & S. Jelgersma (eds.) *Greenhouse Effect, Sea Level and Drought* NATO ASI Series, C-325, Dordrecht: Kluwer, pp. 565-574.

Issar, A.S., A. Bein and A. Michaeli (1972) On the ancient water of the upper Nubian Sandstone Aqufer in Central Sinai and Southern Israel. *Journal of Hydrology* 17 (4): 353-374.

Issar, A.S. and H.J. Bruins (1983) Special climatological conditions in the deserts of Sinai and the Negev during the latest Pleistocene. *Palaeogeography, Palaeoclimatology, Palaeoecology* 43:63-72.

Issar, A.S., with Zhang Peiyuan, H.J. Bruins, M. Wolf and Z. Ofer (1995) Impacts of climate variations on water management and related socio-economic systems. Paris: Unesco, International Hydrological Programme IHP-IV Project H-2.1.

Kally, E. (1973) Israel's water economy and its problems in the early seventies. In J. Bonné, S. Grossman-Pines & Z. Grinwald (eds.) Water in Israel - Part A, Selected Articles, Tel-Aviv: Ministry of Agriculture, Water Commission, Water Allocation Department, pp. 105-120.

Kally, E. (1979) Water supply to arid areas: the Israeli lessons. In G. Golany (ed.) *Arid Zone Settlement Planning: The Israeli Experience*. Pergamon Press, New York, pp. 393-410.

Katsnelson, J. (1964) The variability of annual precipitation in Palestine. *Archiv für Meteorologie, Geophysik und Bioklimatologie, Serie B,* 13(2):163-172.

Kedem, O. (1996) Desalinating in arid regions. International Conference of Water resource Management Strategies in the Middle East. Association of Engineers and Architects in Israel, Society of Water Engineers, Herzlia, Book of Abstracts.

Lee, K.S., Sadeghipour, J. and Dracup, J.A. (1986) An approach to frequency analysis of multiyear drought durations. *Water Resources Research* 22(5):655-662.

Levite, G.A. (1973) Towards the era of desalination. In J. Bonné, S. Grossman-Pines & Z. Grinwald (eds.) Water in Israel - Part A, Selected Articles, Tel-Aviv: Ministry of Agriculture, Water Commission, Water Allocation Department, pp. 135-142.

Melloul, A. and M. Bibas (1990) General and Regional Hydrological Situation in the Coastal Plain Aquifer of Israel and Water Distribution according to Quality Standards (Chlorides and Nitrates) in 1987/88 and expected to 1992. Ministry of Agriculture, Water Commission, Hydrological Service. Jerusalem: Rep Hydro 1990/3 (in Hebrew, abstract in English).

Ministry of Agriculture (1973) Israel's Water Economy. Reprints of published papers. Tel Aviv: Ministry of Agriculture, Water Commission.

Ministries of Foreign Affairs & Finance (1996) Development Options for Cooperation: The Middle East / East Mediterranean Region, Jerusalem: Government of Israel.

Palmer, W.C. (1965) Meteorological drought, *Research Paper No 45*, US Weather Bureau, Washington D.C.

Penman, H.L. (1948) Natural evaporation from open water, bare soil and grass. *Proceedings of the Royal Society, Section A*, 193:120-145.

Schwarz, J. (1990) Management of the water resources of Israel. *Israel Journal of Earth Sciences* 39:57-65.

Shanan, L. (1997) Irrigation development: proactive planning and interactive management. In H.J. Bruins and H. Lithwick (eds.) *The Arid Frontier: Interactive Management of Environment and Development*, Dordrecht: Kluwer Academic Publishers, Chapter 1, pp. 251-276.

Shanan, L. and Berkowicz, S. (1995) The context of locally managed irrigation in Israel: policies, planning and performance. Report No. 10, Short Report Series on Locally Managed Irrigation, Colombo: International Irrigation Management Institute.

Sharma, T.C. (1994) Stochastic features of drought in Kenya, East Africa. In K.W. Hipel (ed.) *Stochastic and Statistical Methods in Hydrology and Environmental Engineering* Vol.1:125-137. Dordrecht: Kluwer Academic Publishers.

Sperber, D. (1978) *Roman Palestine 200-400 - The Land: Crisis and Change in Agrarian Society as reflected in Rabbinic Sources.* Bar-Ilan University, Ramat Gan.

State Comptroller (1990) Report on the Management of Water Resources in Israel. State Comptroller, Jerusalem.

Statistical Abstracts of Israel (1993) Volume 44, Central Bureau of Statistics, Israel.

Thorthwaite, C.W. (1948) An approach towards a rational classification of climate. *Geographical Review* 38:55-94.

Trewartha, G.T. (1961) *The Earth's Problem Climates.* Madison: The University of Wisconsin Press.

UNEP (1992) *World Atlas of Desertification*, Edward Arnold, London.

UNESCO (1979) Map of the world distribution of arid regions. Explanatory note. Paris: Unesco, Man and the Biosphere (MAB) Technical Notes 7.

Warren, A. (1997) Environmental science and desertification at the Arid Frontier. In H.J. Bruins and H. Lithwick (eds.) *The Arid Frontier: Interactive Management of Environment and Development,* Dordrecht: Kluwer Academic Publishers, Chapter 6, pp. 117-127.

Wilhite, D.A. (1993) The Enigma of Drought. In D.A. Wilhite (ed.) *Drought Assessment, Management and Planning: Theory and Case Studies* Boston: Kluwer Academic Publishers, Chapter 1, pp. 3-15.

Wilhite, D.A. and M.H. Glantz (1985) Understanding the drought phenomenon: The role of definitions. *Water International* 10:111-120.

Wilhite, D.A. and W.E. Easterling (1987) Drought policy: toward a plan of action. In D.A.Wilhite and W.E. Easterling (eds.) *Planning for Drought - Toward a Reduction of Societal Vulnerability,* Boulder: Westview Press, pp. 573-589.

World Health Organization (1988) Guidelines for Drinking Water Quality. Geneva, W.H.O.

Zangvil, A. (1979) Temporal fluctuations of seasonal precipitation in Jerusalem. *Tellus* 31:413-420.

# CALIFORNIA'S APPROACH TO MANAGING WATER SUPPLIES DURING DROUGHTS

Paul Boulos, Ph.D., Russell Mau, Ph.D., David Ringel, Harold Glaser
*Montgomery Watson*
*300 North Lake Avenue. Suite 1200. Pasadena, California 91101.*
*United States*

ABSTRACT

The State of California has a unique water resources setting — approximately 75 percent of the natural water sources are located in northern third of the state while an estimated 80 percent of the demand occurs in the southern two-thirds. To mitigate this imbalance, massive water facilities were constructed to convey water from the north to the south. However, these facilities generally were designed to transfer water and did not necessarily eliminate drought risk as evidenced by water shortages experienced during the state-wide drought of 1987 to 1992. As this particular drought persisted, available water resources were strained. At the same time, increased environmental legislation enacted to protect endangered species in the Sacramento-San Joaquin River Delta estuary restricted water exports. In previous droughts, community and regional drought management techniques focused primarily on increased water conservation, mandatory rationing, or other reactive mechanisms to balance demands with limited supplies. With steadily increasing demands, conservation measures alone were observed to be inadequate to balance supplies and demands. Instead, a combination of proactive measures including increased water conservation, drought contingency plans, fallowing of farmland, water rationing, use of recycled water, water transfers, conjunctive use of stored groundwater, and desalination were employed to satisfy demands. The measures that were implemented actually allowed southern California to shift water back to northern California, extending their limited supplies. In this chapter, the development and implementation of these approaches to managing shrinking water supplies during droughts are detailed.

## 1. Introduction

California presents a unique setting for assessing water supply and drought management. A majority of the southwestern United States, where California is an integral component, corresponds to desert, semi-arid, or chaparral land types with notable exceptions being the Coastal and Sierra Nevada mountain ranges and the Mediterranean coastal re-



*E. Cabrera and J. García-Serra (eds.), Drought Management Planning in Water Supply Systems,* 322–360.

gion. As such, the nature of managing California's water supplies during any supply or demand scenario (drought or normal) can be extremely involved. **Figure 1** shows key areas of interest in the southwestern United States with a particular focus on California.

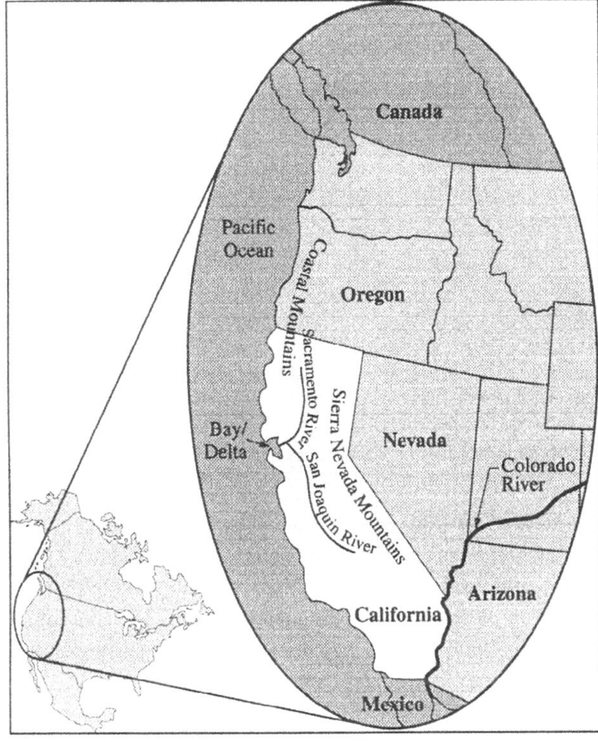


**Figure 1. Location Map Showing Key Areas of Case Study**

Not only are natural conditions non-conducive to lush plant growth such as truck farming or animal foodstuffs, but the region is not capable of providing a local, renewable water supply for continued regional population growth or additional development.

A map depicting May through October average historical rainfall totals for selected U.S. cities [1] is shown in **Figure 2**. While a grass lawn needs at least 12 inches of water to survive, the data presented in this map indicate that California receives less than 2 inches (50.8 mm) in the summer. This figure clearly demonstrates that conditions in California are not favorable for growing water-thirsty vegetation.

Water surplus and deficiency, where the delineation between surplus and deficiency is based on a comparison of the amount of precipitation required for well-watered vegetation, not just total rainfall, and the amount of local rainfall historically recorded across the United States [2] are also shown in Figure 2. Most notable in this figure is that California falls in the categories of deficient to extremely deficient. In fact for California, the ratio of consumptive use to renewable supply was estimated as approximately 35 percent in 1980 [2] while the same ratio for the Colorado River region was estimated as 105 percent (i.e., use actually exceeded available, renewable supply). These figures indicate that the desert southwest operates under a continuous drought-type condition where demand nearly matches supply and any consequential, yet predictable, decrease in supply or increase in demand could push demand beyond supply. These conditions have only worsened since the western United States has been the country's fastest growing region. For example, a 2°C increase in temperature coupled with a 10 percent

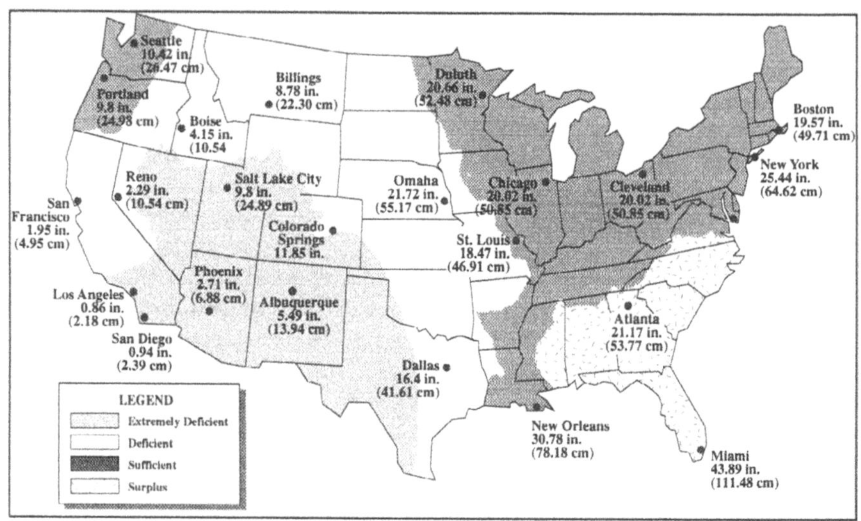


**Figure 2. Map Indicating May Through October Rainfalls in inches (cm) and Water Surplus and Deficiency Across the United States (Adapted from [1] and [2])**

reduction in rainfall has been estimated to result in a 53 percent reduction in water supply for the western United States [2]. This reduction in water supply resulting from events of reasonable probability would nearly erase the 65 percent supply surplus that has been approximated for California. With imminent population growth and additional development envisioned for the state, supply prognosticators predict the occurrence of a water supply crisis for California before 2020, unless major changes transpire in the management of California water supplies and water use [3,4,5].

Although the state surfacially presents a microcosm of current demographic trends of population migration to cities, the state also fosters a burgeoning agricultural industry, currently the largest in the nation, that accounts for over 10 percent of the state's gross revenue [6]. This dichotomy of both urban and rural development in a state whose population and areal extent ranks are 28 and 54 [7], respectively, among *nations* worldwide, has continually stressed the state's indigenous water supplies and water conveyance infrastructure. Further growth has only exacerbated the problem because approximately 75 percent of the natural water sources are located in northern portion of the state while an estimated 80 percent of the demand occurs in the southern portion [8,9].

For California, the issue of state water supply management has crossed regional, state (including Arizona, New Mexico, Nevada, and Oregon), and even national (including both Canada and Mexico) boundaries or jurisdictions; has evolved into a complex interplay of economic, social, environmental, and political considerations and stakeholders; and has transitioned toward more centralized control with an increasingly contentious relationship between state and federal entities. A discussion of each of these areas follows:

- **Multiple Jurisdictions.** The distributions of population and water supply are greatly out of balance in California — half of the population lives in the southernmost fifth of the state and 75 percent of the population reside where only one-quarter of the precipitation occurs [10]. Water distribution in the state involves more than just regional rivalries but entails core concerns regarding ethical and legal constraints over water rights and environmental protection. In addition, California withdraws water from the Colorado River at a flow rate exceeding its entitlement, and this supplemental water supply can be terminated whenever the state of Arizona exercises its full allocation [10]. Internationally, a long-running feud between Mexico and the United States exists over the quality and quantity of Colorado River water delivered to its mouth. Meanwhile, the National Democratic government of Canada recently voted to prohibit water exports or river diversions from British Columbia to the United States [11].

- **Multiple Stakeholders.** Traditionally, agriculture has enjoyed a highly subsidized and unchallenged status in California's economy and society. The recent drought, the worst since 1929 to 1934, forced politician and resident alike to rethink this arrangement. Agriculture uses about 80 percent of the state's developed water while only generating about 10 percent of the state's gross product, and four alfalfa-type crops use 45 percent of the available developed water while generating approximately a mere 1 percent of the state's gross product [10]. Through intense pressures exerted by environmental groups, the Clean Water Act, Endangered Species Act, and Miller-Bradley Bill were enacted that set-aside substantial quantities of water to environmental purposes, and these quantities need to be satisfied prior to any diversions for human consumption or use [8]. The implications of these mandates range from basin-wide water quality requirements in the Sacramento-San Joaquin River Delta to highly localized concerns. Recently, State Water Project (SWP) pumps located in the Sacramento-San Joaquin Delta were required to shut down during the height of the drought due to substantial kills of juvenile winter-run Chinook salmon (1,500 by March) migrating through the Delta to the sea. Under a federal-state pact, no more than 2,700 winter-run Chinook may be killed through the end of May [12].

- **Separation of State and Federal Control.** The construction and management of the water conveyance systems in California has been administered by local, state, and federal entities. As such, conflict has arisen among these governing bodies over control of the state's water. In 1992, the Miller-Bradley Bill, which authorized federal guidelines for water projects in the western United States, was signed into law [8]. The governor of California objected to this increasing intervention by the federal government on the grounds that the federal government was trespassing on the state's water rights prerogatives [13,14].

While administering the allocation of supply to demand is daunting during normal conditions in California, water supply management can and nearly did become catastrophic during a true drought. This chapter has been divided into two principal parts to place structure on the purported complexity in managing water supplies in California:

1. **History of Water Use in California.** The growth of regional water demands is outlined, water supplies and water conveyance systems are described, and legislation intended to regulate water supply management are identified. This part provides the requisite background regarding the critical nature of water supply issues in California.

2. **Water Supply Management in California.** Methods used by water purveyors to forecast water supplies and demands and to optimize water system operations are reported, conventional drought management practices are presented, and innovative approaches to manage water supplies during droughts are detailed.

## 2. Part 1 — History of Water Use in California

To fully appreciate the gravity of water supply management in California, the presentation of the history of water use in the state consists of the following: 1) growth and projection of regional water demands for different water use categories; 2) development of water collection, storage, and conveyance infrastructure to transport water from northern California to southern California; and 3) enactment of political legislation at the local, state, and federal levels governing water allocations, water and wastewater treatment, protection of the environment, and conservation measures. The authors would have felt reticent if these background issues had been ignored.

### 2.1. WATER DEMANDS

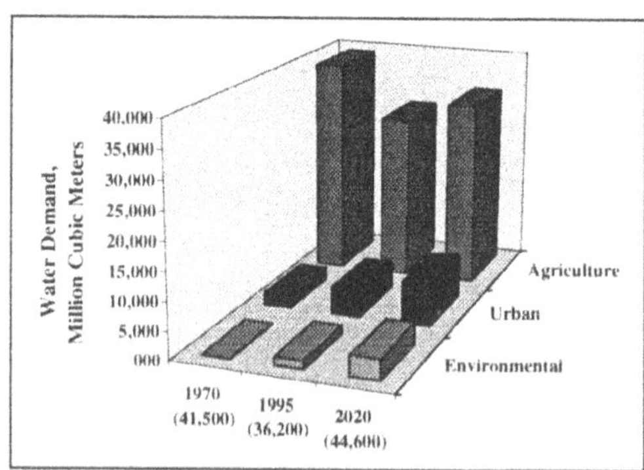


**Figure 3. Water Demands by Use Category for California (Yearly Totals in Parentheses)**

Consumers of California water can be grouped into three broad categories: urban, agricultural, and environmental. **Figure 3** shows annual demands for each water use category ranging from historical water use through existing demands to projected needs. Historically, agricultural interests and urban inhabitants peacefully coexisted while environmental uses were largely ignored. Accelerated by explosive urban growth in southern California and exacerbated by the crippling drought, the face of water use across the state dramatically changed, pitting central Californian agriculture against urban southern California, northern versus southern California, and urban and agricultural users against the environment. These transformations and the order of magnitude of water use within the state are detailed in this subsection.

Globally, mankind consumes 26 percent of total terrestrial evapotranspiration and 54 percent of accessible runoff [15]. California mirrors this apportionment, thus it is poised to have water shortages when runoff declines. Across California, the northern third of the state exhibits a surplus of natural water with respect to local demands while the southern two-thirds display moderate to substantial deficiencies where local demands greatly exceed supply [16]. This distribution of surplus and deficiency across the state by hydrologic region is shown in **Figure 4**.

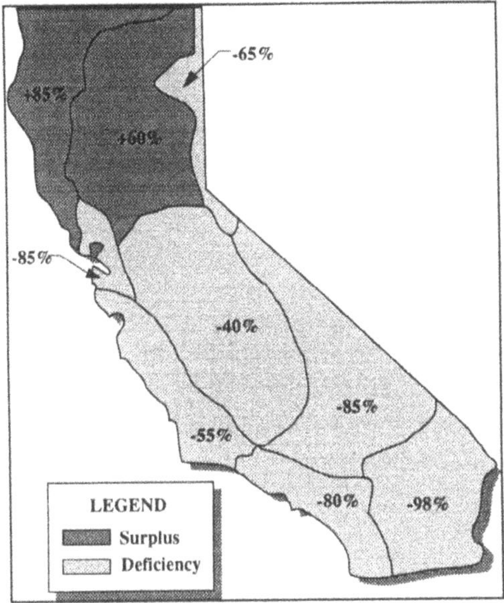


**Figure 4. Surplus and Deficiency of Water Supply Versus Demand in California (Adapted from [16])**

*2.1.1. Agriculture*

Agricultural irrigation is the primary user of California water with its current use being estimated as approximately 80 to 85 percent of the available developed water; however, agribusiness interests face an uncertain future having lost significant political clout [17,18]. For example, with reservoirs at 90 percent of capacity in 1994, a year after the end of California's most recent and most devastating drought, farmers were only receiving as low as one-third of the water they normally receive [17], and water officials did not plan to promise full deliveries in the near future [19]. In addition, the cost of agricultural water has always been subsidized by state and federal agencies, with prices ranging from $15 to $55 per acre-foot (af) (1.24 to 4.54¢ per $m^3$) for federal water in the Central Valley Project (CVP) and for state water from the SWP, respectively. Contrast these costs with the $485 per af (39.4¢ per $m^3$) paid by residential consumers in San Francisco.

Circa 1975, agriculture used 31.6 million acre-feet (maf) (39,000 million $m^3$ - mcm) of water while only generating a fraction of the production achieved today [20,21]. This historical quantity can be compared to the theoretical quantity of water needed to produce adequate food for the 32 million state residents - 20 maf (25,000 mcm). By 1994, significant improvements in water use efficiency lowered total farm irrigation water use to between 24 and 26.6 maf (29,600 and 32,800 mcm) [3,10,22] while agriculture's contribution to the state's income has reached 10 percent [6].

*2.1.2. Urban*

Historically, urban water use, composed of residential, commercial, and industrial users, has been a small percentage of total state water use. As of 1994, urban water use was approximately 4 maf (4,900 mcm) per year for a state-wide population of 32 million [22]. With a population projected to increase to 36 million by 2000 [23] and 49 million by 2020 [8,10] with estimates of SWP service area population increases of 400,000 people per year [24], the projected increase in urban water demands will be approximately 1.2 to 2.1 maf (1,500 to 2,600 mcm) per year over 1994 demands.

Accompanying future population increases, yearly water demands in the City of Los Angeles are expected to rise from 0.63 maf (780 mcm) in 1995 to 0.75 maf (930 mcm) by 2015, with a population increase of 900,000 [25]. This 20 percent increase may be dramatic; however, it only represents a change of less than 0.5 percent in state-wide demand. Demand in the Metropolitan Water District (MWD) service area is projected to increase from 4.0 to 4.9 maf (4,950 to 6,050 mcm) between 1990 and 2020.

#### *2.1.3. Environmental*

In 1993, a landmark year for environmentalists, federal legislation was passed to protect the environment through water set-asides. California was especially susceptible to repercussions from this legislation because it followed a six-year drought and decades of water diversions had not only severely damaged affected ecosystems but also caused consumers to become increasingly reliant on these water sources. Examples of the effects of these legislative mandates on water supplies follow with more on political activities presented later in Part 1 of this chapter:

- In the CVP, federal reforms have mandated the set-aside of 0.8 maf (1,000 mcm) of water per year for restoration of fish and wildlife populations, with this set-aside being approximately 20 percent of total CVP water [10,18,26].
- In the Delta, water deliveries are to be trimmed by 9 and 21 percent during normal and drought years, respectively [27], with these percentages corresponding to annual flows of 0.5 and 1.4 maf (620 and 1,700 mcm), respectively [3,28].
- By 2020, the state estimates the allocation of 1.0 to 3.0 maf (1,200 to 3,700 mcm) per year to conservation and environmental restoration [10].

Although dramatic increases in population yield a relatively inconsequential change in water demand across the state, this assessment needs to be tempered by the fact that environmental water allocations are rising to the same order of magnitude as urban use with the net effect of increased demands on supply being quite substantial.

### 2.2. WATER SUPPLIES AND WATER STORAGE AND DELIVERY SYSTEMS

California has four water supply sources: 1) developed water, 2) groundwater, 3) desalination, and 4) recycled water that represents a derived water supply used to reduce demand for the other three water supply sources. The average annual precipitation in the state is approximately 193 maf (240,000 mcm) with 121 maf (150,000 mcm) consumed by evapotranspiration and the remaining 72 maf (90,000 mcm) forming runoff; however, the volume of runoff can vary from a low of 15 maf (18,500 mcm) during droughts to 135 maf (170,000 mcm) during wet years [8]. The current state-wide distribution of water supplies for consumptive use is shown in **Figure 5**.

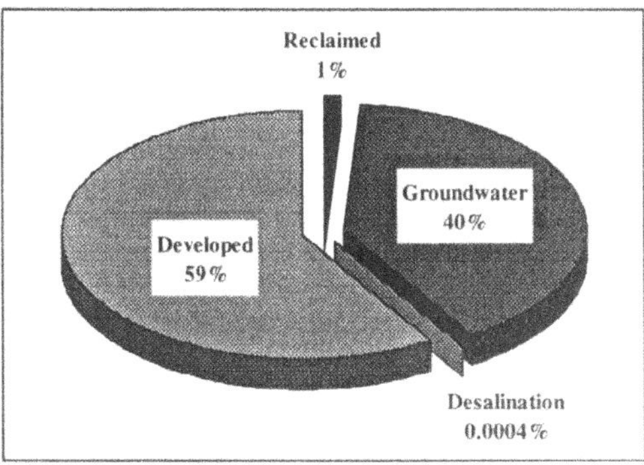


**Figure 5. Distribution of Water Supplies for California**

*2.2.1. Developed Water*

Water that is collected, stored, diverted from rivers, or otherwise managed unnaturally is defined as "developed" water, developed for alternative consumption rather than terrestrial runoff or evapotranspiration [8]. In California, where water supplies are widely separated from demands, developed water is transported through massive conveyance infrastructure to consumers.

Available state storage capacity totals about 36 maf (44,400 mcm) located in 1,313 local, state, and federal reservoirs [8], where 155 are major reservoirs, with total annual rainfall runoff, 72 maf (90,000 mcm), exceeding storage capacity by a factor of two [4,8,10]. In contrast, the Colorado River basin receives less than one-fourth of its storage capacity, 65 maf (80,000 mcm), in total annual runoff [10].

The conveyance facilities for California are shown in **Figure 6**, and a development chronology of these systems is provided in **Table 1** [29]. The authors were fortunate to have participated in the design of a portion of the SWP, Coastal Branch of the California Aqueduct. Two governmental agencies have been predominantly responsible for the major development of the conveyance systems in the state: the U.S. Bureau of Reclamation for the CVP and All-American Canal and the California Department of Water Resources (DWR) for the SWP. Other significant water projects are administered by other public entities: the City of San Francisco and the Hetch Hetchy Reservoir, MWD of Southern California and the Colorado Aqueduct, and the City of Los Angeles and the Los Angeles Aqueducts, with other smaller projects developed by local agencies [8,30]. The principal dis-

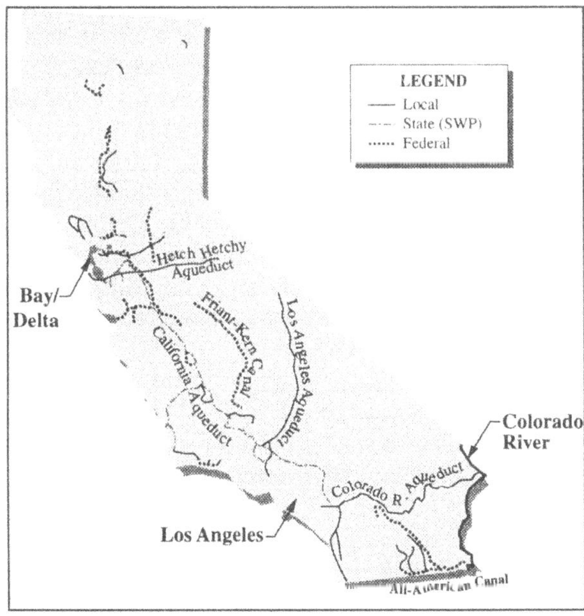


**Figure 6. Water Conveyance Infrastructure in California [29]**

| | |
|---|---|
| 1901 | First Colorado River deliveries to Imperial Valley for agriculture irrigation |
| 1910 | First Orland Project deliveries to Sacramento Valley |
| 1913 | First Los Angeles Aqueduct (Owens River Aqueduct) begins service |
| 1928 | Hetch Hetchy Valley placed in service to provide water supply for San Francisco |
| 1933 | Central Valley Project authorized |
| 1941 | First MWD Colorado River Aqueduct deliveries |
| 1941 | First All-American Canal deliveries to Imperial Valley |
| 1941 | First Contra Costa Canal deliveries, first unit of Central Valley Project |
| 1947 | San Diego completes connection to Colorado River Aqueduct |
| 1951 | First Central Valley Project deliveries from Shasta Dam to San Joaquin Valley |
| 1959 | State Water Project authorized |
| 1967 | First State Water Project deliveries |
| 1970 | Second Los Angeles Aqueduct begins service from Lower Owens Valley |
| 1973 | First State Water Project deliveries to southern California |
| 1997 | Coastal Branch Phase II begins operation to deliver water to Santa Barbara |

**Table 1. Chronology of Development of Water Conveyance Systems in California [29]**

tinction between the SWP and the CVP is the method of financing where interest costs for the CVP are borne exclusively by the federal government; whereas, all costs associated with the SWP are paid by SWP water contractors [8].

Diversions from the Sacramento-San Joaquin River Delta form a major source of supply for both the CVP and the SWP [8,31] and represent approximately 80 percent of the state's fresh water for approximately 18 to 20 million urban users, of the state's approximately 30 million residents, and 2.7 million acres (1.1 million ha) of agricultural lands [4,10,27,31]; however, recent legislation has significantly curtailed these diversions.

The federally funded and operated CVP has four major northern California reservoirs, with a combined capacity of 8.25 maf (10,200 mcm) [32]. Southern facilities of the CVP are utilized to convey water around the Delta to the Central Valley and southern California. The CVP system can transport 7.3 maf (9,000 mcm) per year throughout the state [8].

The SWP has one major northern California reservoir with significantly less storage capacity than the CVP [8]. Approximately 70 percent of the SWP deliveries are intended for urban users, mostly in southern California, with the remaining 30 percent for agriculture [8]. The SWP has contracted to sell 4.2 maf (5,200 mcm) of water per year throughout the state, but the annual dependable yield is only approximately 2.4 maf (2,900 mcm) per year [24].

Twenty-two compacts or agreements for allocating river water among states in the western United States were in effect in 1994, with two agreements involving California: Colorado River and Klamath River [33]. Under these compacts, California is entitled to import annually 4.4 maf (5,400 mcm) from the Colorado River [8], southern California's agricultural entitlement being 3.85 maf (4,750 mcm) [34], and to receive 1.4 maf (1,700 mcm) from the Klamath River [8]. However, MWD's maximum supply of Colorado River water is limited to 1.2 maf (1,500 mcm), the Colorado River Aqueduct capacity [8,35].

Southern California has continually relied on conveyance systems to maintain adequate water supplies for urban users. Less than 2.0 maf (2,500 mcm) of storage capacity is currently available in southern California compared to 35 to 37.7 maf (43,200 to 46,500 mcm) of storage capacity available across the entire state [8,10]. Although California has more dams than any other state, the last major dam was constructed in 1979 [10]. Since then, explosive growth in southern California and its corresponding increase in political power have not been able to offset environmental opposition to the development of new storage facilities, with new dam facilities being the subject of intense scrutiny [8,36]; however, recently three new storage facilities have been approved and are in various stages of design and construction.

For the City of Los Angeles, the Los Angeles Aqueduct satisfies about half of the City's water needs; local groundwater provides 15 percent of total demand; and purchases of water from the MWD, a water wholesaler, account for the remaining demand [25]. Previously, the City was able to provide more than two-thirds of its demand with aqueduct water, but recent court decisions have critically limited diversion of water from Mono Lake tributaries [8,37,38,39,40]. The combined storage capacity of seven Los Angeles Aqueduct reservoirs is approximately 0.325 maf (400 mcm), representing on the order of six months of current water demands [25].

*2.2.2. Groundwater*

As shown in Figure 5, the state currently takes 40 percent of its water from groundwater resulting in an annual overdraft of approximately 2.0 maf (2,500 mcm) [10,41,42,43]. Southern California urban users only extract about 15 percent of their demand from groundwater while a considerably greater percentage of agricultural and urban demand in the low and high deserts of southern California use groundwater. In 1991, the Imperial Irrigation District of California submitted a request to the DWR seeking to drill wells along the All-American Canal to recover seepage that has leaked from the canal [44].

*2.2.3. Desalination*

Bordered on one boundary by the Pacific Ocean, California has a virtually limitless supply of seawater. Prior to the recent drought, the state did not have a single major desalination facility in operation, matching the number of desalting plants for domestic use in the remainder of the United States — zero. In 1991, the first major desalination plant began operation generating a flow of approximately 13 af (16,000 $m^3$) per year [45]. During the height of the drought, a number of desalination facilities were considered or envisioned [46]. For the City of Santa Barbara, desalination was shown to be less expensive than the construction of water conveyance infrastructure to deliver agricultural water transfers to the City [47]. More on desalination is presented later in this chapter.

*2.2.4. Recycled Water*

Reclaimed water, or "recycled" being the latest moniker, in the form of tertiary treated wastewater is a water source derived from consumed potable water and represents a novel alternative water source that can offset a portion of the demands exerted on existing water supplies. The recycled water can be used to irrigate landscapes or agriculture [48], provide toilet flushing or industrial water [49,50,51,52], provide groundwater recharge [53], establish seawater intrusion barriers [54,55], or even supplement existing drinking water reservoir supplies [56,57]. Historically in California, over 70 percent of treated wastewater, 1.8 maf (2,200 mcm) [58] and 2.5 to 3.0 maf (3,100 to 3,700 mcm) per year in 1975 and 1994, respectively [59], was discharged directly into saline waters and lost for any reuse. More on recycled water is presented later in this chapter.

## 2.3. POLITICAL REGULATIONS: FEDERAL, STATE, AND LOCAL

Political impacts ranging from local water audits, through state-level requirements for recycled water treatment, to state and federal legislation, e.g., installation of low-flow flush toilets in new construction and allocation of water for environmental purposes, have profoundly altered the way that water is managed within California [20]. In the political arena, old enemies have become allies and vice versa while the scramble for water management authority has pitted the state against the federal government [13,14,60]. For instance, subsidies provided by the state and federal governments originally enacted to protect the fledgling agricultural industry have been mercilessly attacked by urban users and environmentalists as water has become more scarce [10,17]; whereas, urban users and agribusiness interests have lobbied against environmentalists who have pushed for the set-aside of water to restore natural habitats that have been damaged by decades of water diversions [28]. The information presented in this subsection is not intended to condone nor condemn the use of the legislative process, rather, it provides a narrative of key legislation and their influences on water supply management in California.

|  |  |
|---|---|
| 1850 | Arkansas Act: granted state flood and swamp lands on condition of reclamation |
| 1922 | Colorado River Compact apportions water to river's two basins |
| 1937 | Rivers and Harbors Act approved authorizing construction of CVP by Army Corps |
| 1944 | Mexican-American Water Treaty guaranteeing Mexico allocation of Colorado River water |
| 1959 | Delta Protection Act to resolve delta issues |
| 1963 | Supreme Court allocates Arizona Colorado River water |
| 1968 | Wild and Scenic Rivers Act to preserve Middle Fork of Feather River |
| 1972 | Clean Water Act to clean polluted waters |
| 1974 | Safe Drinking Water Act to assure drinking water safety |
| 1982 | Reclamation Reform Act limits amount of land a family can own and receive low cost federal water |
| 1986 | Safe Drinking Water Act Amendment |
| 1992 | Energy Policy Act to promote resource use efficiency |
| 1992 | CVPIA for habitat restoration |
| 1994 | Bay/Delta Protection Plan to set water quality standards for the Delta |
| 1996 | Water Resources Development Act to restore and preserve wetland habitats |

**Table 2. Chronology of Federal Legislation [29]**

### *2.3.1. Federal*

**Table 2** provides a chronology of federal legislation affecting California water management [29]. As a condition of statehood bestowed in 1850, California was granted 2.1 million acres (0.85 million ha) of swamps and flooded lands by the federal government contingent on the use of revenue gained through real estate sales to provide drainage and flood control facilities to reclaim these lands [61]. At the time, California was virtually unsettled, and the federal government was keenly interested in developing the region. California gratefully consented to this governmental decree. Over a century later, the same federal government is now requiring the restoration of the swamplands and other water habitats [27,62,63], while public opinion supports protection of western states' natural waterways and wetlands [63,64].

Through various federal acts and legislation, the federal government has ventured into the traditional realm of the state regarding authority for controlling local and regional water supplies and resources with the effect of federal intervention exemplified by two specific cases, Bay/Delta and Central Valley, as noted in the following:

- **Bay/Delta.** One of the most controversial areas in the state has been the Bay/Delta region where upstream diversions of fresh water have resulted in seawater intrusion into the Bay/Delta, rendering salinity levels that threaten existing water quality [28]. Four federal agencies, the U.S. Fish and Wildlife Service, the National Marine Fishery Service, the U.S. Environmental Protection Agency (EPA), and the U.S. Bureau of Reclamation, consorted to study the Bay/Delta to ensure that agreement on public policy was pursued [8]. Over a period of three years from 1992 through 1994, this compendium shaped the Bay/Delta Protection Plan, eventually in agreement with the state of California, that established water quality standards (i.e., set volumes of water required to mitigate Bay/Delta water quality problems) for the Bay/Delta [8], under the combined guise of the Endangered Species Act, Miller-Bradley Bill (a.k.a., Central Valley Project Improvement Act - CVPIA), and Clean Water Act [3,8,13,27, 28,65,66,67,68]. Specifically, four standards were established: 1) salinity criteria in Suisun Bay; 2) survival targets for juvenile, migrating Chinook salmon; 3) salinity criteria for lower San Joaquin River; and 4) descriptive criteria for maintaining Suisun Marsh tidal wetlands [66]. Under the Plan, water diversions or exports are required to be reduced by 0.5 to 1.4 maf (620 and 1,700 mcm) per year for normal and drought years [3,28], respectively, with respective reductions in water supplies of 9 to 21 percent [13].

- **Central Valley.** Through the CVPIA, an annual budget of $50 million, administered by the U.S. Bureau of Reclamation, was established for habitat restoration [8,17,26, 62,69]. The immediate impact in California was the annual set-aside of 0.8 maf (1,000 mcm) of CVP water to fish and wildlife purposes [8,26,62,69] in addition to 0.34 maf (420 mcm) already required for maintaining the Trinity River system [8], even as the state was negotiating to gain control of the CVP [70].

At the same time, other federal actions have been promulgated. Through the federal Clean Water Act, the U.S. EPA authorized the use of federal money, normally used to finance state revolving fund loans for treatment facility construction, to acquire agricultural water rights along the Truckee River in California and Nevada to maintain minimum flow rates in the river [60]. Additionally, the Water Resources Development Act of 1996 was drafted to provide federal monies for the development and preservation of United States water infrastructure [71]. Finally, the passage of the Energy Policy Act mandated that new residential toilets reduce water consumption by 70 percent and hold no more than 1.6 gallons (6.06 liters) of water [72,73,74,75].

*2.3.2. State*

The evolution of California water supply management has embodied the divisive issue of riparian (property) versus appropriative (eminent domain) rights, resulting in diametric alignment of northern and southern California regarding control of water in the state [16,30,61,76]. Under the doctrine of riparian water rights, a property owner has first rights to all water within or contiguous to the property, including groundwater, streamflow, or rainfall runoff. Under appropriative rights, riparian rights are always subordinate to the needs of the greater community, or state in this case. Implementation of appropriative rights may be as obvious as water entitlements or may incorporate subtle externalities such as state-wide support for water infrastructure development with negative externalities felt by northern Californians who realize a cost without benefit or positive externalities for southern Californians who do not bear the full cost but still receive the full benefit. Northern Californians staunchly support riparian rights while southern Californians, by necessity with growing population, are displaying increasing hegemonic attitudes toward their entitlement to water and support appropriative rights.

The state's foray into the management and control of water supply and resources has been manifested in three fronts: 1) direct water supply control, 2) institution of water recycling treatment guidelines, and 3) establishment of water conservation practices. Under direct water supply control, state-level activities either have been legislation enacted by California to both mirror and typically respond to federal actions or have been coordinated maneuvers intended to provide more efficient water supply management. The chronology of state legislation that has affected the management of California water is provided in **Table 3** [29].

In general, the state has supported export reductions for protection of the Bay/Delta [8]; however, it objected to the intervention of the federal government into state water rights prerogatives [13,14]. To satisfy federal agencies, the state passed Decision 1630, which would have mandated the reallocation of diverted waters to the Delta [8,77], but withdrew these interim standards in an effort to assert state control over water resource management [13]. In the end, the federal government authored the Bay/Delta Protection Plan to provide protection and set water quality standards for the Delta because the proposed standards given in Decision 1630 were considered too weak [8,67,68]. The state then agreed to the standards given in the Plan.

Additionally, water transfers in which any entity can sell their entitlement of water without forfeiting their water rights were first investigated by the MWD. Under the CVPIA, water transfers were legally sanctioned for CVP water rights holders, where any purchase of water by an entity outside of the CVP was charged an additional $25 per af (2¢ per $m^3$) that directly funds the CVPIA Restoration Fund [8,26,62]. The use of

|  |  |
|---|---|
| 1850 | Statehood granted |
| 1931 | State Water Plan published |
| 1931 | County of Origin Law to define water rights |
| 1933 | CVP to utilize water in Central Valley |
| 1937 | Colorado River Board of California formed to ensure rights to river water |
| 1945 | State Water Resources Control Board to oversee water rights and water quality |
| 1951 | Feather River Project Act is authorized, becomes State Water Project |
| 1956 | Department of Water Resources created to combine independent state agencies |
| 1957 | First California Water Plan published |
| 1959 | Delta Protection Act for water quality in Delta |
| 1960 | State Water Resources Development Act to authorize construction of SWP facilities |
| 1972 | Wild and Scenic Rivers Act to protect natural waterways |
| 1974 | Title 22 of California Administration Code as companion to federal SDW Act |
| 1978 | Water Rights Decision 1485 issued by SWRCB regarding Delta water quality |
| 1980 | State Wild and Scenic Rivers Act placed under federal Wild and Scenic Rivers Act |
| 1980 | SWRCB provides financial assistance for water reclamation projects |
| 1983 | Urban Water Management Planning Act to enforce water efficiency for large users |
| 1985 | Water transfers are investigated by MWD |
| 1991 | Water Conservation in Landscaping Act to enforce conservation in landscaping |
| 1992 | Decision 1630 to set water quality for Delta |
| 1994 | SWRCB ends Mono Lake water diversions |

**Table 3. Chronology of State Legislation [29]**

water transfers has since been applied to any water entitlement holder throughout the state. More on water transfers is presented later in the chapter.

As mentioned, prior to 1994 the City of Los Angeles had received approximately 15 percent of its water supply from the Mono Lake area. However, the State Water Resources Control Board (SWRCB) amended the City's water rights licenses to Mono Lake tributary waters, and the City is not allowed to divert tributary water from the lake until its level has risen by 17 feet (5.18 m), which may take more than 20 years [8,38,39,40].

For water conservation, the Water Conservation in Landscaping Act mandated: 1) all cities and counties in California adopt a model ordinance enforcing water conservative landscaping for new development, or 2) present findings that indicate an ordinance is not required [78]. During the height of the recent drought, the DWR adopted policies allowing for the limited use of graywater in single family residences [79], but a bill to extend graywater use to multi-family dwellings and commercial buildings [80] was vetoed. The state also adopted the Urban Water Management Planning Act that requires large water suppliers to prepare water plans addressing efficient use of water supplies and development of alternative water supplies [81].

For wastewater recycling, in the absence of federal regulations, the state has promulgated wastewater treatment regulations for health protection that have been incorporated into the State of California, Code of Regulations, Title 22 Environ-mental Health, which was first adopted in 1974 and administered by the California Department of Health Services (DHS) [53,82,83,84] while environmental protection has fallen under the auspices of the State Regional Water Quality Control Board (RWQCB) .

*2.3.3. Local*

Local agencies have followed the lead of the state and have adopted water conservation plans, enacted xeriscaping ordinances for new developments [78], investigated the operation of eight graywater systems (Los Angeles [79]), banned regenerative water softeners because of their high brine content discharge that would restrict recycling of the treated wastewater [85], and implemented dual water systems in commercial buildings where local guidelines were developed jointly with local health departments [50].

### 2.4. CUMULATIVE IMPACTS ON WATER MANAGEMENT

Water scarcity is becoming a world-wide problem [86] where one-fifth of the world's population has no reliable source of fresh water [87]. Although new dam construction has the potential to increase the capture of runoff by about 10 percent globally over the next 30 years, the world's population is projected to increase by more than 45 percent during the same time frame [15]. Case studies on rivers such as the Gila and Salt in Arizona, the Colorado in California, the Yellow in China, and the Ganges in India show that diversions of streamflows in these rivers to irrigation and urban uses have so greatly reduced their flows that these rivers no longer flow to their natural discharge locations [88,89]. In California, the aqueduct systems have helped to solve the need for water but have generated two related problems: 1) continued expansion and development across the state and 2) adverse environmental impacts [90,91].

Locally, the Pacific Coast ecosystem has been damaged by unsound water management [92,93]. Diversions of water from the Bay/Delta have drastically impacted aquatic flora and fauna in the bay that thrive on the fresh water/salt water interface within the estuarine zone as salt water intrusions creep upstream [94]. Considerable public interest in protection of the environment has surfaced recently [5,93].

With both state-wide population growth — approximately 400,000 people per year in areas served by the SWP [24] — and expanded environmental protection, a total average annual water demand of approximately 36 maf (44,400 mcm) is estimated for California by 2020. Estimates of annual shortages [3,22,26] range from 3 to 9 maf (3,700 to 11,100 mcm) [3,4]. The DWR states in their recent California State Water Plan Update that severe water shortages are imminent if short-term drought management and long-term demand management and water supply development options are not implemented [95]. However, environmental constraints and public opposition have curtailed efforts to develop new water supplies or augment water storage [10]. With the development of new water supplies and storage essentially at a standstill for approximately two decades (with very recent designs and construction underway), urban southern California will need to rely on water transfers to supply increased demands [23].

Even with such doomsday reports and predictions of impending droughts and shortages, individual residents have expressed little support for conservation or water-wise efficiency programs such as growth moratoriums, mandatory water conservation programs, water transfers, water marketing, or reallocation of water from agriculture to urban users [96]. Nonetheless, California has embarked on plans for alternative water supply development such as desalination and recycling [10].

The recent drought was an epiphany for the state with glaring revelation of the underlying mismanagement of water supplies evidenced by water shortages and mandatory rationing. In Part 2 of this chapter, the innovative and progressive measures invoked by water agencies in the state to administer water supplies and water use, which guided the state through the most severe drought in modern state history, are presented.

## 3. Part 2 — Water Supply Management In California

This section details specific water supply management approaches implemented in California during the recent drought to mitigate water supply management problems. These management practices provide witness to the evolution of water management in California from crisis management to risk management. The specific approaches employed by California are: 1) forecast modeling of water supplies and demands and optimization modeling of system operation, 2) conventional drought management practices, and 3) proactive drought management approaches.

### 3.1. FORECAST MODELING OF WATER SUPPLIES AND DEMANDS AND OPTIMIZATION MODELING OF SYSTEM OPERATION

Astute management of water supply systems entails basically two components: 1) defining or forecasting water supplies and demands and 2) planning system operations based on anticipated demands and available supplies. Currently, relatively sophisticated analytical techniques to model each of these elements are available as discussed below.

#### *3.1.1. Forecast Modeling of Water Supplies and Demands*

Forecasting serves both short- and long-term purposes by facilitating financial planning and management by projecting earned revenues, estimating costs of service, setting rates, planning water system infrastructure, and, of course, matching demands with available supplies (2). In our context, it is vital that forecasting models predict the impacts of drought conditions on demand where the effect is predicated on an intricate relationship between water supplies and demands. For example, droughts can impair local supplies *and* subsequently lead to higher levels of water consumption, e.g., landscape irrigation [2]. Methods for proper water supply and system management are available that incorporate the effects of conservation on demand [97,98,99,100,101]. Incorporating conservation measures into water master planning strategies provides incentives to the water utility through reduced capital investment, operation, and maintenance costs, and extension of existing water supplies as well as disincentives including revenue shortfalls, more frequent rate adjustments to mitigate these shortfalls, and unreliable future demand predictions [99,102,103]. For the utility, the key in establishing economic incentives for conservation is to link increased water efficiency with equivalent water supply capacity, thereby, shifting from supply-side to demand-side management. This approach is necessarily long-term because existing rate structures are based on providing debt-service coverage for current capital expenditures and may require many years for retirement.

Although numerous forecasting methods are available, three approaches are described herein that encompass the most common methods, are used in California, and can incorporate the effects of conservation [2]: 1) extrapolation of time series data; 2) statistical, econometric, or stochastic models; and 3) end-use aggregate demand.

*Extrapolation of Time Series Data.* The simplest approach to forecasting future water use or supply is time series models where historical patterns are projected indefinitely. The general daily time series models [104] have been enhanced to include trends such as seasonal climatic effects [105], conservation [101], and transfer functions [106]; however, even with these enhancements, the time series models can be limited in their capability to predict future conditions because of possible changes in socioeconomic conditions, water allocation system structures, and limits to growth [107]. For the typically water-rich southeastern United States, demand has been shown to be more dependent on climatic factors such as rainfall events than on any conservation measure, and serious overestimation of the effects of conservation would have occurred with the use of simple before-after comparisons [97].

*Statistical, Econometric, and Stochastic Models.* These models are designed to incorporate a randomness based on statistically correlated variables into the forecast of demand with six basic types of models of this form having been identified by the U.S. Army Corps of Engineers [2]. While historical methods cannot predict the mercurial behavior of major demographic or climatic events, a statistical or stochastic model can assign a recurrence frequency for weather phenomena and assign a risk-based probability to any demand or supply estimate [2].

*End-Use Aggregate Demand.* For utilities, future water requirements traditionally have been based on gross per capita needs [2]. This approach has been shown to be grievously deficient for forecasting regional aggregate water demands in growing urban areas because growth is rapid, income levels are high (relatively elastic demands with respect to price when sufficient discretionary income is available [108]), seasonal uses are high, and industrial uses are unusually low when evaluated on a per-employee basis [109].

*3.1.2. Optimization Modeling of System Operation.*

Optimizing the operation of water systems entails the linking of forecasts of water supplies and demands with ongoing allocations of available supplies to existing demands allowing for sound decisions regarding management of water systems to mitigate adverse effects and extents of potential water supply shortages. Through modeling, water agencies can assign event occurrence probabilities and define acceptable levels of risk.

*3.1.3. Models Applied in California*

A number of models have been implemented in California and have proven successful in simulating water supply, use, and system operation, and **Table 4** summarizes published uses in California. This table does not delineate between models used as forecasting or optimization tools, because several of the listed models can serve as both types.

**Table 4. Summary of Models Used in California**

| Model | Purpose | Reference |
|---|---|---|
| IWR-MAIN (Institute for Water Resources Municipal And Industrial Needs) | • Provides two distinct improvements over the per capita approach: 1) disaggregates urban water use into homogeneous categories and 2) links water use to factors for need for and intensity of use<br>• Simulated water use for Los Angeles and San Diego within 3 % accounting for weather conditions, conservation, and price<br>• Estimated water savings from selected conservation practices in 57 water demand areas of the MWD | 109,110 |
| LAST (Lane's Applied Stochastic Techniques) | • For operation of complex reservoir systems where stochastically generated sequences or traces of flow occurrences are simulated<br>• Applied for managing the CVP and was especially effective in providing objective estimates of nonexceedance probabilities for low flows during drought periods | 111 |
| Optimization model solved using MINOS | • Applied to the SWP and portions of the CVP<br>• Accounts for major hydrologic, regulatory, and operational features of the water conveyance systems<br>• Results indicated that yield from the SWP could potentially increase by 13 percent | 112 |
| DSPM (Distribution System Planning Model) | • Determined system conveyance and storage capacities<br>• Identified system bottlenecks and recommended system upgrades<br>• Evaluated the timing and amount of additional transfer water required to meet demands during droughts<br>• Estimated system reliability | 113 |
| IGSM (Integrated Ground-water Surface Water Model) | • Models surface water and groundwater components of the hydrologic cycle | 114 |
| Optimization model solved using MINOS | • Conjunctive use management model for Mad River Basin planning and operational strategies | 115 |
| Computer-aided Negotiation | • Analyze water supply problems and resolve disputes among customers in two portions of the CVP system | 116 |
| NEOSYS | • Simulates and optimizes five main SWP reservoirs on the California Aqueduct on a monthly basis | 117 |
| Position analysis model | • Assess risk of CVP reservoir storage falling below target levels | 118 |
| SANTCUM (San Joaquin-Tulare Conjunctive Use Model) | • Evaluated the long-term management of drainage and drainage related problems in the western San Joaquin Valley and uniquely coupled surface water deliveries with regional groundwater model | 119 |

## 3.2. CONVENTIONAL DROUGHT MANAGEMENT PRACTICES

A medley of conservation measures, denoted herein as conventional drought management practices, has been both encouraged and mandated in California in response to dwindling water supplies. Such measures should not be considered a panacea when implemented as short-term reactive solutions, but they can have lasting impacts when incorporated into development planning and achieve cultural acceptance. An audit of the

City of Los Angeles Department of Water and Power (DWP) concluded that existing and planned water supplies are inadequate to meet future City water needs while estimates of the benefits of conservation may be overly optimistic, and the City's planned reliance on such savings may not be warranted [120]. The state has grouped conservation methods into three categories of best management practices (BMPs) [78]: 1) reserved for conservation programs that provide definitive savings, 2) includes conservation measures that are worth implementing but water savings cannot be directly computed, and 3) more esoteric and embody conservation programs that might be worth pursuing. Conservation measures can be further categorized according to their targeted method of reducing water use [2]: 1) conservation of supplies by water suppliers, 2) demand management by water suppliers, and 3) demand management by water users. The remainder of this subsection is organized according to this second set of categories.

*3.2.1. Conservation of Supplies by Water Suppliers*

This category of water conservation is defined by actions intended to provide efficient use of supplies [2]. The specific methods outlined by the American Water Works Association (AWWA) for conserving water by suppliers are water loss reductions, pressure reduction, and resource management. **Table 5**, located at the end of this subsection, lists typical water savings corresponding to these methods.

*3.2.2. Demand Management by Water Suppliers*

The water conservation methods included in this category are intended to reduce end user demands through actions of the water suppliers. As discussed in Part 1 of this chapter, federal, state, and local laws and ordinances have been enacted to enforce conservation at the end-user level including the federal Energy Policy Act requiring new residential toilets to reduce water consumption by 70 percent and hold no more than 1.6 gallons (6.06 liters) of water [72,73,74,75], and the state Water Conservation in Landscaping Act requiring water conservative landscaping for new developments [78]. The three general areas recognized by the AWWA for reducing end-user demands through supplier actions are pricing, public education, and user restrictions [2]. An example for each of these areas is provided, in order, below, and summarized in Table 5.

During the recent drought, the City of Los Angeles enforced mandatory conservation by fining users who failed to reduce water use by at least 10 percent [120].

The Marin Municipal Water District (MMWD), in the San Francisco area, receives 80 percent of its water from an 80,000 acre-foot (100 mcm) reservoir. This relatively small storage facility is susceptible to draining during droughts; therefore, the MMWD has had substantial experience in developing and implementing rationing programs that are perceived as fair and equitable by the public [121].

Due to the sustained drought, several water utilities in California, namely, Palo Alto, Los Angeles DWP, San Jose, and a partnership of several agencies in Ventura County, formed programs to evaluate nonresidential water conservation [122]. The conservation techniques that were investigated included low flow plumbing equipment that matched residential devices and recycling of either cooling or process water. Such conservation measures were shown to be successful in both producing significant reductions in water usage and providing cost-effective retrofits.

*3.2.3. Demand Management by Water Users*

The water conservation methods comprising this category are targeted to reduce end-user demands through end-user actions [2,123]. The methods for reducing demand include changing water use practices, installing more efficient appliances and plumbing fixtures, and xeriscaping — conservation through the use of water-wise plants. The demand management methods employed by urban, industrial, and agricultural users are detailed, in order, below, and summarized in Table 5.

Urban water users currently consume approximately 15 percent of the developed water in California. This demand belies the need for conservation at the residential level because estimates indicate only 20 percent of residential water is used efficiently [124] and dramatic increases in the state's population are expected. The three largest users in the home, hence the largest potential water wasters, are toilets, showers, and landscape irrigation. Monetary savings due to reductions in combined water and sewer fees corresponding to use of high efficiency toilets provided a rate of return on investment of more than 70 percent for residential users and up to 200 percent for office buildings [125]. To assist residential consumers in conserving water in the home, the MWD funded a report that identified alternative flushing and retrofit devices for the toilet [126], while the Los Angeles DWP outlined a number of conservation measures that can be applied at home or in the workplace including xeriscaping, public information programs, and pricing [127]. The MWD and other water agencies have provided rebates to customers who have installed low flow flush toilets or operated toilet distribution programs with estimated replacement of more than 900,000 toilets.

California has pioneered the use of xeriscaping by requiring water-wise planting for all state buildings [128]. Xeriscaping includes the use of drip irrigation in lieu of spray irrigation, mulches to hold moisture, preservation of existing trees and indigenous plants, and automatic timing devices for controlling irrigation [129]. San Diego prepared a manual that outlines seven fundamental steps for implementing a xeriscape program and includes pictorial examples of water-saving landscapes with cited flora [130]. A listing of 92 drought-tolerant plants suitable for use in California has been compiled by Sunset Magazine [131] while an extensive listing of drought-tolerant trees, shrubs, and ground covers are compiled in a conservation handbook along with 40 methods, devices, and products to conserve landscape water [132]. These plants are native either to California or Mediterranean climates. In addition, Sunset identified ways to redevelop existing residential landscapes to be water-wise [1]. A study conducted by the North Marin Water District (NMWD) in California indicated that significant monetary savings related to reductions in water use and decreased gardening labor, about $75 per dwelling unit per year, could be realized by using water conserving landscapes [133].

In California, water demands of commercial services and manufacturing constitute approximately 12 and 18 percent of urban water usage, respectively [134]. Nationally, industrial conservation programs resulted in savings of 40 percent in water use between 1977 and 1982. In San Francisco, 15 independent industries significantly reduced water consumption from 25 to 90 percent [134]. The conservation techniques that were implemented ranged from employee education to major equipment retrofits.

Due to the wide disparity in the quantity of water used for agriculture, the state can expect to save considerably more water through reductions in agricultural water use, where a modest 20 percent reduction would nearly equal the total quantity of water consumed by all other use categories [135]. Water-saving methods for agricultural irrigation include sprinklers and gated piping systems, where correctly operated sprinkler systems provide better leaching of salts while generating less subsurface drain water than traditional surface irrigation. However, these methods involve higher labor and energy costs which may actually exceed the monetary value of the water savings [136].

#### *3.2.4. Summary of Conservation Measures*

Table 5 lists typical conservation methods and associated water savings percentages. It should be noted that this listing only includes conservation measures that have specifically referenced water savings. The reader is encouraged to consult other publications that provide more thorough lists of conservation measures [137,138,139,140,141].

### 3.3. PROACTIVE APPROACHES TO DROUGHT MANAGEMENT

Traditional water management has focused on water supply — developing water collection, storage, and conveyance systems to meet demands. However, the era of massive infrastructure construction is ending as water supplies are becoming scarce with the future direction being more efficient use of water through conservation and recycling, better management through water transfers and conjunctive use, and innovative supply development through desalination [142]. In striking contrast to its surfer image, California has been extremely aggressive in the application of both new technologies and forward-thinking methods to administer these technologies for the management of water resources. The scenery of the state is dotted with explorations of new ways to extend or augment water supplies to meet ever growing demands. In the following, these proactive approaches for managing water supplies, particularly during droughts, are catalogued.

#### *3.3.1. Technological Approaches*

As recently as five years ago, the technological approaches of water recycling and desalination were quixotic; however, improvements in technologies engendered by drought and water supply shortage exigencies have rendered them practical and cost-effective.

*Water Recycling.* Globally, communities are facing problems related to both limited water supplies and disposal of treated wastewater effluent. Water recycling can provide a majority of a city's nonpotable needs while reducing the quantity of effluent requiring

**Table 5. Summary Of Water Conservation Methods**

| Water Conservation Method | Water Savings Percent | Reference |
|---|---|---|
| **Conservation of Supplies by Water Suppliers:** | | |
| • Reducing Leaks | 15-30 | 2 |
| • Metering | 30,20 | 2,81 |
| • Pressure Reduction | 3 | 2 |
| **Demand Management by Water Suppliers:** | | |
| • Pricing | 22-40 | 2 |
| • User Restrictions | 22-25 | 100 |
| • Public Education | 22-25 | 100 |
| **Demand Management by Water Users:** | | |
| Residential | | |
| • Low Flow Flush Toilets | > 300,57-78 | 123,124,143 |
| • Fixing Leaking Plumbing | 10-50 | 124 |
| • Low Flow Shower Head | 25-75,7-11,17-70 | 123,124,143 |
| • Faucet Aerators | 3-5,27-70 | 123,124 |
| • Personal Habits | 200-2,000 | 2 |
| • Landscape Irrigation | 10,30-60 | 81,123 |
| Industrial | | |
| • General | 25-90 | 52,134,144 |
| • Water Reuse in Food Processing | 50-70 | 145 |
| Agricultural | | |
| • Irrigation | 40 | 123 |

disposal [83]. Water recycling is defined by two categories: 1) graywater reuse and 2) recycled treated wastewater. This subsection will primarily detail treated wastewater recycling, while graywater reuse will be briefly discussed. The interrelationship of water supply, graywater reuse, and domestic sewage recycling is illustrated in **Figure 7**.

Graywater refers to potable water that has been used within the home but is not tainted by any pathogenic contaminants. It is specifically defined in the 1994 Uniform Plumbing Code as, "Graywater is untreated household waste water which has not come in contact with toilet waste. Graywater includes used water from bathtubs, showers, bathroom wash basins, and water from clotheswashing [sic] machines and laundry tubs. It shall not include waste water from kitchen sinks or dishwashers." [79]. The City of Los Angeles sponsored the evaluation of graywater use in eight different graywater systems in 1992 [79]. The successful completion of this study ultimately led to the legalization of residential graywater use in California. In 1994, a bill to extend the use of graywater to multi-family dwellings and commercial buildings was vetoed because a clause in the bill would have allowed local areas to alter state graywater standards [80].

The remainder of this subsection is dedicated exclusively to water recycling. In our context, water recycling is the use of tertiary treated wastewater effluent for municipal reuse, construction water, groundwater recharge, agricultural irrigation, in-

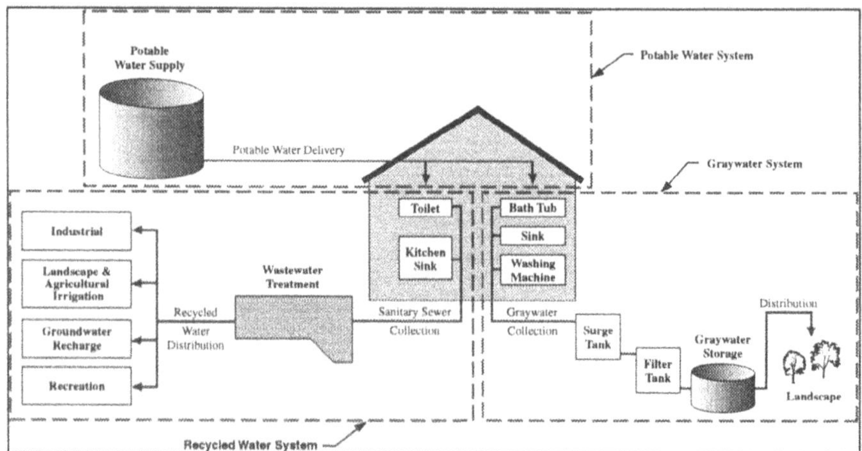


**Figure 7. Relationship Between Domestic Water Use and Water Recycling (Adapted From [79])**

dustrial reuse and recycling, and recreation [58,146,147,148]. Because no federal regulations governing water recycling exist [84], California has been a leader in this field with uniform state-wide criteria to regulate the use of recycled water for the protection of human health having been established under the guise of Title 22. For example, for groundwater recharge, the requirements are intended to control the migration of pathogens, nitrogen, and organics through the aquifer [82]. The regulatory and implementation framework established in California is considered a benchmark for other agencies in which water recycling provides both pollution control and water supply needs [149].

The use of recycled water in California increased 45 percent, from 0.175 to 0.267 maf (215 to 330 mcm) per year, between 1970 and 1987 [150]. In southern California, water recycling increased by more than 123 percent between 1987 and 1993, from 0.12 to 0.27 maf (150 to 330 mcm) per year [59]. Plans and projections for future water recycling across the state indicate an additional 2.5 to 3.0 maf (3,100 to 3,700 mcm) per year of treated wastewater is envisioned for recycling [59].

The County Sanitation Districts of Los Angeles County have participated in recycling wastewater since 1949 [146]. As early as 1976, public attitude surveys conducted in California indicated acceptance for low or non-body contact uses such as groundwater recharge, irrigation, and industrial applications; however, higher contact uses including human consumption met considerable resistance [151]. Pilot studies conducted in the state circa 1977 tested the effects of recycled water on soil, groundwater, and crops [152]. In 1989, urban user assessments of the merits of innovative wastewater treatment, recycling, and reuse were surveyed with alternatives employing high degrees of recycling and reuse receiving the most favorable evaluations [153].

Historically, the California SWRCB administered funding to local agencies for design and construction of water recycling facilities, with 19 in operation as of

1995 [150]. Under legislation adopted in California in 1977, the SWRCB promoted the use of recycled water as a water supply substitute through the Clean Water Grant Program; however, a change in the federal Clean Water Act in 1979 restricted public funding of recycling projects to the proviso of water pollution control and not water supply. These grant programs were eliminated starting in 1984 and replaced with the State Revolving Fund that provides low interest loans with loan repayments returned to the fund to finance new loans [150]. The severity of the drought in the entire western United States during the late 1980s and early 1990s coupled with restrictions on the construction of new water supplies prompted a renewed interest in the investigation of innovative conservation and water management approaches including recycling. To promote the reuse of treated wastewater, a federal bill was approved in 1991 that authorized a program to investigate and identify opportunities for recycling and reuse of domestic, industrial, and agricultural wastewaters [154]. Under this federal program, feasibility studies were conducted to determine the potential for recycling facilities, and California headed the list with studies in the southern counties of Imperial, Los Angeles, Orange, San Bernardino, Riverside, San Diego, and Ventura and the City of San Jose in the north.

High total dissolved solids (TDS) in recycled water is a typical problem in southern California due to generally high TDS in the original potable water [155]. Water with TDS greater than 1,000 mg/L will not be accepted by recycled water users. To lessen the impacts of TDS input from normal municipal water use, two water agencies in California banned regenerative water softeners because the operation of these units inputs brine into the sewage stream [85,156]. However, the Water Quality Association filed a lawsuit against the City of San Diego and the San Diego County Water Authority, pointing to previous laws that have guaranteed the right to own a water softener [156].

As a pioneer in the water recycling industry, California has a number of success stories regarding both the innovative technologies associated with, and forward-thinking management of, recycled water. Three case studies are presented below, accompanied by **Table 6**, a tabulation of a cross-section of recently planned or constructed recycling facilities or projects in California that have been published:

- **Irvine Ranch Water District (IRWD).** The IRWD located in Irvine, California (Orange County), has been practicing water recycling since 1967 [51]. By 1993, the District was recycling approximately 17,000 af (21 mcm) per year [157], nearly double the 9,000 af (11 mcm) per year recycled in 1986 [51]. To facilitate the safe use of recycled water, the IRWD operates dual distribution systems, one for potable water and the second for recycled water [50]. The recycled water is distributed to three end users: 1) landscape irrigation, 2) agricultural irrigation, and 3) sanitary plumbing in commercial high-rise buildings. The IRWD even provides recycled water to toilets in a neighborhood park's public restroom [51].

- **Orange County Water District (OCWD).** In the early 1970s, OCWD constructed Water Factory 21, a water recycling facility, that continues to provide recycled water

Table 6. Published California Water Recycling Projects

| Entity | Type of Recycle | Key Components | Reference |
|---|---|---|---|
| Contra Costa Water District & Central Contra Costa Sanitary District | • Industrial<br>• Landscape Irrigation | • 5,600 af (6.9 mcm) per year | 48,158 |
| Snow Valley Ski Resort | • Artificial Snow | • 80 af (0.10 mcm) | 159 |
| Metropolitan Water District & West Basin Municipal Water District | • Landscape Irrigation<br>• Industrial<br>• Seawater Intrusion Barrier | • 70,000 af (86 mcm) per year<br>• Eliminates disposal to ocean | 160,161 |
| East Bay Municipal Utility District | • Industrial | • 5,600 af (6.9 mcm) per year | 162 |
| City & County of San Francisco Department of Public Works | • Industrial<br>• Landscape and Agricultural Irrigation | • 16,000 af (20 mcm) per year<br>• Selected in-line filtration and ultraviolet disinfection | 163,164 |
| City of San Jose | • Landscape and Agricultural Irrigation | • 38,000 af (47 mcm ) per year<br>• Evaluation of positive and negative impacts of recycling water | 165 |
| Salinas Valley | • Seawater Intrusion Barrier | • 22,400 af (28 mcm) per year | 166 |
| City of Los Angeles | • Landscape Irrigation | • 3,000 af (4 mcm) per year | 25 |
| Capistrano Beach Water District | • Landscape Irrigation | • 500 af (0.6 mcm) per year | 167 |
| City of Escondido | • Landscape and Agricultural Irrigation<br>• Groundwater Recharge | • 27,000 af (33 mcm) per year | 168 |

to a series of injection wells that are used to create a seawater barrier [54]. This project consisted of 7 extraction and 23 multi-point injection wells [55]. Evaluation of the facility showed that recycled water costs less than imported water in southern California [169], and when appropriately priced, recycled water can provide a favorable cost/benefit ratio with respect to capital and operation and maintenance costs [170].

- **City of San Diego.** San Diego has embarked on an aggressive Clean Water Program to mitigate severe water supply and treatment problems due to rapid population growth, limited treatment system capacity, and continued drought [171]. The authors are intimately familiar with this work, being part of a Project Team assigned to manage the City's Clean Water Program. San Diego is especially at risk because it imports up to 90 percent of its water [35]. A groundwater management plan was being developed for the San Pasqual Valley with initial groundwater recharge of 8,000 af (10 mcm) per year [53] and plans to recycle 70,000 af (86 mcm) per year by 2010 [172]. The recharge project was subject to regulations from the Drinking Water Division of the California DHS and the San Diego RWQCB who stipulated that a two-dimensional model be utilized to analyze water quality and hydraulic impacts on the groundwater basin. Through the Total Resource Recovery Project, the City of San Diego and the San Diego County Water Authority proposed to recycle water through the San Vicente Reservoir, a primary raw water supply source [35,57]. A health effects study was performed to estimate the health risks associated with the introduction of recycled water into the reservoir [173,174]. A tracer study of the reservoir was used to determine the repurified water retention time in the reservoir and compare this retention time to the requirement established by the DHS [57,35]. Related to this work, the application of membrane systems for production of highly treated reclaimed water suitable for introduction into a domestic water supply reservoir was examined [56].

*Desalination.* Due to the energy intensive nature of desalination technology, desalted water can cost nearly an order of magnitude greater than developed water. For example, for the City of Santa Barbara, the cost of desalted water was estimated as approximately $1,900 per acre-foot ($1.54 per m$^3$) [175,176]; whereas, production costs for the City of Oceanside Water Utilities Department were $387 per acre-foot ($0.31 per m$^3$) [175]. These costs can be contrasted with those of developed water which are on the order of $350 to $500 per acre-foot ($0.28 to $0.41 per m$^3$) in southern California [176]. Although desalination previously had been considered too expensive, drought conditions coupled with prevailing sorely over-taxed groundwater and surface water supplies have provided California with the impetus to aggressively investigate or consider desalination to meet or supplement their potable water demands [46,176,177,178]. As of 1991, four California communities were either operating or constructing desalting facilities [45]. Federal legislation enacted in 1991, where $1 million in federal funds were allocated to desalination research and development [179], should further the investigation and use of desalination on a large-scale basis. With increased emphasis on improving the efficiencies and economies of desalination, this technology can cost less than the allocated capital cost associated with the construction required to deliver surface water, as in the case for Santa Barbara [47]. Currently, the largest producer of desalinated water in the world is Saudi Arabia, accounting for 30 percent of total world output.

As a leader in the domestic use of desalination in the United States, California is at the cusp of a pivotal period where dependence on naturally occurring fresh water may be greatly limited. Two case studies are presented below, along with **Table 7**, a tabulation of desalination facilities or projects in California that have been planned or constructed recently and have a published reference:

- **Metropolitan Water District.** Although desalination initially was considered too costly, the MWD has approved funding for demonstration plants [180,181,182,183] and has teamed with the San Diego County Water Authority and the Los Angeles DWP to assess the feasibility of a desalination plant near the Mexican border [184].

- **San Diego County Water Authority.** The San Diego County Water Authority partnered with the San Diego Gas & Electric Company to investigate the development of

**Table 7. Published California Desalination Projects**

| Entity | Key Components | Capacity | Reference |
|---|---|---|---|
| Monterey Peninsula Water Management District | Desalination alone or with new surface water reservoirs | • 3,400 af (4.2 mcm) per year<br>• 7,800 af (9.6 mcm) per year | 185 |
| City of Santa Barbara | Bolster water supply and reduce dependence on local surface water | • 5,000 af (6 mcm) per year | 186 |
| Marin County | Feasibility Report: determined siting, processes, off-site facilities, and costs | • 5,600 af (6.9 mcm) per year<br>• 11,200 af (13.8 mcm) per year | 187 |
| City of Oceanside | Operational reverse osmosis plant | • 2,200 af (2.7 mcm ) per year | 175 |
| Sweetwater Authority | Evaluated disposal or use of desalted residual from desalination facilities | • 16,000 af (20 mcm) per year | 188 |
| City of Santa Barbara and<br>City of Morro Bay | Recent legal developments:<br>• California Environmental Quality Act<br>• Public Contract Code<br>• California Coastal Act<br>• SWRCB<br>• RWQCB<br>• State Lands Commission | • 5,000 af (6 mcm) per year<br>• 7,000 af (1 mcm) per year | 176,189 |
| City of Ventura | Voted for desalination facility rather than hook-up to SWP | • 7,000 af (1 mcm) per year | 190 |

a dual purpose facility for generating electrical power and desalted water [191]. The use of an integrated pumping and energy recovery method for desalting was shown to be less energy intensive than importing SWP water [192,193].

Dual purpose or co-located facilities provide a collateral benefit inherent to the synergistic coupling of power generation and desalination where the desalting process is more efficient when applied to higher temperature waters and power generating facilities require substantial quantities of process water. California has once again stepped to the forefront of innovative implementation of technology transfer and has begun investigating the retrofit of existing power facilities along the Pacific Coast to direct the elevated-temperature process water to desalination facilities [194].

### *3.3.2. Administrative Approaches*

The administrative approaches invoked in California were innovative programs designed to mitigate water supply shortages during the 1987 to 1992 drought and included: 1) best management practices, 2) water transfers, 3) conjunctive use, and 4) system modeling methods. The utilization of these approaches has persisted beyond the cessation of the drought due to their efficacy in providing forward-thinking management of California's water. Of these, the first three are presented herein while the fourth was discussed previously in this subsection. As testimony to the success of these approaches, a drought management plan for urban water supply incorporating major elements of California's drought management approaches was prepared for the U.S. Army Corps of Engineers [140] and applied to a midwestern United States city [141].

*Best Management Practices.* This approach forms the framework used by water agencies to develop alternatives, formulate evaluations, and guide decisions to administer water resources. To foster better management of water, the state enacted the Urban Water Management Planning Act, where all large water suppliers are required to prepare and update a water plan that outlines efforts to efficiently use and develop alternative water supplies [81]. This Act requires drought contingency plans where agencies state methods to mitigate water shortages of up to 50 percent with most plans including water rationing to conserve supplies in extreme droughts. Other BMP examples are given below.

Drought mitigation initiatives employed by the California DWR, as summarized by the U.S. Natural Resource Conservation Service in 1995, include [195]:

- Conducted workshops on drought contingency planning.
- Enacted new legislation to facilitate water recycling.
- Strengthened the Urban Water Management Planning Act and passed legislation requiring water agencies to prepare drought contingency plans.
- Established a State Water Bank to remove legal and institutional barriers to water transfers while still protecting water rights and environment.
- Initiated groundwater management plans.
- Developed and marketed innovative technologies for agricultural and urban use and studied the design and effectiveness of urban and agricultural conservation programs.
- Formed a Drought Information Center.
- Evaluated statewide rationing but supported water conservation programs.
- Developed and distributed computer programs to water user categories: AgWater for farmers to improve irrigation practices [196], CIMIS for citrus growers to plan irrigation schedules [197], and WaterPlan for urban water retailers to set price.
- Established statewide standards for safe use of residential graywater.
- Contributed prominently in establishment of the 1992 National Energy Policy Act.
- Promoted public input in designing and implementing conservation programs.

The CVP enacted the following progression of reactive water management approaches during the drought in an attempt to stretch dwindling supplies [198]: 1) years one and two, reservoir storage was used to meet demands with no reductions in deliveries; 2) year three, some types of deliveries were eliminated or restricted; 3) year four, up to 50 percent reductions occurred; 4) year five, up to 75 percent reductions occurred. A hardship allocation process was instituted in years four and five to permit water users to appeal for additional deliveries. Even with the enactment of these extensive management measures to conserve supply, it was serendipitous that the drought ended in the sixth year because the CVP was relegated to the last of its reservoir storage. Due to the severity of the drought and the nearly catastrophic water shortages that resulted, the CVP has developed new plans to proactively manage their supplies and deliveries as follows [198]: 1) perform runoff forecasts to predict aberrant behavior such as droughts versus simple time series projections and 2) make conservative annual water allocations in February that will not require reduction regardless of ensuing hydrologic conditions.

In 1993, the MWD launched an unprecedented effort to develop an integrated water resources plan through collaboration with dozens of retail water providers [199]. This coordinated effort confirmed the benefits of a multi-faceted program that optimizes the combination of imported water supplies coupled with conservation and development of other water sources including groundwater, recycled water, and desalination. Specifically, the MWD initiated the following series of innovative water management programs [23,24,43,200]: 1) a seasonal storage program where substantial price reductions were afforded to water purchases in the winter during lower demand and greater flows; 2) an irrigation efficiency conservation program for the Imperial Irrigation District (IID) where a portion of the water savings would be provided to the MWD, estimated at 0.106 maf (130 mcm) per year; 3) a joint program with the Bureau of Reclamation and the IID to line the All-American Canal and portions of the Coachella Canal where the MWD would cover the costs of lining and receive an estimated 0.100 maf (123 mcm) per year; 4) a groundwater storage program in conjunction with the Arvin-Edison Water Storage District that would direct unneeded entitlements to the Storage District for storage and receive 0.100 maf (123 mcm) per year from the Storage District's surface waters; and 5) a farmland fallowing program with the Palo Verde Irrigation District with 0.200 maf (250 mcm) per year of water assigned to the MWD, where the authors assisted the MWD.

In 1994, the MWD and U.S. Bureau of Reclamation formed a partnership to encourage and fund innovative urban water conservation projects with the following recommended programs [201]: 1) system leak detection and repair pilot program (losses for MWD water agencies ranged from 5 to 15 percent), 2) low flow flush toilets, 3) moisture sensors for irrigation control, and 4) a school showerhead distribution program.

The OCWD has developed a water resources management program that involves innovative control strategies, proven management policies, and current technology to form a groundwater management model that can be used as a benchmark for other water agencies. The approaches that have been employed include [202]: 1) control of extractions from the groundwater, 2) development of a reliable revenue base to fund operations and projects, 3) evaluation of new local water supplies to reduce reliance on the SWP, and 4) formulation of a groundwater quality control plan. The principle program method identified to augment water supplies was recycled water for groundwater recharge, seawater intrusion barriers, and landscape irrigation.

Residential water audit programs conducted by the Contra Costa Water District in California during the recent drought were estimated to yield aggregate water savings of about 6 percent, matching similar audit programs in the NMWD [203] and the cities of Concord, Pasadena, and Novato [204], all of California. These water audits represented moderately intensive conservation programs where water purveyors provided complimentary examinations of both internal and external water use and made recommendations to improve the water use efficiency.

*Water Transfers.* In California, water transfers, conducted in a water market that resembles a standard commodity market [205], involve the selling of water without loss of water rights. Historically in the United States, water rights have been considered to be usufructuary rather than possessory, i.e., a right to use the water as opposed to ownership of the water [206]. This stance has been relaxed by recent court cases that have begun to recognize water rights as property rights, thus allowing sale or transfer of the water. The passage of the federal CVPIA legalized and stipulated conditions for water transfers, thereby reversing precedent that prohibited water transfers [23,206]. The state is drafting its own legislation through the Model Water Transfer Act [23,81,205] with proposed standard conditions governing water transfers being [23]: 1) water transfers must be voluntary, 2) water rights cannot be transferred, 3) transfers are subject to review by water rights holders, 4) transfers are limited to quantity historically consumed or stored by transferor, 5) transfers through the Delta shall allow for carriage losses, 6) transferred water shall remain subject to relevant provisions of transferor's water supply contract, 7) water transferees must have a water conservation program, 8) transfers shall not harm any legal user of water or the environment, 9) no transfers from areas with groundwater overdraft unless consistent with approved groundwater management plan, 10) no groundwater shall be transferred, 11) other water users in a given transferor's jurisdiction have first right of refusal, and 12) public review and comment are required.

Water transfers can be implemented in five basic forms [207]: 1) permanent transfers; 2) contingent transfers; 3) spot market transfers; 4) water banks; and 5) recycled, conserved, and surplus water transfers. The emphasis in their implementation is integration of water transfers with traditional supply expansion and water conservation measures. The actual method of accomplishing water transfers and exchanges need not actually involve water and can take any of the following forms [200]: 1) interagency storage, 2) Kern Water Bank, 3) Sacramento Regional Cooperation, 4) pumped storage, 5) transfers of conserved urban water, 6) payments for water conservation, 7) payments for alternative water supplies, and 8) water wheeling. Four sources of water transfers have been cited for agriculture water [208]: 1) fallowing or retirement of cultivated land, 2) conservation of excess irrigation water, 3) reducing evapotranspiration of plants, and 4) development of agricultural groundwater basins.

A Drought Emergency Water Bank, operated by the DWR in 1991 and 1992, completed major water transfers by purchasing and distributing 0.820 and 0.190 maf (1,000 and 230 mcm) of developed water, respectively, in California [23,200]. The State Water Bank was the first large-scale water transfer program in the United States with a state serving as the predominant water broker [200]. Government sponsorship and centralized control is credited with the success of this widely lauded program [200,207]. An economic analysis of the 1991 Water Bank provided evidence that agricultural supply and urban demands were more elastic than presumed, and the water transfers generated direct benefits for the state by creating a net gain in income and employment by trading water from lower to higher value uses [209].

Water Bank success has been refuted where the use of groundwater and the finite quantities of actual water transfer have been criticized [210]. One suggested improvement in the Water Bank operations would be to restructure the financial transactions to reflect market-determined prices [211].

In response to a critically dry year in 1994, the DWR planned the 1995 California Drought Water Bank Program by purchasing water supply options from willing sellers and signing tentative agreements with purchasers [9]. If drought conditions had prevailed in 1995, the Water Bank would have been raised to active status, and the DWR would have exercised its options and begun physical transfers. Eventually, drought conditions did not materialize and the Water Bank options were canceled; however, this experience indicated that the state could quickly organize and administer water transfers in response to potentially disastrous water supply shortages.

In addition to the State Water Bank, several other independent water transfers were operated in the state [200]: 1) MWD agreements with the Imperial Irrigation District, Coachella Valley Water District, Desert Water Agency, and Arvin-Edison Water Storage District; 2) transfers within the CVP; 3) transfers within East Bay Municipal Utility District; 4) transfers in Solano County; 5) transfers involving the San Francisco Water Department; and 6) Yuba County Water Agency sales.

*Conjunctive Use.* "Conjunctive use" refers to the coupled beneficial use of surface water and groundwater, where the source of the water used for supply depends on climate conditions. For wet conditions, developed or surface water that may be in excess is used for both consumption and groundwater recharge rather than allowed to form runoff. During dry conditions, with possible shortages in surface water, the stored groundwater is pumped for supply. Several southern California water agencies have practiced conjunctive use since the 1970s. The largest conjunctive use project in California is the Kern Water Bank, which is designed to increase groundwater storage up to 1.0 maf (1,230 mcm) per year [24,42,200]. State-wide, the potential for groundwater storage through conjunctive use is estimated to range from 2 to 4 maf (2,500 to 5,000 mcm) per year [24]. An example of a potential conjunctive use project is an intensive water resources management study that was performed to assess methods to increase water supplies in five Central Valley counties where water demand models predicted yearly water shortages of 0.521 maf (650 mcm) [212]. The study reviewed two options: 1) conjunctive use with diversions of storm water flow for recharge during wet years and groundwater pumping during dry years, or 2) expansion of existing surface water storage. The conjunctive use approach was critiqued more favorably.

## 4. Conclusion

Although California has relied on developed water to satisfy regional needs, this basic approach cannot be a long-term solution [135] because the function of the state's various water systems are vulnerable to three factors [213]: 1) systems are neither hydrologically nor administratively unified; 2) major facilities and agencies are multi-purpose, for instance, being responsible for both water supply and flood control; and 3) design and operation of the water systems are politically sensitive. These conclusions are supported by the data discussed in this chapter and presented in **Figure 8** that demonstrate that California uses on the order of 35 percent of the available, renewable supply [2,29].

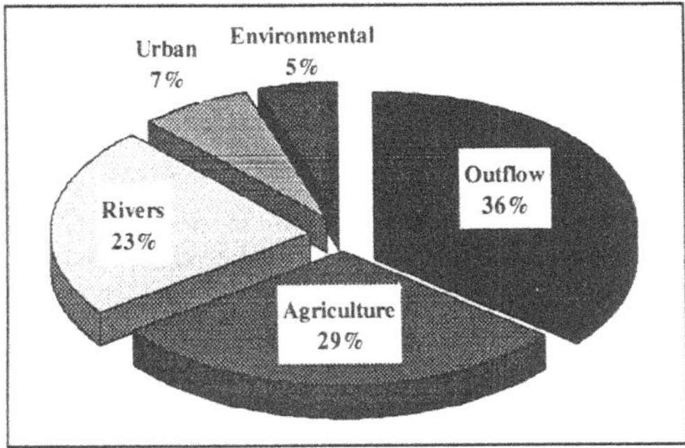


**Figure 8. California's Annual Total Runoff Distribution**

One of the key elements in implementing programs for managing water resources is effective communication where successful water conservation campaigns provide clear message content and delivery [214]. For San Francisco, the degree to which water users curtailed their consumption was purported to be influenced more by the extent to which they believed a shortage existed than by any other factor [215]. A Quick Response project, which studied the impacts of news media and public information on water conservation, showed that the most highly publicized conservation programs achieved the greatest water use reductions [216]. This assessment also was upheld by experiences in Los Angeles and San Diego where reductions of 22 to 25 percent in water use were achieved in 1991 through contrasting drought response measures [100]: Los Angeles ordered mandatory rationing, while San Diego encouraged voluntary conservation through cooperation, publicity, and timely information dissemination. Support for water recycling continues to require public education regarding the advantages and inherent safety of water recycling techniques. The current success of water recycling projects will be sustained if the public perceives recycling as desirable and not adversely affecting public health or the environment [217].

The most significant contribution of conservation programs and innovative approaches to drought management used by California has been the paradigm shift away from water source development toward water supply and demand management with increased focus on optimizing the efficiency of use. In meeting future water supplies, dams are not the answer; instead, the real cost of transporting water should be charged and more efficient use of water should be enforced [218]. The innovative methods employed by California during the latest drought revealed that a water shortage crisis could be solved only by better management of water supplies through the implementation of forward-thinking approaches and application of state-of-the-art technologies.

**Acknowledgments**

The authors would like to express our most gracious thanks to Mary Apostol for her tireless efforts in transforming our engineering prattle into eloquent verbiage, Mike Surace for his personal dedication to quality and accuracy, and the PAS-1 library for the enormous assistance as evidenced by the following reference list.

## References

1. Anon (June 1990) Time to ask serious questions about lawns and water, *Sunset*, 175-180.
2. Beecher, J.A. and Laubach, A.P. (1989) Compendium on Water Supply, Drought, and Conservation, *National Regulatory Research Institute Report, NRRI 89-15*, pp. 388.
3. Anon (1994) Warring Factions Coalesce, *ENR* 233, 21.
4. Anon (1997) Serious Water Shortage Could Hit California by 2020, *Christian Science Monitor*, 89, 13.
5. Quinn, N.W. (1991) Environmentally Sound Irrigated Agriculture in the Arid West: New Challenges for Water Resources Planners and Environmental Scientists, *Proc Conf. Envir. Sound Ag.*, 9 pages.
6. Oldham, J. (September 19 1997) California: Agricultural Giant, *LA Times* Part D, p. 2.
7. Anon (July 1996) CIA World Fact Book.
8. Herzog, S.J. (1996) California Water: The New Gold, *Appraisal Journal* 64, 134-147.
9. Jercich, S.A. (1997) California's 1995 Water Bank Program: Purchasing Water Supply Options, *J. Water Resources Planning and Management* 123, 59-68.
10. Rosenbaum, D.B. (1993) California Faces Growing Conflicts: Environmental Restrictions Take Toll on Already Overly Taxed Supplies, *ENR* 231, 26-28,30-31.
11. Anon (1995) It's British Columbia's Water, *The Economist* 335, 49.
12. Anon (1993) Calif. Project Shut Down for Fish, *U.S. Water News* 9, 4.
13. Anon (April 1994) Federal Foray Into State Water Rights?, *State Legislatures* 20, 6.
14. Anon (1993) Calif. Governor Says Back Off: Feds Force Issue of Delta Standards, *U.S. W. Nws* 10, 11.
15. Postel, S.L., et al. (1996) Human Appropriation of Renewable Fresh Water, *Science* 271, 785-788.
16. Kaufman, R.F. (1995) The Anatomy of the Great California Water Debate, *J. of the West* 34, 77-87.
17. Margolick, D. (1994) As Drought Looms, Farmers in California Blame Politics, *NY Times* 143, A1.
18. Wood, D.B. (1993) Drought Leaves a Legacy in Golden State: Farmers Will No Longer Get Special Treatment in Allocation of California's Water Supplies, *Christian Science Monitor* 85, 8.
19. Anon (1993) With Drought Officially Over, Calif. Office Shifts Focus to Water-saving, *U.S. Water News* 9, 8.
20. Reinhold, R. (Feb. 25, 1993) Drought Ends, Having Altered Political Landscape, *NY Times* 142, A14.
21. State of California (1978) *The California Water Atlas*, General Services Publications Section, pp. 117.
22. Wallace, A. (1994) Water Recycling and the Urban Vs. Farm Battle for Water: The California Example, *Communications in Soil Science and Plant Analysis* 25, 119-124.
23. Baumli, G.R. (1993) Without Water Transfers, Cities Will Thirst, *Proc. Manage. Irrig. Drain. Syst. Integr. Perspect. - ASCE*, 85-92.
24. Baumli, G.R. (1990) Innovative Ways of Meeting California's Water Needs, *Proc. 1990 Natl. Irrig. Drain. Conf.*, 411-418.
25. Anon (1997) *Water Supply Fact Sheet, Los Angeles*, www.ladwp.com/water/supply/facts/index.htm.
26. Anon (1993) Future of CVP is Pondered, *U.S. Water News* 9, 12.
27. McCoy, C. (Dec, 16, 1993) U.S. Proposals Would Force California to Redirect Water Resources to Wildlife, *Wall Street Journal* 222, B8.
28. Wood, D.B. (1994) Clock Starts Ticking on California's Water Wars, *Christ. Sci. Monitor* 86, 12-13.
29. Water Education Foundation (1997) *Layperson's Guide to California Water*, pp. 24.

30. Kahrl, W. (1993) Acquisitions and Aqueducts, How California's Water System Evolved, *Pacific Discovery* 46, 21-25.
31. Mitchell, M.D. (1996) The Sacramento-San Joaquin Delta, California: Initial Transformation Into a Water Supply and Conveyance Node, 1900-1955, *J. of the West* 35, 44-53.
32. Jackson, M. (1990) Central Valley Project Operations, *Proc. Int. Symp. Hydraul. Hydrol. Arid Land Natl. Conf.*, 188-193.
33. McCormick, Z.L. (1994) Interstate Water Allocation Compacts in the Western United States - Some Suggestions, *Water Resources Bulletin* 30, 385-395.
34. Anon (1993) Colorado R. Users Seek to Divvy Up Flow, *U.S. Water News* 9, 4.
35. Gagliardo, P. (1997) Southern California Water History and Policy, *Proceedings of the 24th Annual Water Resources Planning and Management Conference* 110-113.
36. Egan, T. (June 9, 1996) Great Dam: Life Saver, Or a Big Boondoggle? Proposal Is Criticized as Hurting Nature, *New York Times* 145, 22.
37. Kondolf, G.M. and Vorster, P. (1989) Changing Water Balance Over Time in Rush Creek, Eastern California, 1860-1992, *Biol. Bull. Mar. Biol. Lab. Woods Hole* 176, 823-832.
38. Anon (1994) Want Some More? California's Water, *The Economist* 333, A30-31.
39. Roos-Collins, R. (1993) The Mono Lake Cases, *Rivers* 4, 328-336.
40. Stine, S. (1991) Geomorphic, Geographic, and Hydrographic Basis for Resolving the Mono Lake Controversy, *Environ. Geol. Water Sci.* 17, 67-83.
41. Smith, Z.A. (1989) *Groundwater in the West*, Academic Press, Troy, Mo., pp. 308.
42. Easton, J.L. (1989) Water Marketing in California, *Proc. Legal, Institutional, Financial, and Env. Aspects of Water Issues - ASCE*, 16-23.
43. Brown, J.W. and Schild, N.W. (1990) Government Cooperation in Water Resources Planning and Management, *Proc. Optimizing Resources for Wat. Man., 17th Annual Nat. Conf. - ASCE*, 698-702.
44. Anon (1991) Wells Could Tap Seepage from Calif. Canal, *U.S. Water News* 7, 3.
45. Wood, D.B. (1991) California Desalination: Thirsty State Looks to Sea for Water, *Chr. Sc. Mon.* 83, 3.
46. Anon (1991) California Is Set to Spend Millions on Pilot Desalination Plants, *U.S. Water News* 7, 4.
47. Fernandez, L.M. (1991) Environmental and Economic Feasibility of Water Supply Alternatives in Hawaii and California. Desalination, Water Marketing and Conservation, *Proc. Symposium on Coastal and Ocean Management - ASCE* 3, 2618-2632.
48. Requa, D.A., et al. (1991) Successful Implementation of an Urban Landscaping Recycled Water Program, *Proc. Water Supply and Water Reuse - AWRA*, 279-290.
49. Fahrer, N.D. and Massey, S.W. (1984) Sewer Plant Effluent - A Usable Source of Cooling Tower Makeup Water for Power Plant Operations, *Proc. Intl. Water Conf.*, 349-359.
50. Parsons, J. (1990) Irvine Ranch's Approach to Water Reclamation, *Wat. Env. & Tech.* 2, 68-71.
51. Young, R. (October 1988) Planning for Water Reclamation in Southern California, *Proc. Symposium III, Association of Water Reclamation Agencies*, 11-18.
52. Trussell, R.R., et al. (1980) Evaluation of Industrial Cooling Systems Using Reclaimed Municipal Wastewater, Application for Potential Users, *California SWRCB Report*, pp. 205.
53. Arakaki, G. and Otte, G. (1997) Conjunctive Use With Reclaimed Water: Modeling Water Quantity and Quality, *Proc. 24th Annual Water Res. Plan. and Man. Con. - ASCE*, 134-139.
54. Mills, W.R. (1993) Groundwater Recharge Success, *Water Environment & Tech.* 5, 40-44.
55. Rigby, M.G. and Mills, W.R. (1991) Direct Injection of Recycled Water Into Potable Aquifers, *Proc. AWWA Annual Conference*, 251-265.
56. Martella, S. (1997) Water Reuse Research Needs Assessment, *Sum. of Bureau of Reclam. Sem.*, pp. 14.
57. Williams, R.T. (1996) Tracer Study of San Vicente Reservoir, *Proc. 1995 Water Quality Technology Conference - AWWA*, 921-933.
58. Anon (1975) Inventory of Waste Water Production and Waste Water Reclamation in California 1973, *California Dept. of Water Resources Bulletin*, 68-73, pp. 32.
59. Sheikh, B. (1994) Water Reclamation Potential in Southern California to the Year 2020, *Proc. 21st Annual Conf. On Water Policy Man. Solving Prob. - ASCE*, 661-664.
60. Anon (1996) Novel Use of Clean-Water Loans Brightens Outlook for a River, *NY Times* 145, A25.

61. Mitchell, M.D. (1994) Land and Water Policies in the Sacramento-San Joaquin Delta, *The Geographical Review* 84, 411-423.
62. Anon (1993) Legislation Gives Water Rights to California Ecosystem, *Water Env. Tech.* 5, 27-28.
63. MacDonnell, L.J. (1991) Water Rights for Wetlands Protection, *Rivers* 2, 277-284.
64. Tarlock, A.D. (1991) Global Climate Change and Instream Values: Why Growth Control Must Be Placed on the Western Flow Protection Agenda, *Rivers* 2, 185-189.
65. Grunwald, L. (1994) Bay-Delta Management Plan Signed by EPA Administrator, *W. Env. Tech.* 6, 35.
66. Anon (1995) Agencies Announce Environmental Protection Plan, *Water Env. Tech.* 7, 21.
67. Anon (1993) San Joaquin Delta Plan Disappoints Both Sides of Issues, *U.S. Water News* 9, 18.
68. Anon (1993) EPA Calls Calif. Bay-Delta Plan Too Weak, *U.S. Water News* 9, 1.
69. Patterson, R.K. (1994) The Impact of the Central Valley Project Improvement Act, Responses to Changing Multiple-use Demands: New Directions for Water Resource Planning and Management, *AWWA Report*, pp. 13-15.
70. Anon (1992) Initial Details Being Fleshed Out for California Valley Project Transfer, *Wat. News* 9, 5.
71. Anon (Oct. 21, 1996) Statement on Signing the Water Resources Development Act of 1996, *Weekly Compilation of Presidential Documents* 32, 2062-2063.
72. Sullivan, R.L. (1994) The Scoop on Poop, *Forbes* 154, 20.
73. Kiernan, V. (1994) Water-wise Toilets, *Technology Review* 97, 15-16.
74. Murdoch, G. (1994) Outhouse Blues, *Consumer's Research Magazine* 77, 2.
75. Blatterman, J.F. (1996) Plumbing Designs Meet New Rules, *Architectural Record* 184, 58.
76. Staples, M.A. (1992) How to Promote Water Reclamation and Reuse Through the California Water Rights System, *Desalination* 88, 189-199.
77. Huntley, E.F. and Winkler, E.D. (1993) California Water Issues. Bay-Delta Hearings: Impacts on State Water Project, *Proc. Nat. Conf. On Hydraul. Eng.*, 1-7.
78. Minton, J. (December 1990) Agreement Reached on Urban BMP's, *Water Con. News*, 1-4.
79. Pope, T. (1993) Greywater: A Recyclable Resource, *Water Cond. & Purif.* 35, 66,68,70,72.
80. Myers, C. (1996) Water Recycling in California, *Worldwater Environ. Eng.* 19, 14-15.
81. Anon (1994, 1996) *Memorandum of Understanding for Urban Water Management Conservation* .
82. Hultquist, R.H. et al. (1991) Proposed California Regulations for Groundwater Recharge with Reclaimed Municipal Wastewater, *Proc. 91 Spec. Conf. Environ. Eng. - ASCE*, 759-764.
83. Okun, D.A. (1990) Realizing the Benefits of Water Reuse in Developing Countries, *Water Environment & Technology* 2, 78-82.
84. Crook, J. and Surampalli, R.Y. (1996) Water Reclamation and Reuse Criteria in the U.S., *Proc. Water Reclamation and Reuse 1995* 33, 451-462.
85. Anon (1990) To Reduce Salt Load in Sewers Calif. District Invokes Ban on Regenerative Water Softeners, *U.S. Water News* 6, 18.
86. Hinrichsen, D. (1996) The World's Water Woes: Pollution, Population and Waste Are Creating Dangerous Water Scarcities Throughout the Globe, *International Wildlife* 26, 22-27.
87. Anon (1994) The Big Thirst, Commonwealth 121, 3-5.
88. Luoma, J.R. (1996) The Drying of the Rivers, *Audubon* 98, 26-28.
89. Postel, S. (1995) Earth's Rivers Are Running Dry, *USA Today (Magazine)* 124, 74-76.
90. Long, D.R. (1995) Pipe Dreams: Hetch Hetchy, the Urban West, and the Hydraulic Society Revisited, *J. of the West* 34, 19-31.
91. Williams, T. (1994) Death in a Black Desert: in California's Fields, Toxic Runoff Is Poisoning the Land and Killing Birds by the Thousands, *Audubon* 96, 24-29.
92. Daniel, J. (1994) The Limits of Paradise, *Sierra* 79, 64-74.
93. Kattelmann, R.C. and Dozier, J. (1991) Environmental Hydrology of the Sierra Nevada, California, *First USA/USSR Joint Conference on Environmental Hydrology and Hydr ogeology*, pp. 49-56.
94. Stuller, J. (1994) The Fragile Mix, *Sea Frontiers* 40, 28-32.
95. Anon (1994) California Water Plan Update Reveals Shortages, Solutions, *Citrograph* 80, 12.
96. Krannich, R.S., et al. (1995) Social Implications of Severe Sustained Drought: Case Studies in California and Colorado, *Water Resources Bulletin*, 31, 851-865.

97. Little, K.W. and Moreau, D.H. (1991) Estimating the Effects of Conservation on Demand During Droughts, *J. AWWA - Management and Operations* 83, 48-54.
98. Behling, P.J. and Bartilucci, N.J. (1992) Potential Impact of Water-Efficient Plumbing Fixtures on Office Water Consumption, *J AWWA - Management and Operations* 84, 74-78.
99. Macy, P.P. (1991) Integrating Conservation and Water Master Planning, *J. AWWA - Man. and Op.* 83, 44-47.
100. Shaw, D.T. et al. (1992) Urban Drought Response in Southern California, *J. AWWA - Management and Operations* 84, 34-41.
101. Shaw, D.T. and Maidment, D.R. (1988) Effects of Conservation on Daily Water Use, *J. AWWA - Management and Operations* 80, 71-77.
102. Madduas, W.O., et al. (1996) Integrating Conservation Into Water Supply Planning, *J. AWWA* 88, 57-67.
103. Vickers, A. and Markus, E.J. (1992) Creating Economic Incentives for Conservation, *J. AWWA - Management and Operations* 84, 42-45.
104. Smith, J.A. (1988) A Model of Daily Municipal Water Use for Short-Term Forecasting, *Water Resources Research* 24, 201-206.
105. Miaou, S. (1990) A Class of Time Series Urban Water Demand Models With Nonlinear Climatic Effects, *Water Resources Research* 26, 169-178.
106. Maidment, D.R., and Miaou, S. (1985) Transfer Function Models of Daily Urban Water Use, *Water Resources Research* 21, 425-432.
107. Domokos, M. et al. (1976) Problems in Forecasting W ater Requirements, *Wat. Res. Bul.* 12, 263-275.
108. Howe, C.W. and Linaweaver, F.P. (1967) The Impact of Price on Residential Water Demand and Its Relation to System Design and Price Structure, *Water Resources Research* 3, 13-32.
109. Dziegielewski, B. and Boland, J.J. (1989) Forecasting Urban Water Use: The IWR-MAIN Model, *Water Resources Bulletin* 25, 101-109.
110. Madduas, W.O., et al. (1991) Water Savings from Water Conservation Best Management Practices in Southern California, *Proc. Of 1991 AWWA Annual Conference*, 755-762.
111. Frevert, D.K. and Lane, W.L. (1989) Use of Stochastic Hydrology in Reservoir Operation, *J. of Irrigation and Drainage Engineering* 115, 334-343.
112. Lefkoff, L.J. and Kendall, D.R. (1996) Optimization Modeling of a New Facility for the California State Water Project, *Water Resources Bulletin* 32, 451-463.
113. Anon (1993) Optimal Utilization of a Multi-reservoir Water Distribution System, *Proc. of the 20th Anniversary Conference on Water Management in the 90s - ASCE*, 205-211.
114. Futter, M. (1995) Integrated Water Resources Management, *World Water Env. Eng.* 18, 24.
115. Matsukawa, J. et al. (1992) Conjunctive-use Planning in Mad River Basin, California, *J. Water Resources Planning and Management* 118, 115-132.
116. Randall, D. et al. (1988) Computer-aided Negotiation of Water Resources Issues in California's Central Valley Project, *Proc. 3rd Water Res. Oper. Manage. Workshop - ASCE*, 88-96.
117. Coe, J.Q. and Rankin, A.W. (1988) California's Adaptable Model for Operations Planning for the State Water Project, *Proc. 3rd Water Res. Oper. Manage. Workshop - ASCE*, 119-130.
118. Randall, D. and Sheer, D.P. (1990) Old Ideas, New Applications. Expanding the State of the Practice, *Proc. 17th Ann. Natl. Conf. Optim. Resour. Water Management - ASCE*, 413-417
119. Quinn, N.W. (1992) Conjunctive Management of Groundwater and Surface Water Resources in the San Joaquin Valley of California, *Report Proc. Availability of Water Resources - AWRA*, pp. 15.
120. Nichols, A.B. (1990) Southern California's Water Crisis Deepens as Drought Continues, *Water Environment & Technology* 2, 17,19.
121. Fryer, J. (1995) Rationing the Plans You Need - But Dread to Use, *Proc. Of Conserv '96*, 677-680.
122. Ploeser, J.H., et al. (1992) Nonresidential Water Conservation: A Good Investment, *J. AWWA - Management and Operations* 84, 65-73.
123. Vickers, A. (1991) The Emerging Demand-Side Era in Water Management, *J. AWWA - Man. and Operations* 83, 38-43.
124. Stone, L. and Weiss, J. (1995) Water Efficiency, *Mother Earth News* 149, 20.

125. Preston, B.L. (June 1987) High-Efficiency Water Closet Analysis, *Plumbing Engineer*, 19-23.
126. Honen, T.P., et al. (1996) Alternative Flushing and Retrofit Devices for the Toilet, *Report No 304 for MWD*, pp. 94.
127. Harasick, R. (1990) Water Conservation in Los Angeles, *Proc. Hydraul./Hydrol. Arid Lands*, 131-136.
128. Murphy, K. (1996) Lush Landscapes That Don't Waste Water, *Business Week* 3486, 80-81.
129. Anon, (1997) *Landscaping and Water Conservation*, www.calwater/conservation/landscaping.shtm.
130. Stewart, K. (ed.) (1990) Xeriscape · San Diego Style, San Diego County Water Authority R eport.
131. Anon (Oct. 1976) Good looking ... unthirsty, *Sunset*, 78-86.
132. Smith, L.K. (1977) *40 Ways to Save Water in Your Yard & Garden*, pp.14.
133. Nelson, J.O. (1987) Water Conserving Landscapes Show Impressive Savings, *J. AWWA - Management and Operations* 79, 35-42.
134. Manzione, M., et al. (1991) California Industries Cut Water Use, *J. AWWA - Man. and Op.* 83,55-61.
135. Farvolden, R. (1995) Water Crisis: Inevitable or Preventable?, *Geotimes* 40, 4.
136. Wichelns, D., et al. (1996) Labor Costs May Offset Water Savings of Sprinkler Systems, *California Agriculture* 50, 11-18.
137. ASCE, AWRA, and AWWA (1996) Responsible Water Stewardship, *Proc. of Conserv 96*, pp. 1047.
138. ASCE, AWRA, and AWWA (1993) The New Water Agenda, *Proc. of Conserv 93*, pp. 2025.
139. Maddaus, W.O., (1987) Water Conservation, *AWWA Report*, pp. 93.
140. Army Corps of Engineers (1983) Evaluation of Drought Management Measures for Municipal and Industrial Water Supply, *Contract Report 83-C-3*, pp. 136.
141. Army Corps of Engineers (1983) Prototypal Application of a Drought Management Optimization Procedure to an Urban Water Supply System, *Contract Report 83-C-4*, pp. 75.
142. Solley, W.B. and Pierce, R.R. (1995) Transitions in Water Resources Management and Changing Water-use Patterns in the United States, *Proc. Wat. Res. And Env. Hazards - Am. Wat. Res. Assoc*.
143. Vickers, A. (1990) Water-use Efficiency Standards for Plumbing Fixtures: Benefits of National Legislation, *Journal AWWA - Management and Operations* 82, 51-54.
144. O'Rourke, J.T. (1982) Water Recycling in the Pulp and Paper Industry in California, *California. SWRCB Report*, pp. 188.
145. Russell, L.L., et al. (1981) Water Recycling in the Fruit and Vegetable Processing Industry, *California SWRCB Report*, pp. 158.
146. Garrison, W.E. and Miele, R.P. (1977) Current Trends in Water Reclamation Technology, *J. AWWA - Water Technology/Quality* 69, 364-369.
147. Smith, R.G. and Walker, M.R. (1991) Water Reclamation and Reuse, *Res. J. WPCF* 63, 428-431.
148. Asano, T. et al. (1992) Water Reuse: Evolution of Tertiary Treatment Requirements in California, *Water Environment & Technology* 4, 36-41.
149. Miller, K.J. (1990) U.S. Water Reuse: Current Status and Future Trends, *Wat. Env. & Tech.* 2, 83-89.
150. Mills, R.A. and Asano, T. (1996) A Retrospective Assessment of Water Reclamation Projects, *Water Science Technology* 33, 59-70.
151. Stone, R. (June 1976) Water Reclamation: Technology and Public Acceptance, *J. Environmental Engineering Division ASCE*, 581-594.
152. Blanton, M. (May 1977) California's Water Reclamation and Desalinization Projects, *Water and Sewage Works*, 60-62.
153. Bruvold, W.H. (1991) Public Evaluation of Municipal Water Reuse Options, *Proc. Water Suppy and Water Reuse - American Water Resources Association*, 363-369.
154. Anon (February 26, 1991) S. 485: This Act May Be Referred to as the Reclamation Waste Water and Ground Water Study Act, *First Session, 102nd Congress Record*, pp. 12.
155. Hon, K. (1997) Controlling Brine to Improve Reclaimed Water Quality, *Proc. 24th Annual Water Res. Plan. and Man. Con. - ASCE*, 125-129.
156. Anon (1992) POU and Reclamation: Effect on California Projects Site-specific, *Wat. Tech.* 15, 22,24.
157. Anon (1993) Irvine Ranch, Harvey Mudd College Team-up in Study: Improvement of Water Reuse Is Focus of Joint District, College Project, *U.S. Water News* 9, 25.
158. Dolan, R. (1990) Contra Costa Water Reclamation Project, *Proc. Symposium III, AWRA*, 5-10.

159. Anon (1993) California Ski Resort to Make Snow From Treated Wastewater, *ENR* 230, 24.
160. Anon (1992) Reclamation Project Will Serve Triple Duty, *ENR* 228, 16.
161. Anon (1991) West Coast Project Will Be Largest Reuse Plant in Calif., *U.S. Water News* 8, 11.
162. Sundberg, S.R. et al. (1991) Reclaimed Wastewater for Industrial Utilization, *Proc. Water Supply and Water Reuse - AWRA*, 311-320.
163. Jolis, D. et al. (1996) Assessment of Tertiary Treatment Technology for Water Reclamation in San Francisco, California, *Water Science & Technology* 33, 181-192.
164. Jolis, D. et al. (1995) Desalination of Municipal Wastewater for Horticultural Reuse: Process Description and Evaluation, *Desalination* 103, 1-10.
165. Anon (1995) City of San Jose South Bay Water Recycling Program, Santa Clara County, California, *Department of the Interior, Bureau of Reclamation Report*, pp. 265.
166. Anon (1993) Salinas Water Threatened, *ENR* 231, 23.
167. Everest, W.R. (1995) Groundwater Conservation by Desalting in South Orange County, *Proc. Conserv '96: Responsible Water Stewardship - AWWA*, 1017-1021.
168. Wohlgemuth, G. et al. (1991) Developing a Water Reclamation System for Drought Stricken Community, *Proc. Water Supply and Water Reuse - Am. Water Res. Assoc.*, 355-361.
169. Argo, D.G. (1980) Cost of Water Reclamation by Advanced Waste Water Treatment, *J. of the Water Pollution Control Federation* 52, 750-759.
170. Ashcraft, J.G. and Hoover, M.G. (1991) Water Reuse – Implementation and Costs in Southern California, *Proc. Water Supply and Water Reuse*, 345-353.
171. Findley, P.L. and Hamilton, S.C. (1990) San Diego Devises Wastewater System to Reduce Water Needs, *Water Environment & Technology* 2, 17-18.
172. Bailey, H.E. et al. (1992) Analysis of Changing Constraints and Planning for Flexibility in a Water Reclamation Program, *Water Science and Technology* 26, 1525-1535.
173. Thompson, K. et al. (1992) City of San Diego Study of Potable Reuse of Reclaimed Wastewater: Final Results, *Proc. Water Forum '92 - ASCE National Conference*, 133-138.
174. Thompson, K. et al. (1992) City of San Diego Potable Reuse of Reclaimed Water: Final Results, *Desalination* 88, 201-214.
175. Szymborski, S.E. (1995) Early Action Plan - San Luis Rey Basin Desalting Facility, City of Oceanside, California, *Desalination* 103, 147-153.
176. Smith, N.C. (1992) Developing and Financing New Water Facilities: Alternatives for Desalination and Reclamation Plants, *Desalination* 87, 85-95.
177. Anon (July 17, 1991) Desalination Research, *First Session, 102nd Congress Record*, pp. 170.
178. Kartinen, E.O. (1993) Is There a Future for Desalting in Meeting California's Water Needs?, *Proc. National Conference on Hydraulic Engineering - ASCE*, 162-167.
179. Anon (1991) Congress Hears Merits of Desalination, *U.S. Water News* 8, 8.
180. Anon (1990) Drought Provokes Desalting Plan, *U.S. Water News* 7, 4.
181. Davenport, W.F. (1994) Desalination? Decision Making, *Proc. U.S./Middle East Joint Seminar on Innovative Desalination Tech.* 99, 245-255.
182. Hammond, R.P. et al. (1992) Large-scale Sea-water Distillation for Southern California, *Desalination* 87, 69-83.
183. Hammond, R.P. et al. (1994) Seawater Desalination Plant for Southern California, *Desal.* 99, 459-481.
184. Anon (1991) Drought Forces Desalting in California, *U.S. Water News* 7, 8.
185. Anon (1993) Monterey Peninsula Water Supply Project Supplemental Draft Environmental Impact Report/Statement II. Volume 1 and 2 and Appendices, *Army Corps of Engineers Report*, pp. 513.
186. Katz, W.E. (1992) Drought-proofing in Southern California by Seawater Desalting Is Economic Today, *Desalination* 86, 1-8.
187. Young, R.F. (1995) Proposed Marin Desalination Plant - Siting Criteria, Evaluation and Results, *Desalination* 102, 171-178.
188. Everest, W.R. and Murphree, T. (1995) Desalting Residuals: A Problem Or a Beneficial Resource?, *Desalination* 102, 107-117.
189. Ossiff, J. (1992) Emerging Legal Issues in Siting Desalination Facilities in California, *Desal.* 83, 1-36.

190. Anon (1993) Ventura Seeks Desalted Independence, *U.S. Water News* 9, 8.
191. Hess,G. and Morin, O.J. (1992) Seawater Desalting for Southern California: Technical and Economic Considerations, *Desalination* 87, 55-68.
192. Childs, W.D. and Dabiri, A.E. (1995) VARI-RO 'Low Energy' Desalting for the San Diego Region, *Final Technical Report, Water Treatment Technology Program Report No. 4* , pp. 49.
193. Childs, W.D. and Dabiri, A.E. (1995) VARI-RO Super 'Low Energy' Desalting for the San Diego Region, *Desalination* 103, 49-59.
194. Yamada, R.R. et al. (1995) Co-located Seawater Desalination/Power Facilities: Practical and Institutional Issues, *Desalination* 102, 279-286.
195. California Department of Water Resources (1995) *Drought Mitigation Tools for States*, enso.unl.ed/ndmc/mitigate/policy/tools.htm.
196. Hawkins, T. and Burt, C.M. (1990) AGWATER. Irrigation Management and Planning Expert System, *Proc. Third National Irrigation Symposium - ASAE*, 64-68.
197. Cehrs, D. (1996) Irrigation in the SJV: Evapotranspiration Data Aids Scheduling, *Citrogr.* 6, 4, 6-7.
198. Burke, J.R. (1992) Operation of the Central Valley Project During California's Drought, *Proc. Irrigation and Drainage Sessions at Water Forum '92 - ASCE*, 348-353.
199. Georgeson, D.L. (1995) How to Reduce Conflict in Water Management: Employ New Resources or Change Existing Habits, *Water Supply* 14, 203-204.
200. Israel, M. and Lund, J.R. (1995) Recent California Water Transfers: Implications for Water Management, *Natural Resources Journal* 35, 1-32.
201. Anon, (1996) *U.S. Bureau of Reclamation Annual Report for Federal Grant No. 4-FG-30-00120*, www.lc.usbr.gov/~wtrconsv/partnership-text/htm.
202. Owen, L.W. and Mills, W.R. (1991) California's Orange County Water District A Model for Comprehensive Water Resources Management, *Proc. 18th Annual Conf. and Symposium - ASCE*, 1-5.
203. Nelson, J.O. (1992) Water Audit Encourages Residents to Reduce Consumption, *J. AWWA - Management and Operations* 84, 59-64.
204. Bruvold, W.H. and Mitchell, P.R. (1993) Evaluating the Effect of Residential Water Audits, *J. AWWA - Management and Operations* 85, 79-84.
205. Wood, D.B. (1995) California Water War's Moment of Truce, *Christian Science Monitor* 87, 4.
206. McCormick, Z. (1994) Institutional Barriers to Water Marketing in the West, *W. Res. Bul.* 30, 953-961.
207. Lund, J.R. and Israel, M. (1993) Water Supply Systems Planning with Water Transfers, *Proc. 20th Anniversary Conference: Water Management in the '90s - ASCE*, 336-339.
208. Jacobi, L.A. and Carley, R.L. (1993) Ag-to-Urban Water Transfer in California: Win-Win Solutions, *Proc. 20th Anniversary Conference: Water Management in the '90s - ASCE*, 332-335.
209. Howitt, R.E. (1994) Empirical Analysis of Water Market Institutions: the 1991 California Water Market, *Resource and Energy Economics* 16, 357-371.
210. Frederiksen, H.D. (1996) Water Crisis in Developing World: Misconceptions About Solutions, *J. of Water Resources Planning and Management* 122, 79-87.
211. Anon (1993) Water Markets Seen as Answer to California's Regulatory Drought, *U.S. Wat. N.* 10, 18.
212. Sacramento Metropolitan Water Auth. (1996) American River Water Resources Investigation; El Dorado, Placer, Sacramento, San Joaquin, and Sutter Counties, California, *EPA Rep. #960054D*, pp. 532.
213. Knox, J.B. and Buddemeier, R.W. (1989) Impacts of Climate Change on California Water Resources, *Proc. Coping with Climate Change - Lawrence Livermore National Lab.* , 469-473.
214. Dziegielewski, B. (1979) *The Drought Is Real Designing a Successful Water Conservation Campaign*, unesco.org.uy/phi/libros/efficient_water/wbendzieg.htm.
215. Hoffman, M. et al. (1979) Urban Drought in the San Francisco Bay Area: A Study of Institutional and Social Resiliency, *J AWWA - Management and Operations* 71, 356-363.
216. Shaw, D.T. (1991) Responses to Municipal Water Supply Cutbacks in San Diego County, California, Spring 1991: Field Observations, *Quick Response Research Report #47*, pp. 39.
217. Van Riper, C. and Geselbracht, J. (1996) Water Reclamation and Reuse, *Wat. Env. Res.* 68, 516-520.
218. Anon (1995) Flowing Uphill: Water, *The Economist* 336, 36.

# THE MANAGEMENT OF WATER RESOURCES DURING DROUGHT IN SOUTHERN ITALY

M.R. MAZZOLA, C. ARENA, V. DI LEONARDO
*University of Palermo, Department of Hydraulics and Environmental Applications*
*Viale delle Scienze, 90128 Palermo*

## 1. Introduction

The aim of this paper is to describe how one of the most severe droughts of the century in Sicily has brought awareness of the insufficiency of the water resources system supplying Palermo, due not only to the lack of timely infrastructural measures, but also of an appropriate decisional framework. The experienced shortage has directed research efforts to overcome this problem. As a matter of fact, the 1989-90 drought gave clear evidence of how an heavy hydrologic failure can result in a real crisis depending on the system which suffers from it. Unplanned urban expansion and increased demands levels, mixed with the lack of a drought-oriented water managenment, have produced considerable discomfort to civil population and relevant damages to economic activities, especially agriculture. From this point of view this is the story of how to learn from the past in order to plan the future. Section 2 is devoted to the description of the Palermo water resources system and of its evolution; the following sections will deal with the research developments carried out in order to gain a deeper insight of drought hydrology and management.

## 2. The Palermo water resources system

### 2.1. A HISTORIC BACKGROUND

With a population of over 700.000 inhabitants, Palermo is the fifth Italian city, the second in southern Italy and by far the largest in Sicily. Its civil water supply is based on many sources supplying other users, including not only smaller coastal towns, but also hydropower plants, industrial and agricultural districts. Such a context clearly produces conflicts among users, which will be treated in detail in section 5.

The water resources system consists of water sources different in quantity and quality, including springs, multipurpose reservoirs, well-fields and weirs (see Figure 1).



*E. Cabrera and J. García-Serra (eds.), Drought Management Planning in Water Supply Systems,* 361–387.

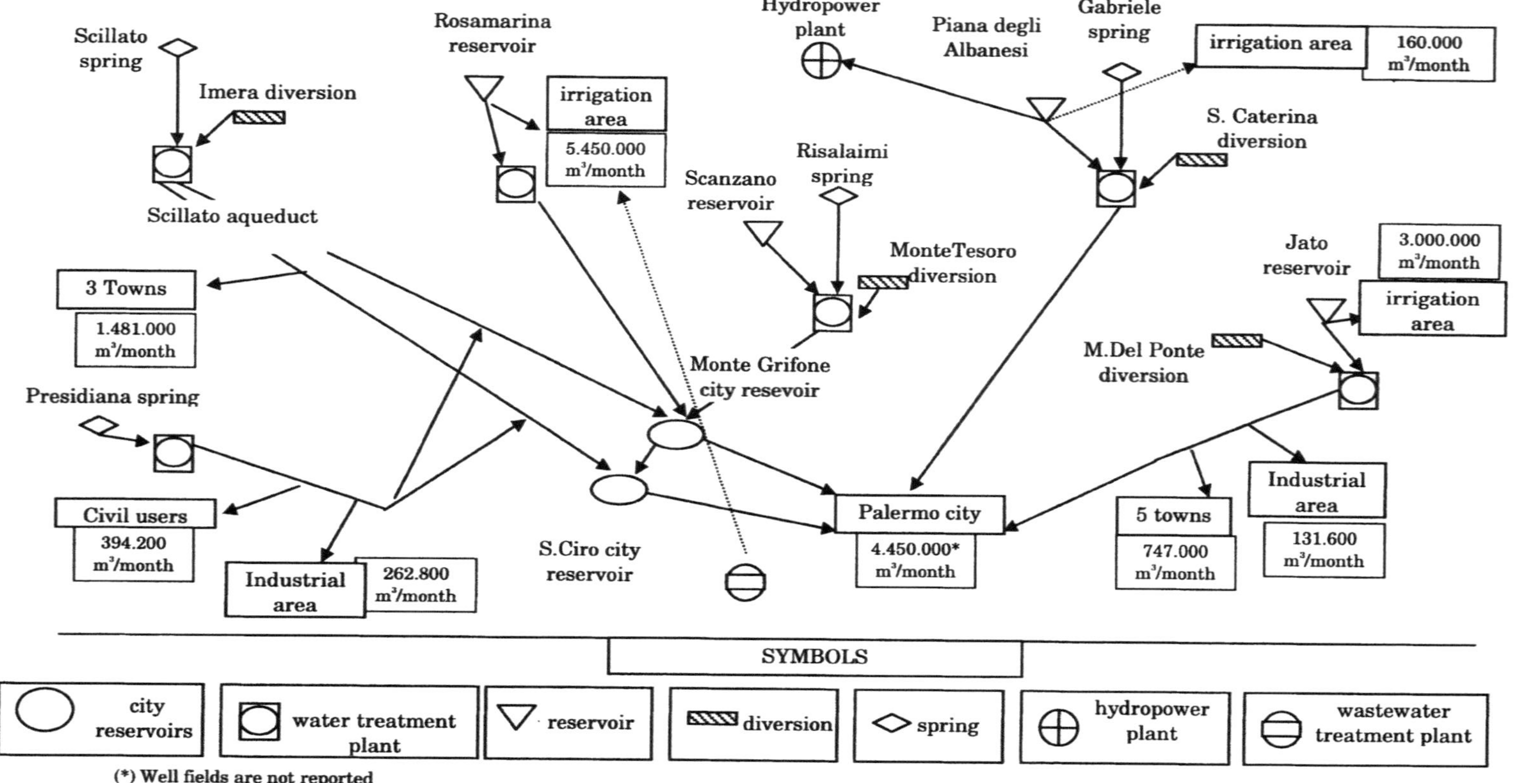


*Figure* 1. Palermo water system

Palermo water supply has not always been a problematic issue: in the pre-industrial period water supply was granted by local sources, as small springs placed in the mountains surrounding the city. Water towers are a typical feature of Palermo landscape and *qanat* (free-surface underground aqueducts), dating back to the times of Arab domination in Sicily, are still a functioning evidence of this civilised water culture. The first example of planned waterworks to meet increased demand dates back to the very first years of the present century, when the Scillato aqueduct, a 70 Km-long channel, was built, conveying a flow of 400 l/s from the homonymous springs to two city reservoirs with a total capacity of 36500 $m^3$. The water volumes supplied from Scillato springs actually exceeded the total demand of Palermo, whose population at that time was about 300.000 inhabitants. Before the second world war the supply capacity of the water system was increased by other springs and some city reservoirs were built supplying the north western part of the city, subject to greater expansion from the early sixties of this century: in the meantime the distribution network was extended without a masterplan. The consequence of limited supply and high level of network leaks was the rationing of water distribution service for some hours a day and low pressure in most of the city districts. In order to have continuous distribution within the households, private pumps, tanks and boosters have been installed in all the city, with a very high increment of supply costs suffered by the consumers. In the years after the second world war the search of different supply sources began: they were identified in surface water to be stored in new reservoirs and wells. Interest was turned to this kind of sources in order to avoid overexploitation of groundwater resources supplying agricultural areas surrounding the city; agriculture was in fact still the major economic activity. The first reservoir built exclusively for Palermo civil water demand was the Scanzano reservoir on Eleuterio river with a capacity of 16 $Mm^3$; its operation started in 1970 together with a new aqueduct conveying the stored waters and the flows of Risalaimi springs to the Monte Grifone city reservoir. Water volumes coming from the Piana degli Albanesi reservoir had actually been used for civil supply since 1958. This reservoir, with a capacity of 24 $Mm^3$ and operating since 1923, had been initially designed for hydropower generation and agricultural uses only. Its exploitation for civil use became possible when Gabriele treatment plant went into operation: two years later, at the same plant, treatment of the water withdrawn from Oreto river at Santa Caterina began. As mentioned above, wells exploitation as source for civil supply increased in the years after the second world war to such an extent that some of them were excluded from operation owing to progressive quality depletion chiefly caused by sea water intrusion.

Another important step in the definition of the present layout of Palermo water resources system was taken when the Jato scheme was completed in 1978: it comprises Poma reservoir on Jato river, with a capacity of 68 $Mm^3$, and an aqueduct supplying irrigation districts and towns (mainly Palermo, but also other minor coastal towns). Later on the same scheme has been completed by some weirs on nearby watersheds in order to increase water resources for agriculture.

The last works have been completed after the severe drought of 1989-90 as measures to reduce shortages in those years: the Scillato intake on river Imera has been operating since 1991, while Presidiana spring on the western coast near Cefalù has been exploited during the 1989-90 drought. Together with Scillato springs and some wells they complete the Scillato scheme conveying water to Monte Grifone city by the new Scillato pressure aqueduct.

## 2.2. OPERATION ANALYSIS OF THE WATER SYSTEM IN THE PAST THIRTY YEARS

The water volumes supplied by the different water sources in the last three decades have increased at first, doubling from year 1958, when 45 $Mm^3$ were supplied, to year 1979 with a supply of 92 $Mm^3$. After 1979 they decreased, settling to 73,7 $Mm^3$ in 1993. The increase of water supplied is mainly due to the operation of reservoirs from which water began to be withdrawn for civil uses (as in the case of Piana degli Albanesi in 1958 and Poma in 1979). In the same three decades population has increased of about 40%, namely from 578.000 in 1958 to 700.000 in 1993; as a consequence the gross daily per capita consumption has grown from 200 l/day to 360 l/day in 1979, settling to 290 l/day in 1993 (Figure 2). The reason for the decreasing water supply after the peak of the late seventies is the diversion of water resources from Poma reservoir to irrigation districts, for which it had been originally designed. The volumes withdrawn for civil supply, which reached 24,5 $Mm^3$ in 1980, settled after this year to values never exceeding 10 $Mm^3$. The minimum volumes in the last years (table 1) have been supplied in 1989 (69 $Mm^3$) and 1990 (61 $Mm^3$) with a minimum gross daily consumption of 240 l/p.c. It should be noticed that the metered consumption is less than sixty percent of the gross one, basically because of the deficient maintenance of the distribution network (Mazzola, 1990). The consequences of water leaks are great lack of gains for the Palermo water agency and social discomfort and hygienical risk for the population.

It is interesting to analyse how the different kinds of water supply sources have contributed to the annual water yield in these last thirty years. Wells contribution varies from 27% to 49% of the total annual civil supply: it is remarkable that supply from wells has reached the peak in 1990 which shows that they must be considered as a reliable resource during water shortages to be carefully safeguarded, both in quantity and in quality. Spring vary largerly than wells in their monthly yields with a peak flow in March and April. In the whole, groundwater resources range from 55% to 80% of the total water supply.

Surface water comprises water intakes operating only in winter months, owing to the torrential nature of most Sicilian rivers, and reservoirs. The first ones supply a percentage of the total annual volume varying from 0,4 to 7%, the second from 8% (in 1990) to 43%. In the whole, surface resources have contributed to Palermo water supply with percentages fluctuating in the range 20% - 45%.

## 2.3. WHAT HAPPENED IN 1989 AND 1990

The most severe drought for the city of Palermo in the last thirty years occurred in the period 1989-90. A previous serious shortage for the city occurred in 1975, when the gross supply was 63,5 $Mm^3$. The water rationing was reduced to very few hours per day in all the city and to alleviate social discomfort the new aqueduct from Poma reservoir was built in few years. It started to operate in 1978 and in the period 1979-88 the mean gross water supply has been 83,3 $Mm^3$/yr., with a minimum value of 74,8 $Mm^3$ in 1983 and a maximum of 92,3 $Mm^3$in 1980. The mean percentage of leaks in the period has been 36%, and it proves the inefficiency of the distribution network. The policy of using all the resources available in the reservoirs in order to reduce, but actually failing to eliminate, rationing in the city has been followed by the Water Agency. The availability of new resources from the large Poma reservoir convinced the Water Agency managers of the impossibility of another shortage of the same intensity of that of year 1975 to occur.

In Table 1 the 1989-90 historical series of monthly gross supply and the metered volumes in the city show how this belief was wrong. Owing to the absence of a long-term operating rule, in the second year of drought reservoirs were completely empty, and the gross supply in the city was only 61,2 $Mm^3$ and metered volumes amounted at 45 $Mm^3$. In the period July—November the monthly gross supply was less than 5 $Mm^3$, with a minimum of 4,6 $Mm^3$.The consequence of this very considerable shortage has

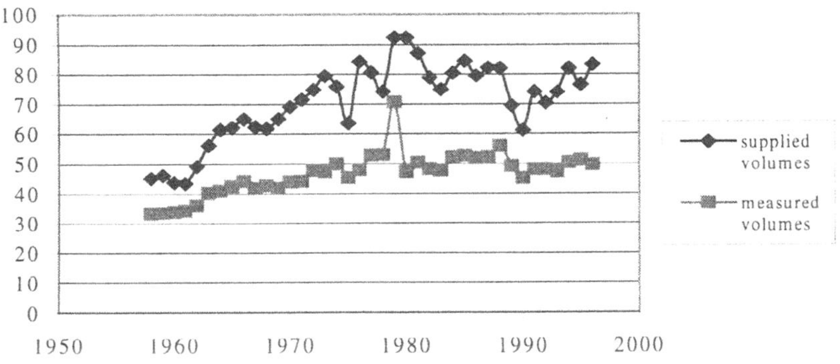


*Figure 2.* Supplied and metered volumes [$Mm^3$] for Palermo city in years 1958-1996

been distribution of water for few hours every two days with very strong sanitary and social problems. In the period 1991-96 the mean yearly gross supply in the city was 76,7 $Mm^3$ and the mean yearly metered volume 49,1 $Mm^3$, with a leak percentage of 36%. These data point out the actual possibility of climate change and the consequent need not only to increase supply by new works and reduce network leaks, but also to define operating rules for the system taking into account the chance of other long droughts.

## 2.4. POSSIBLE DEVELOPMENTS FOR PALERMO WATER RESOURCES SYSTEM

The experience of the two years drought period (1989-90) teaches that a proper management of water resources can greatly alleviate social discomforts and economic losses when a severe multiyear drought occurs. Nonetheless, it is clear that system management cannot substitute the appropriate infrastructural measures to meet fully system requirements. The improvement of the distribution network with losses reduction is preliminary to any other structural measures, as the exploitation of new water sources.

The ongoing projects to ensure continuous water distribution with pressure in all the city, eliminating in-house private disposals, are:

- renewal of distribution network, to reduce leaks percentage to 10%;
- construction of new city reservoirs to supply the most elevate areas of the city;
- control of groundwater exploitation, in order to avoid seawater intrusion;

TABLE 1. Supplied and metered water volumes [$m^3$] for civil uses during years 1989 - 1990

| | YEAR | | | |
|---|---|---|---|---|
| | 1989 | | 1990 | |
| MONTH | Gross supply ($m^3$) | Metered volumes($m^3$) | Gross supply ($m^3$) | Metered volumes($m^3$) |
| January | 6.849.500 | | 5.637.511 | |
| February | 5.898.887 | | 5.089.252 | |
| March | 6.838.282 | | 5.470.989 | |
| 1st Trim. | 19.136.169 | 13.734.589 | 16.197.752 | 11.166.140 |
| April | 6.013.211 | | 5.262.703 | |
| May | 6.125.740 | | 5.317.420 | |
| June | 5.123.575 | | 5.162.140 | |
| 2nd Trim. | 17.262.526 | 12.095.171 | 15.742.263 | 10.995.757 |
| July | 5.660.459 | | 4.983.062 | |
| August | 5.346.917 | | 4.624.618 | |
| September | 5.318.559 | | 4.629.191 | |
| 3rd Trim. | 16.325.935 | 11.049.700 | 14.236.871 | 11.390.930 |
| October | 5.846.707 | | 4.996.700 | |
| November | 5.268.591 | | 4.872.218 | |
| December | 5.527.9440 | | 5.113.208 | |
| 4th Trim. | 16.643.242 | 12.394.596 | 14.982.126 | 11.444.562 |
| TOTAL | 69.368.372 | 49.274.056 | 61.159.012 | 44.997.389 |

- utilisation of stored water volumes in Rosamarina reservoir; it is located 50 km east from Palermo and with its 100 $Mm^3$ of capacity is the largest of the whole system. Even if it is questionable whether this reservoir will be actually able to supply the planned 50 $Mm^3$/year for civil uses, owing to the present negative trend in yearly streamflows, it remains an important source for the whole system, since it will make available for civil and agricultural uses more than 60 $Mm^3$/year.
- Wastewater reuse for agriculture, with consequent diversion to civil uses of surface and groundwater resources.
- Improvement to 800 l/s of the water yield from Presidiana spring, presently supplying 400 l/s. The present constraint in Presidiana spring exploitation is the too high chloride concentration which imposes mixing with Scillato spring waters. The treatment by inverse osmosis of Presidiana flows will permit its complete exploitation.

This brief overview on Palermo water resources system points out that its complexity consists in the large dimension of the main civil demand centre compared to the other civil and agricultural demand centres and in the fact that the ratio between supply and demand in the whole area is close to one, that is on the limit of environmental sustainability. The need of an effective system management becomes evident during drought periods when water conflicts explode chiefly because of Palermo large requirements. Modelling such conflicts and the possible management alternatives, on the basis of correct hydrologic inputs, is the necessary companion to the infrastructural measures described above.

Another interesting issue in drought management is the real time operation of reservoirs which requires specific procedures for streamflows forecasting within a season and optimisation models to define optimal operation rules.

In the following sections these matters are dealt with separately: section 3 is concerned with the effectiveness of real time operation of reservoirs to mitigate the effects of droughts. Section 4 is devoted to the hydrologic characterisation of droughts and to the description of some very simple hydrologic models for drought frequency analysis and water systems long-term planning and management. Section 5 deals with the use of multicriteria analysis for the assessment of the overall effects of selected management alternatives on conflicting interest groups during droughts.

## 3. Real time optimisation model

The problem of water resource allocation over a finite time period can be treated as a sequential decision process and solved by multistage decision-making procedure, as dynamic programming, where the allocation in each step is considered a stage in a sequence of decisions (Loucks *et al.*, 1981). If hydrologic year features a wet season starting in October and a dry one during summer, as often in the Mediterranean area, the inflows Ai to be forecasted for each year are only those of the wet season (i = 1 to 8, from October to May). The releases Xi (i = 1 to 8) to be optimised are those to meet civil uses in the wet season, while in the dry season also agricultural uses must be considered. The release in the dry season is assumed to be aggregated in the value $X_9$.

The objective of the decision process can be assumed as the minimisation of the sum of the squared deviations of standardised deficits in a given period, computed from current month i till the last one (i=8) of the hydrologic year (Mazzola, 1994). Constraints for the optimisation problem are:

- the continuity equations of the reservoir for each time step;
- the nonnegativity of stored volumes and releases;
- the maximum capacity of reservoir.

A further constraint can be introduced on the stored volume at the end of the dry season, denoted withVn, if the occurrence of multiyear droughts is to be allowed for. It states that Vn must be greater than a minimum strategic reserve SFmin whose value depends on the probability that the next year will be a dry one.

The dynamic programming optimisation model should be used as a subroutine of a larger model, which simulates the real time reservoir management. At each time step the optimisation model defines the optimal allocation policy under given reservoir inflow forecast (section 3.1), which determines the reservoir behaviour in the period. The simulation model uses the historical inflow to the reservoir in the first month instead of the forecasted one.

Consequently in the next step new conditions are set based on the previous policy assumptions and on the actual inflows. The optimisation is repeated for a period shorter of one month. This procedure is carried out during the wet season months, with the starting point set at the beginning of October, using the forecasts before mentioned.

In figures 3 and 4 the first two steps of the allocation procedure are drawn. In step 1, the end of September, the initial volume stored in the reservoir is known and equal to V1, while the volumes in the other stages derive from the forecasted inflows $A_1$-$A_8$ and the decision rules inflows $X_1$-$X_9$. The demand $D_9$ is equal to the demands in the dry season (June — October), in which the agricultural demand is concentrated. The volume stored $S_{10}$, which is the predicted volume in the reservoir at the end of September, could be treated as a decision variable, or set equal to a minimum strategic reserve for long duration droughts.

Once the allocation problem has been solved for step 1, the procedure continues taking into account the historical reservoir inflow $I_1$ instead of the forecasted one $A_1$. By the continuity equation in which the real values of inflow and release ($I_1$ and $R_1$, possibly different from the forecasted inflow and the optimal release ) are used, it is possible to calculate the real amount of water stored in the reservoir at the end of step 1 ($V_2$) to be used as starting point for the subsequent step.

## 3.1 THE FORECASTING MODEL

Many inflow forecasting models have been proposed for real-time reservoir operation. A simple model is often effective enough for prediction. In the hydrologic situations with presence of strong memory of the past events an autoregressive model is normally suitable. For instance this kind of model has been chosen for the 10-day streamflow prediction of Tanshui River in Taiwan (Kuo *et al.*, 1990).

When the memory of the past hydrologic events is short, owing to little soil permeability and limited watershed extension, autoregressive model approach does not often give satisfactory results and different models should be developed. Furthermore when reservoir capacity, mean yearly inflows and demands assume similar values, the forecasting horizon has to be extended at least to an hydrologic year to allow deficit smoothing during the whole period. The proposed forecasting procedure has been developed for climate with hydro-meteorological characteristics similar to the Sicilian ones, with a wet season in the period October-May and a dry season normally lasting from June to September.

The preliminary step of the adopted forecasting method (Mazzola, 1992) is the selection and parameter estimation of the cumulative distribution functions $F_Y(y)$ of streamflows for the wet season from the historical record. $F_Y(y)$s are estimated not only for monthly streamflows but also for the cumulative streamflows from September till the end of each month of the wet season, i.e. for flows of period Sept., Sept.-Oct.,...Sept.-May. The normal distribution is often suitable for monthly, cumulative and seasonal streamflows. To start the procedure it is necessary to have a streamflow value for the month previous to the first to be forecasted; therefore it is necessary to estimate $F_Y(y)$sept.

The forecasting model predicts the monthly streamflows for each month of the wet season and updates prediction in each month. In the application the procedure starts at the beginning of October and follows these steps:

a1) determination of the quantile $p_{sept.}$ by the function $[F_Y(y)]_{Sept.}$, given the reservoir inflow in september $y_{sept.}$;

a2) the overall streamflow quantile $p_{PR}$ is assumed equal to $p_{sept.}$ and the value of the streamflow for the whole wet season can be determined from $[F_Y(y)]_{PR}$, whose form and parameters were previously estimated;

a3) disaggregation of the seasonal streamflow value in monthly values $A_i$ for the period October-May by simply multiplying $y_{PR}$ by monthly distribution coefficients Ki obtained from historical data;

a4) repetition of steps a1-a3 at the beginning of November: the quantile whose value is assumed to equal the seasonal one is now drawn from the cumulative distribution of streamflows of period September-October, as the true value of inflow in October is now known.

a5) repetition of step a4) at the beginning of each month.

As previously mentioned, the above described real-time operation model does not specify the best value of the volume stored at the end of the dry season, which is a reserve for the occurrence of future drought years. If the exploitation constraint on the final volume is not introduced, the yearly optimisation model leads to a complete exhaustion of reserve to face shortages without accounting for over-year compensation in case of drought persistence. This constraint could be defined as a function of drought persistence probability.

The phenomenon of persistence in streamflow series has been analysed and modelled as a Markov process by some researchers (Jackson, 1975, Gottschalk, 1976). Dividing the streamflow values in P intervals, the transition probability $d_{sp}$ is the conditioned probability that next state is $q_p$ given a current streamflow state of $q_s$. The transition probability matrix M gives complete information of probability of each initial state to remain in the same state or leaving for another one (Benjamin and Cornell, 1970).

To define normal and severe droughts, two truncation thresholds $P_N \leq P$ and $P_S \leq P$ could be chosen, defining the superior limits of the intervals for which these definitions hold. For instance historical yearly reservoir inflows can be grouped in four classes (P=4), high flows ($q_1 \geq 1.5\ q_m$), normal flows ($1.0\ q_m \geq q_2 \geq 1.5\ q_m$), low flows ($0.5\ q_m \geq q_3 \geq 1.0\ q_m$), very low flows ($0.5\ q_m \geq q_4$), being $q_m$ the mean yearly historical inflow.

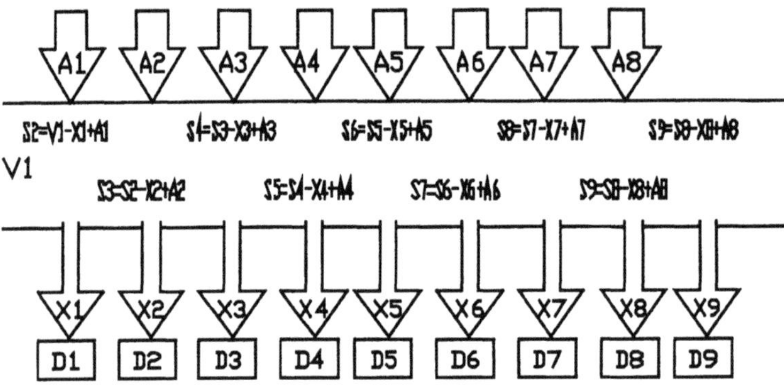


Figure 3. Sequential allocation diagram for step 1.

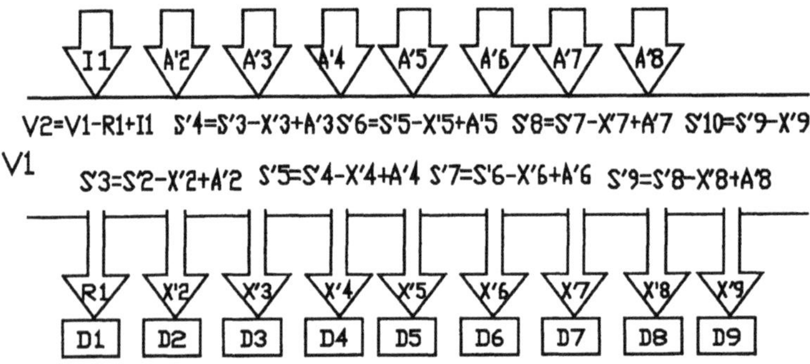


*Figure 4. Sequential allocation diagram for step 2.*

The details regarding the computation of the probability matrices of normal and severe drought duration $A_N$ and $A_S$ and of continuing normal and severe drought, $B_N$ and $B_S$ can be found in Mazzola, 1994. Two different algorithms have been developed for the definition of the minimum value of the final storage constraint $SF_{min}(i, j)$ in the dynamic programming optimisation model. In the first algorithm, identified as N-one, $SF_{min}$ at the beginning of each iteration is function of initial storage $V_i$, current year and month inflow state s(i,j), monthly rank correlation coefficient $r_j$, normal drought duration $L_N(i)$ and value $b_N$. In the second algorithm, identified as S-one, $SF_{min}$ is function also of $L_s(i)$ and $b_s$. The N-algorithm imposes an exploitation limit of volume stored in reservoir to take into account the possibility of future normal droughts, while the S-algorithm increases it even more in order to face more severe droughts.

## 3.2. REAL TIME MANAGEMENT OF ROSAMARINA RESERVOIR

The optimisation model has been applied to the management of Rosamarina reservoir on San Leonardo River, for which the historical streamflow series is 53-years long (1928-1980). The first 43 yearly values of this series have been used to calibrate the statistical functions described in the previous section, while the last 10 years have been used to validate the forecasting method of steps a1-a5. As before stated, the reservoir has a storage capacity of 100 $Mm^3$, and it supplies the city of Palermo and a large agricultural area. The watershed is 500 $km^2$, and it is almost completely impervious. Historical yearly inflow mean over the period 1928-80 is 90 $Mm^3$.

Monthly and yearly streamflow series have been assumed normally distributed according to statistical tests. The analysis of the rank correlation coefficients between the cumulative inflows series at the end of each month and the inflow of the whole wet season (Mazzola, 1992) proves the strong correlation of quantiles of the cumulative streamflows already at the beginning of February, especially for the complete historical series. The real time management has been simulated for nine wet seasons, starting from October 1971.

In the application municipal yearly demand of 50 $Mm^3$ and agricultural yearly demand of 40 $Mm^3$ have been considered. The municipal demand is distributed in the year, while the agricultural one is concentrated in the dry season.

The forecasting procedure described in steps a1-a5 of section 3.1 has been applied also without final storage constraint. The reservoir inflows in the real time optimisation model has been forecasted by the $F_y(y)$ functions estimated by the 1928-1970, 1941-1970 and 1956-1970 historical series.

The model is very effective in the prediction of droughts in the years 1974 and 1975 of the simulated period, characterised by severe droughts. The deficit in the dry season has been chosen as the most important index of system vulnerability, because the demand in this period is equal to 62.93 $Mm^3$, that is 70% of the total yearly demand. The comparison among the results obtained with the forecasting models with different lengths of historical series shows that the prediction ability is almost independent from this length, and a 15-year historical series is sufficient to calibrate a good prediction model.

Applying the optimal policy in the simulation, in 1974 the dry season deficit equals 10.28 $Mm^3$, and in 1975 it equals 48.85 $Mm^3$. In 1974 the proposed forecasting model points out an expected dry season deficit of 18.47 $Mm^3$. In 1975, when the real deficit is very large compared to the total demand in the dry season, the proposed model forecasts dry season deficit in December (19.41 $Mm^3$) and in January it also predicts its intensity very accurately (45.69 $Mm^3$).

To control the importance of the drought storage constraint algorithms described in section 3.1, the performances of the optimal operation model without the finale storage constraint (Model A) have been compared with those of the model which uses the N-algorithm for the definition of $SF_{min}$ (Model B) and of the model using the S-algorithm (Model C). The total deficit (TD in $Mm^3$), the objective function values (FO), the maximum length of consecutive periods with deficit higher of 30% (LD) and the number of periods with deficit equal to 100% (ND) for the three models are reported in Table 2.

The better performance of models B and C in smoothing deficits is proved also by the deficit absolute value in the 1975 dry season, which with real-time operation performed by Model A is 48.8 $Mm^3$ (78 %), by Model B is 45.2 $Mm^3$ (72 %) and by Model C is 43.4 $Mm^3$ (69 %).

TABLE 2. Comparison of the performances of forecasting models A, B and C

| MODEL | A | B | C |
|---|---|---|---|
| TD | 85.06 | 87.79 | 88.68 |
| FO | 61.33 | 57.12 | 53.54 |
| LD | 5 | 3 | 1 |
| ND | 4 | 2 | 2 |

## 4. Drought analysis and modelling

In the common hydrological terminology distinction is usually made between low flows and drought: the first only refer to a particular hydrological variable (i.e. the flows) and to a short time step; droughts, on the contrary, refer to long periods (i.e. months or years) in which a given hydrological variable takes on values that are smaller than a certain threshold. In the framework of engineering applications the two fields of interest are apparently different: while low flow studies are usually required in civil or industrial wastewater discharge design or in navigable river and hydropower plants management, droughts are usually studied because of their often relevant impact on water resource systems; gaining a thorough understanding of the way in which dry and normal periods alternate and their extent should be in fact a prerequisite to the design and the management of water resource systems. On the other hand, much emphasis should be placed on the fact that it hardly makes sense to discuss about droughts if there is no water resource system, which can suffer from it. Hence, the water resources system features contribute to determine the impact of a drought and its consequences.

In the following low flows will be neglected, while attention will be focused on droughts.

As previously stated, drought is a period in which a hydrologic quantity X keeps below a given truncation threshold Xo; the three parameters commonly used to identify a drought are its length D (i.e. the number of consequent years in which X keeps below the long term mean), its severity S (i.e. the sum over D of the deficits) and its magnitude (i.e. the ratio S/D). The study of such periods can be carried out on a single rain or streamflow gauge station (point drought) or over a larger region (regional drought). In this case a critical area Ac should also be defined (Rossi, 1989; Santos,1983) such that, if the areal extent of drought is less than Ac at time i, no regional drought has occurred at time i. Although attention has been mainly devoted to point droughts, regional droughts have certainly the greatest practical interest since complex water resource systems often rely on supply sources spread over a large area, so that the possibility of a contemporary failure of the different sources should be checked. As for the unit time step to be chosen, the year seems to be the most appropriate, since multiyear droughts are the most dangerous for water resource system performances. The choice of the truncation threshold plays an important role in drought analysis and should somehow be related to the target demand which a given water source should meet: when no precise idea about the exploitation level is available or when no target demand is actually connectable to the hydrological variable X (as in the case of meteorological drought, where X is the precipitation depth at a gauge station) the long term average $\mu$, can be conveniently used as truncation threshold, or a fraction of it, or a fixed percentile of the cumulative distribution function of X or the long term average, minus a fixed fraction of the long term standard deviation, $\sigma$, that is:

$$Xo = \mu - a*\sigma \tag{1}$$

One of the main difficulties when studying multiyear droughts is the lack of adequately long and continuous historical records from which to draw a reliable distribution of drought duration and severity. In simulation problems, on the other hand, using the only historical record means neglecting the fact that the historical record is only one of the infinite possible realisations of the flow process which the water system could experience during its life. Studies in operational hydrology have since over thirty years presented a great number of models at different time scales to generate synthetic streamflow sequences from the historical record which preserve its main statistics (as the mean, the standard deviation, lag 1 autocorrelation coefficient, cross correlation between different gauge stations). The underlying idea is that if a model is able to reproduce the variability in time and space of flows it should incorporate droughts as an expression of this variability. Askew *et al.* (1971) have shown that the most common synthetic streamflow generation techniques fail in many cases to reproduce long-term droughts; as a matter of fact the latter are never explicitly considered in the modelling of flows.

There is a simple stochastic model at annual scale in the literature (Jackson, 1975) that explicitly takes into account the length of droughts in its formulation; it assumes that the flow generation system can be in either two states, normal or dry and models its sequence as a homogeneous two-state Markov chain and flow values as a mixture distribution of two normal distributions: one for dry years, another for normal year. In its simplest form the model is built assuming annual flows to be uncorrelated: this hypothesis is in almost all cases fully acceptable for Sicilian rivers, characterised by small watersheds with usually low or medium permeability. Theoretical details can be found in Jackson (1975). The author has also modified the form of the model in order to reproduce annual runoff correlation. In its final layout the model is a six-parameter probability distribution function of the form

$$N(\mu,\sigma^2) = \lambda*N(\mu_1,\sigma^2{}_1)+(1-\lambda)*N(\mu_2,\sigma^2{}_2) \tag{2}$$

in which $\mu_1$, $\sigma^2_1$, $\mu_2$, $\sigma^2_2$ are the mean and variance of flows in dry and normal years respectively and $\lambda$ equals $b/(a+b)$,where $a$ and $b$ are the 1,2 and 2,1 terms of the 2×2 transition probability matrix of the Markov chain. Physically $a$ represents the probability that year i is normal, given that year i-1 is dry and $b$ the probability that year i is dry, given a normal i-1 year,.persistence implies that $a + b < 1$. Parameter estimation is, as usual, a thorny issue, since the size of a streamflow record, long as it may be, is clearly not sufficient to allow reliable applications of likelihood estimation techniques; nonetheless, it can be carried out in less rigorous ways that have the advantage of always keeping an eye on the physical reality of the problem. If model parameters are to be estimated one by one, then a truncation threshold must be introduced; it is remarkable that if parameter estimation could be performed all at one time, no definition for low flows would be required, since the parameter best-fit to the historic record would automatically provide such a definition. For the following application dry years have been defined as those featuring flows less or equal to the 0,25 percentile of the empirical cumulative distribution function. Once the fraction of low flows is identified ($\lambda = 25\%$ for instance), the expected value E[L] and the variance VAR[L] of drought length L can be easily calculated from the historic sample: since $a$ is related to both these quantities (Jackson, 1975) $a$ and $b$ can be determined with a certain degree of subjectivity. Given distribution (2), synthetic flow generation can be performed as follows: state of year 1 is chosen arbitrarily and it can be either state 1 (dry) or 2 (normal). A random value u belonging to [0,1] is then extracted from a uniform distribution: if $u < a$ and state of year 1 has been chosen to be of type 1, the system experiences a transition to state 2, i.e. the following year will be labelled as normal, while if $u > a$, the system remains in its present state . On the other hand, if year 1 has been chosen to be a normal one (state 2), comparison should be made between the values of u and $b$: thus, if $u < b$, a transition is experienced from state 2 to state 1, that is the following year will be dry, otherwise year 2 will be normal as year 1. By running this procedure, any number of dry and normal years can be generated. The value of flow at year i is obtained by extracting a value either from distribution $N(\mu_1,\sigma^2_1)$, if year i is a dry one or from $N(\mu_2,\sigma^2_2)$, if year i is a normal one.

This model has been applied to the annual flow series of S.Leonardo river, that is to the inflows of Rosamarina reservoir. Compared with the common lenght of streamflow records in Sicily, an unusually long record consisting in 53-years (1928-80) of observed data is available for this river; they have been used to calibrate a markov mixture model. 300 series of the same length of the historic one have been generated in order to allow comparison between the features of historical droughts and those extracted by the synthetic series. Moreover, 300 series have been generated using a simple regressive model yielding annual runoffs Di as a function of annual rainfall Ai only; such models have been extensively used in surface-water resources planning and management in Sicily both for their simplicity and because they soundly make allowance for the actual climatic and geomorphic characteristics of Sicilian rivers, usually featuring low permeability, so that autoregressive components make hardly sense, especially at yearly scale. The yearly model for S.Leonardo river:

$$Di = 0.489Ai - 154 + \varepsilon_i \tag{3}$$

has a high $R^2$ (0.58) which makes it a reliable tool for streamflow generation and forecasting; such models are often used to assess streamflow in years when no measures are available. In (3) $\varepsilon_i$ is the error, modelled as an independent random variable, normally distributed with zero mean and standard deviation equal to 55 mm. Synthetic series are generated by drawing random values from the distribution of $\varepsilon_i$ and adding them to the linear relation between Di and Ai. One of the drawbacks of this approach is that forecasts can be inaccurate when the part of streamflow explained by regression (3) is small and no other independent variables can be found to increase R and reduce the standard error. The highest number of consecutive drought years, the worst drought severity and magnitude have been extracted from each of 300 series and averaged to obtain the mean values of the greatest drought lenght, severity and magnitude. Table 3 shows the historic values of the three drought parameters and the average values of the same parameters drawn from the series generated by the Markov mixture model and the regressive model: it shows that both models reproduce satisfactorily the sample mean and standard deviation, moreover, while in the average both models fail to equal the length of the longest historical drought, the regressive model actually overcomes the severity of the worst historical drought. In the whole, both models can be said to be equivalent in drought features reproduction.

A further test for the adequacy of both models has been carried out by using them in a real planning situation. As stated in section 2, Rosamarina reservoir will be operated for civil and agricultural uses: as a component of the Palermo water system, its operation must be viewed in the framework of the whole system. A very simple model to simulate system performances can be constructed (Amico *et al.*, 1996) if the year is chosen as time unit and complete interconnection among all supply sources is assumed. The first assumption is legitimate, since a satisfactory within-year water allocation can be accomplished by management measures only if water availability at yearly scale is sufficient The second is not fully met till now, although in the future system interconnection will be increased. A rough operating rule for the four multipurpose reservoirs in the system could be the following: in those years when annual streamflows equal or exceed their mean value, here representing the truncation level, the planned division between civil and agricultural uses is operated; when annual streamflows are less than their mean value, a fraction not exceeding 30% of the agricultural water demand is diverted to civil uses. Rather than an actual operating rule for reservoirs, this should be seen as a way to represent grossly the effects of water conflicts.

*TABLE 3. Comparison of basic statistics of historical and generated annual flows for S.Leonardo river and historical and generated drought parameters*

| | Mean | Stand.dev. | Max. Duration | Max Severity | Max magnitude |
|---|---|---|---|---|---|
| | [$10^6$ m$^3$] | [$10^6$ m$^3$] | [years] | [$10^6$ m$^3$] | [$10^6$ m$^3$/year] |
| Historic record 1928-1980 | 89.38 | 40.34 | 7.00 | 238.5 | 69.14 |
| Regressive model | 90.27 | 40.05 | 6.44 | 263.32 | 74.82 |
| Markov mixture model | 89.40 | 42.15 | 6.58 | 223.17 | 69.88 |

The choice of carrying out simulation at yearly scale results in some simplifications for the model: there is no need to specify how demand varies along the year and an average yearly demand can be used instead. Evaporation from reservoirs can be evaluated as a fraction of yearly inflows, instead of treating it as a function of the water level in the reservoir. Finally, if total interconnection is assumed and an aggregate yearly value for civil, agricultural and industrial requirements is adopted, consumers will be no more dependent on their actual water supply source: the whole supply source system can be therefore reduced to a equivalent reservoir, whose capacity equals the sum of all reservoirs active capacities. System performances can be studied by writing the continuity constraints and the proposed operating policy for such equivalent reservoir (Amico et *al.*, 1996).

Simulation outputs can be shown by the "f-T rate curves" in which the percentage of years of deficit is plotted versus the percentages of target demand T (Del Treste and Mazzola, 1991). These curves have the remarkable property of providing some synthetic indices of system performance. The area A between the curve and the f=100% line measures the average water deficit in the chosen time step; it can be considered as a measure of the risk due to water deficit (Hashimoto *et al.*, 1980, Rossi *et al.*, 1995). These diagrams also enable to compare the effectiveness of different plans for alleviating water deficit. As well known, such plans can consist either in expanding system capacity by new water resources or in scheduling the exploitation of the already existing resources in a different way. Whilst the first ones are able to increase the area A under the curve, the latter can only reduce system vulnerability by changing the curve shape. As a consequence, the static moment Ms of area A around the vertical line intersecting the x-axis at T = 100% can be introduced as another meaningful index to characterise the system aptitude to reduce water deficit risk. When synthetic hydrology is used as input to simulation, the result will be a set of curves (as many as the series employed) from which an expected curve should be obtained whose points are the expected frequency f of meeting a given target demand percentage T.

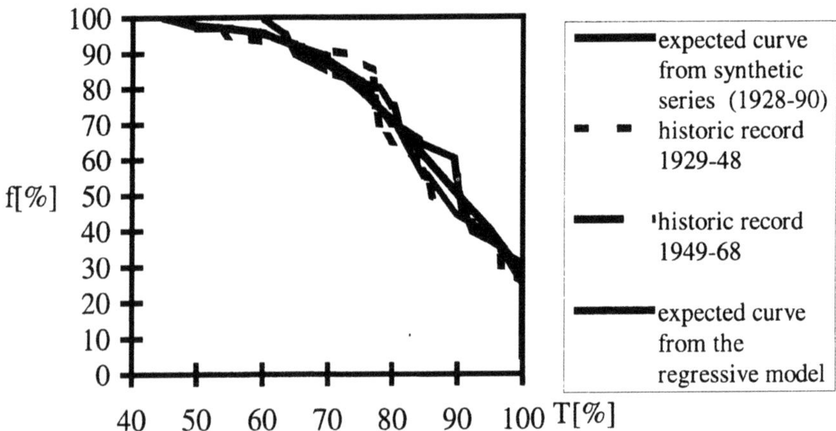


*Figure 5.* Target rate curves from synthetic (full lines) and observed (dotted lines) streamflows of S.Leonardo river at Rosamarina dam

To allow comparison between the two generation techniques only S. Leonardo river flows are generated by means of the regressive model and the Markov mixture model, while historic records are used for the other supply sources. In this application twenty years of water system were simulated with historic data from years 1971-1990. Further comparison can be made by running the simulation model with the 1929-48 and 1949-68 historic record of S.Leonardo river. Figure 5 shows results: it can be clearly seen that the performances of both synthetic streamflow generation techniques are very much and both yield more conservative values of the average annual deficit and moment Ms than the historic inputs do; as a matter of fact both models succeed in predicting years when only very low target percentages could be met. This is probably enough to conclude that synthetic hydrology can be a useful tool for water resource system performance during drought and that no preference can be expressed in terms of better performance between the two models examined: both of them are very easy to use. However, the Markov mixture model can be a valid alternative to the regressive rainfall-runoff model when the results of the regressive analysis are not satisfactory.

## 5. The multicriteria decision analysis for water system during drought

The experience of drought suffered in years 1989-1990 has showed the need of building up a decisional framework to support the decision-maker in the hard task of facing droughts. In this sense, studies have been carried out with the purpose of developing a tool for decision support including a streamflow generation model to supply the synthetic series representing the input data for each supply source of the system and a water allocation model to provide the volumes distributed to each user in the period of simulation. During droughts distribution has to be rationed with consequent shortages suffered by civil users, farmers and industries; reduced availability from natural supply sources also induces consequences for the environment. The impacts of such a situation and the possible conflict among users can be analysed by means of a multicriteria analysis method.

### 5.1. CHOICE OF THE MULTICRITERIA ANALYSIS METHOD

In the last years the decision analysis methods have been highly developed, also in the field of water resources management. In Mazzola and Di Leonardo, 1997 the applicability and effectiveness of a particular multicriteria evaluation method called NAIADE on a problem treating with the management of water resources during droughts is evaluated.

The applications of multicriteria evaluations in the water resources management during droughts and, consequentially, in consumer conflicts, are several. Among these it is possible to distinguish:

- resource allocation linear problems, solvable by means of multiobjective linear programming; the solution of such a problem identifies the unique vector of optimal parameter values (Greis *et al.*, 1983);
- many problems in which judgement factors are neither linearly expressible nor quantifiable (e.g. environmental effects); in these cases it is suitable to use evaluation methods treating explicitly non-quantifiable terms (Zeleny, 1982); they can be grouped in two sub-classes:
  - those starting from an indefinite set of possible alternatives and giving back after some iterations one or more dominant solution;
  - those analysing a finite set of possible alternatives, selected according to the decision maker opinion. The subjective element does represent an advantage and a weakness of this kind of methods at the same time: it is an advantage for

its ability to give elasticity to the process, but it may heavily influence the final choice.

In this paper the second strategy is applied to water system management during droughts. Its main step is the structuration of a proper set of criteria suitable for the description of the effects of water shortages on different classes of water users (Yevjevich *et al.*, 1983).

## 5.2. THE APPLIED METHODOLOGY

The procedure adopted in this study follows the NAIADE evaluation method in which two main steps can be distinguished:

- evaluation of the alternatives,
- analysis of possible conflicts among social parts.

The theoretical basis of the method is contained in Munda, 1995.

The multicriteria evaluation follows the scheme of figure 6. A first input phase can be distinguished in which a matrix is used, containing for each criterion the performances of alternatives. These can be in quantitative, fuzzy, stochastic or qualitative (descriptive) form. The subsequent steps consist in a pairwise comparison of alternatives and in the consequent drafting, by an aggregation process, of the final ranking containing the non-dominated set of alternatives.

In the second part NAIADE implements the conflict analysis showing the possible alliances among the interest groups involved in the process: this is a compulsory step since NAIADE makes use of weights neither for the criteria nor for the interest groups. In fact weights could ascribe more or less significance to particular criteria or groups, thus determining an excessive influence of the decision maker subjectivity on the decision process.

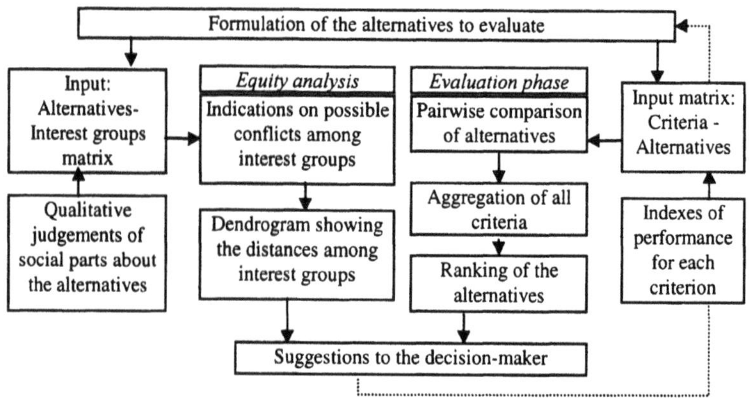


*Figure 6.* NAIADE evaluation procedure scheme

Having selected the evaluation basis, the following step consists in the design and building up of a set of criteria by which a suitable evaluation for the analysed case can be carried out.

The applicable set of criteria in the case of management alternatives for a water system during drought can be divided in four main classes (Yevjevich *et al.*, 1983):

- economic (operating costs, extra costs...);
- social (unmet demands of water, etc..);
- governmental (respect of certain constraints imposed by law);
- ecological (risks for the environment, air pollution level....).

The criteria structuration phase is performed strictly related with and according to the context in which the evaluation is performed. Social parts and environmental factors involved must be considered, so that the two sets on which the evaluation process is based can be identified, namely the *interest groups* and the *judgement criteria*.

## 5.3. THE EVALUATION CRITERIA RELATED TO THE INTEREST GROUPS

The *social groups* considered in this study are: a) civil water consumers; b) farmers; c) industries; d) environmentalists.

The criteria are measures of the performances of each alternative and they are related to the involved interest groups; in fact each single criterion expresses the effects of a single alternative on each interest group. The chosen criteria are:

Operating costs of the water company including a) water treatment costs, b) penalties for missed hydropower production owing to alternative use, c) pumping costs, d) increased network management costs in case of rationing.

Incomes: Assessment of the gross incomes for the water agency.

Farm production: Amount of the total agricultural production in the irrigation districts.

Total costs born by industries: Costs of water supply from alternative sources.

Ecological minimum flow: Total releases for ecological purposes from reservoirs.

Hygienic risk for civil consumers: Qualitative estimation of the hygienic risk that consumers can run during shortages.

Social discomfort: Total discomfort suffered by civil consumers during drought periods; it can be sometimes quantified in monetary terms using Benefit-Cost analysis (Del Treste and Mazzola, 1991).

The features of flexibility of the multicriteria analysis permit the fitting of the above-described criteria to particular cases and also the introduction of new ones, when necessary. Each criterion is simple to compute. In the next sections the evaluation of social discomfort and hygienic risk criteria is described, for which an original approach is presented.

### 5.3.1. *Water shortages effects on civil consumers*

During droughts consumers may suffer to some degree from shortages in water supply, which result in increasing discomfort according to the water shortage intensity. This discomfort can be quantified using Benefit-Cost analysis: consumer benefits, deriving from marginal shifts in the availability of a certain good, are expressed by the demand curve of that good. Recalling that this curve expresses the consumers' willingness to pay, the demand curve of water resource can be plotted under these two assumptions (Mazzola and Del Treste, 1991):

- its central part is linear;

- the characteristics of water market are known, so that the extreme points of the straight line can be determined.

Figure 7 represents the course of the curve in the whole variation domain of 'd' (daily per-capita consumption). As stated before, the part of the curve presenting a practical interest is the approximately linear one ($d_1$-$d_2$) since all variations in d induced by implementation of new infrastructures or by management actions can be reasonably represented in this zone.

The couple ($d_1$ ,$p_1$) can be assessed by appraising costs born by consumers for the minimum water supply (strictly personal and familiar uses) when no water is supplied by the water system and the alternative water supply is represented by private tank-trucks. The value of $d_1$ can be set approximately equal to 0.080 $m^3$/(per capita*day).

The second extreme of the straight line is identified by $d_2$, the optimal per-capita consumption supplied in normal conditions and $p_2$,the price corresponding to the $d_2$ per-capita consumption. It depends on the price policy of each water agency.

In case of free market price coincides with marginal social cost and benefit and a shifting in the amount of available good (from $d_i$ to $d_f$ in Figure 7) produces a gross benefit for consumers equal to area $d_i d_f$LM.

The amount of annual benefit B for N consumers consequent to a variation in the per-capita water consumption can be then computed as

$$B = [(d_f - d_i)*\varepsilon*(N*365)]*[(p_i + p_f)/2] \tag{4}$$

where $\varepsilon$ is the efficiency of the water distribution network (in modern supply systems $\varepsilon$ can be set equal to 0.85).

If an aggregate demand curve is used, this evaluation can be extended to a whole of communities, each having different values of per-capita consumption but equal type of demand curve: the aggregate demand curve will link the values of resource consumed in a given time T ($m^3$/T) with the unit price (£/$m^3$) of water resource.

Social costs of discomfort consequent to a deficit in water supply depend on the selected time step T and on the implemented management strategy. In the case of civil water demand the deficit is equal to $\psi D_2(T)$, being $\psi$ a scaling factor ranging from 0 to

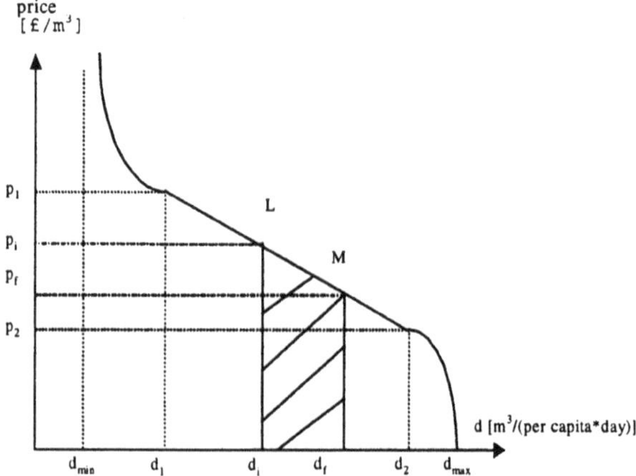


*Figure 7.* Civil water demand curve

1 and $D_2(T)$ the target of water supply in period T.

The monetary value of discomfort is given by (5) where the per-capita consumption does not appear, while it contains the value of the total demand for civil use D:

$$B_{(T)} = [(D_f - D_i)*\varepsilon*(p_i + p_f)]/2 \quad (5)$$

$D_i$ and $D_f$ respectively represent the initial and final values of overall demand with $D_i = D_2$ and $D_f = \psi D_2$, $p_i$ is the initial price, depending on the adopted price policy and the target civil demand $D_2$ and $p_f$ is the final equilibrium price consequent to a variation in the availability of resource during T, expressed by:

$$p_f = \frac{(1-\psi)\cdot p_2 D_2 + \psi \cdot p_1 D_2 - p_2 D_1}{D_2 - D_1} \quad (6)$$

When using a decision tool as NAIADE, water deficits should be computed by a simulation model at the given time step T (e.g. monthly model). The values of deficits for each demand centre will be referred to the chosen time unit and the sum of discomforts during time units T (monthly intervals in our application) over the whole time horizon P of simulation will measure the global discomfort under an assigned system management hypothesis ip:

$$B^{ip} = \sum_{T \in P} B_T^{ip} \quad (7)$$

Consequently, when droughts occur in the simulation period, formula (7) provides measurement of the social discomfort criterion, depending on the applied management alternative.

The criterion described, together with the others already mentioned, gives a first idca of the multidimensional effects in the performances of a set of alternatives, which are the basis for the multicriteria evaluation.

5.3.2. *The collective hygienic risk during drought periods*

The reduction of water supply to civil consumers could sometimes provoke hygienic-sanitary risks. During drought periods such a situation could occur if the available supply is not enough to face the normal demand and if the distribution is rationed. In this latter case pressure may drastically reduce in some area during each supply turn. Under these conditions, infiltration from the surrounding ground could take place, causing the pollution of the water distributed. This is often the case when sewerage does not exist or is in poor repair.

The first step of the procedure herein proposed for the evaluation of hygenic risk for the population of an urban centre is to zone the examined area and to analyse:

- the groundwater levels in each area,

- the maintenance of the sewerage with regard to possible deterioration and risk situations,

- the operation of the water network, with characterisation of all the possible situations of low-pressure functioning. This part of the analysis can be simplified by assuming two situations of operation only: normal supply (no risk) and rationed supply (maximum risk).

These analyses should allow to draw up two thematic maps: one for groundwater levels, the other for sewerage network.

The second phase of the study consists in assigning to the zones within each of the thematic maps a certain degree of risk; therefore each zone will be charachterised by a score varying from 1 to 3 (1: reduced risk, 2: medium risk, 3: high risk).

The third phase consists in superimposing the two risk distribution maps (the 'groundwater level' and the 'sewerage network' one) and summing the scores. The result is a map containing subareas "j" characterised by a common degree of potential unit risk Kj, which is independent of the population density.

The function of absolute risk can be now obtained by linking Kj with the distribution of the population density for each subarea. Vj, the coefficient of absolute risk for the $j^{th}$ sub-area is expressed by formula (8):

$$\mathrm{Vj} = \sum_{i=1}^{Nd} E_j \cdot d_{ij} \cdot A_{ij} K_j \tag{8}$$

with the following meaning of symbols:

- Ej: coefficient of supply, which takes on either value 1 if erogation is continous or value 0 if erogation is rationed,
- Aij: zones contained in the subarea "j "with population density dij;
- _i": index for density $d_i$;
- $N_d$: number of population density levels in which each zone "j" is divided.

The knowledge of the distribution of the degree of risk allows to define distribution strategies during events of serious water shortage: a method to allocate the available resource is to minimise the function of global risk R*tot*:

$$R^{tot} = \int_{Atot} Vj(x,y) \cdot dA = \int_{Acont.} Vj(x,y) \cdot dA + \int_{Aturn} Vj(x,y) \cdot dA \tag{9}$$

subject to the constraints:

$$\int_{Atot} F(x,y) \cdot dA = \int_{Acont.} F(x,y) \cdot dA + \int_{Aturn} F(x,y) \cdot dA = \text{Available Resource} \tag{10}$$

$$\text{Atot.=Acont.+Aturn.} \tag{11}$$

The first constraint expresses the balance between the available water volume for civil consumers and the corresponding distributed volume. *F (x, y)* represents the assigned water requirement in a time unit for unity of area. In the second constraint, the sum of the two areas, the one in which the distribution is continuous and the other with a turned distribution, must be equal to the total town area.

The optimisation problem may be solved by one of the several models available for network simulation. A simplified approach assumes a fixed distribution of the turned areas, obtaining an average monthly degree of risk in the whole critical zone; as a consequence the total risk depends only on the duration of the rationing period.

## 5.4. APPLICATION AND RESULTS

Using the described criteria, a set of management alternatives can be fixed and applied during drought periods. Management alternatives which can be undertaken without planning infrastructural measures are essentially of two types (Rossi *et al.*, 1995): the ones limiting demand satisfaction (type $\alpha$) and those overexploiting certain supply sources for limited periods (type $\beta$). It is also useful to include in the set a zero-option alternative (non-intervention) to control the effectiveness of the other proposed measures. In the case of matter we evaluated seven different management alternatives for the system, three belonging to category '$\alpha$', three to category '$\beta$', while the seventh is the zero" option:

- A: no intervention;
- B: restrictions on water use for irrigation ($\alpha$);
- C: restrictions on water use for irrigation and industrial areas ($\alpha$);
- D: restrictions on water water use for irrigation, industrial areas and civil population ($\alpha$);
- E: exploitation for civil use of a reservoir normally used for hydropower generation ($\beta$);
- F: relaxation of the constraint on the minimum ecological release for all reservoirs ($\beta$);
- G: exploitation of treated wastewater ($\beta$).

It is clear that the set of alternatives has to be adapted by the decision maker to the specific environment which is the object of the analysis; the case study is represented by the water system of Palermo city whose general features are:

Total civil water demand : 156 $Mm^3$/year
Total demand for irrigation use : ~ 33 $Mm^3$/year
Total demand for industrial use : 15 $Mm^3$/year
Total demand : 200 $Mm^3$/year

The use of water supply sources is conflicting because the reservoirs are used both for civil and agricultural purposes.

### 5.4.1. *The process structure*

The structure of the evaluation process is represented in the block diagram illustrated in figure 6; it joins:

- a water system simulation model providing the water allocation in a given period;
- a streamflow generation model (or historical series of streamflows).

They are used to simulate the operation of the given water system in a period of time T in which 's' water crises occur. Drought identification can be pursued by the methods outlined in section 4 or in Rossi (1995). The results of simulation represent thel performances of each alternative and are used in the evaluation of the coefficients of the Criteria-Alternatives matrix which is subsequently used as input for the multicriteria analysis process (Table 4).

5.4.2 *Effects of the application of the alternatives*

Simulation results are summarised in table 4 and figure 9; table 4 reports the Criteria - Alternatives matrix and figure 9 the ranking of the alternatives from the best to the worse as obtained by the application of the NAIADE program (Munda, 1995). In table 4 the criteria performances of each alternative are referred to a four years drought period. The results of the evaluation are expressed at first in terms of dominance of each alternative respect to the others and then, by means of an aggregation process, in absolute terms as showed in figure 9. The same figure reports the set of dominant alternatives which proved stable to sensitivity analysis.

It is worth noticing the importance of the possible reuse of treated wastewater ("F" alternative). Another result is that possible accords among the social parts involved have been pointed out, which allows the decision maker to analyse possible conflicts among water users and the consequent acts aimed at their resolution (Zeleny, 1982). The "dendrogram" of figure 8 shows the degree of similarity and consequent possible alliances among social parts involved.

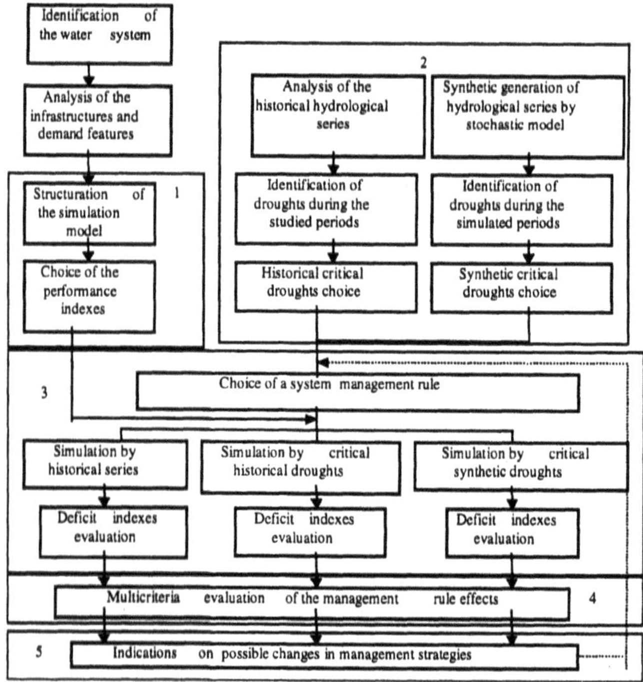


*Figure 8.* Multicriteria evaluation scheme for a water system

TABLE 4. Criteria-Alternatives matrix

| Criterion / Alternative | Nrm. (A) | b-1 (B) | b-2 (C) | b-3 (D) | a-1 (E) | a-2 (F) | a-4 (G) |
|---|---|---|---|---|---|---|---|
| Operating costs (M£) | 109357 | 111219 | 110843 | 105618 | 109357 | 107802 | 105853 |
| Water distribution gross gains (M£) | 483788 | 495678 | 504998 | 491985 | 519182 | 532370 | 529841 |
| Social discomfort (M£) | 312461 | 262903 | 227717 | 215602 | 194854 | 202762 | 178021 |
| Hygienic risk (number of months at risk) | 17 | 17 | 14 | 13 | 14 | 11 | 10 |
| Extra costs for the industries (M£) | 0 | 6437 | 6437 | 6437 | 0 | 0 | 0 |
| Farm production (M£) | 864610 | 737021 | 560476 | 673317 | 868513 | 1086638 | 864610 |
| Environmental flow (M $m^3$) | 48 | 49 | 51 | 57 | 48 | 51 | 0 |

## 6. Conclusions

The severe shortages faced by the city of Palermo in the years 1989-90 showed how unexpected long droughts produce very strong social problems and hygienic risks. In geographic areas where the balance between water demand and supply is jeopardised by environmental sustainability limits, shortages must be faced not only by infrastructural measures, but also by the application of operating rules of water systems based on the possibility of long droughts.

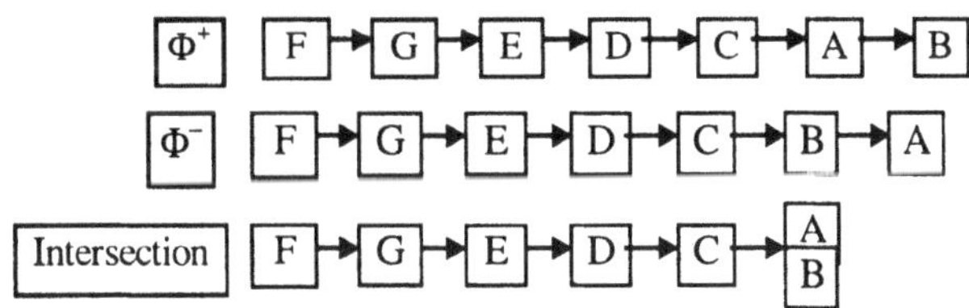


*Figure 9.* Ranking of the alternatives

Many applications have demonstrated how multicriteria evaluation processes, along with drought forecasting models, can be useful decision tools. In the case of water resources management, the NAIADE program, easily treating with qualitative, quantitative and stochastic data has a strong applicability and adaptability to different cases. The application presented in this study shows the possible interesting developments in the field of drought management using a complete approach joining a synthetic streamflow model for the reproduction of drought charachteristics, a simulation model of the water system and a support decision program as the NAIADE used here. The effective use of the multicriteria evaluation tool is strictly related to the structuration of criteria that have to be well fitted in each different application.

In the paper are described the approaches followed for the computation of two out of the seven criteria used in this study, social discomfort and collective hygienic risk, which could be profitably used in other real applications.

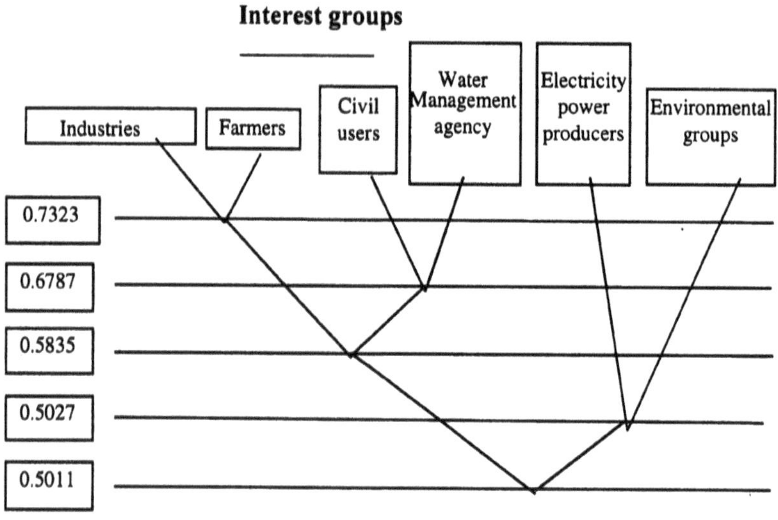


*Figure 10.* Possible alliances among interest groups

## 7. References

Amico, C., Arena, C., Cannarozzo, M. and Mazzola, M.R. (1996) The use of qualitative different water resources to normalise supply in areas exposed to water shortage, *Proceedings of the I international meeting on* "The impact of industry on groundwater resources", Cernobbio, Italy.

Askew, A. J., Yeh, W. W-G. and Hall, W.A. (1971) A comparative study of critical drought simulation, *Water resources research* 7(1), 52-62.

Benjamin, J.R. and Cornell, C. Allin (1970) *Probability, statistics and decision for civil engineers*, Mc Graw-Hill, New York.

Del Treste, A. and Mazzola, M.R. (1991) Valutazione degli effetti economici indotti dal trasferimento di risorse idriche nell'ambito di un sistema ad usi multipli, *Proceedings of the Congress on*: "Grandi trasferimenti d'acqua", Associazione Idrotecnica Italiana.

Gottschalk, L. (1976) Frequency of dry years, *Bullettin series A*, 55, University of Lund - Tenth Anniversary, Lund, Sweden.

Greis, N.P., Steuer, R.E. and Wood, E.F. (1983) Multicriteria analysis of water allocation in a river basin: the Tchebycheff approach, *Water Resources Research* 19(4), 865-875.

Hashimoto, T., Loucks, D.P. and Stedinger, J.R. (1980) Reliability, resiliency and vulnerability criteria for water resource system performance evaluation, *Water resources research* 18(1), 14-20.

Jackson, B.B. (1975) Markov mixture models for drought length, *Water resources research* 11(1), 64-74.
Kuo, J.T., Hsu, N.S., Chu, W.S., Wan, S. and Lin, Y.J. (1990) Real time operation of Tanshui river reservoir, *ASCE WR* 116 (3) 349-361.

La Loggia, G. and Mazzola, M.R. (1992) Real time reservoir management models, *Excerpta*, 6, 185-205.

Loucks, D.P., Stedinger, J.R., Haith, D.A. (1981) *Water resources systems planning and analysis*, Prentice-Hall, New Jersey.

Mazzola, M.R. and Di Leonardo, V. (1997) Multicriteria evaluation for water systems management alternatives during droughts *Proceedings of the International conference on* "Water in the Mediterranean", Istanbul, Turkey.

Mazzola, M.R. (1994) Multipurpose reservoir real time management by stochastic interannual forecasting model *EWRA: Advances in Water Resources Technology and Management*, Balkema, Rotterdam.

Mazzola, M.R. (1990) Analisi dei consumi idropotabili e taratura della curva della domanda *Proceedings of the convention on* "La conoscenza dei consumi per una migliore gestione delle infrastrutture acquedottistiche", CUEN, Sorrento, Italy.

Munda, G. (1995) *Multicriteria evaluation in a fuzzy environment*, Physica-Verlag, Heidelberg.

Rossi, G., Ancarani, A. and Cancelliere, A. (1995) *Gestione dei sistemi idrici durante i periodi di siccità: il ruolo dei modelli*, Istituto di Idraulica,. Idrologia e Gestione delle Acque della Facoltà di Ingegneria dell'Università degli Studi di Catania, Catania.

Rossi, G. (1989) On the identification of regional droughts, *Excerpta*, 3, 145-166.

Santos, M. A. (1983) Regional droughts: a stochastic characterization, *J. of Hydrology*, 66, 183-211.

Yevjevich, V., Da Cunha, G. and Vlachos, E. (1983) *Coping with droughts*, WRP, Littleton.

# EXPERIENCES & CONCLUSIONS AFTER A LONG DROUGHT IN THE METROPOLITAN AREA OF BILBAO
*(August 1988 - November 1990)*

ANGEL L. SILVEIRO Gª-ALZORRIZ
*Assistant Manager. Water Supplies. Water Treatment Plant.*
*(Consorcio de Aguas del Gran Bilbao).*
*Associate Lecturer on Hydroelectric Power Stations*
*Higher School of Industrial Engineering, University of the Basque Country*
*(Dept. of Nuclear Engineering & Fluid Mechanics).*

**Abstract**
From October 1989 till February 1991 Bilbao, the metropolitan area around the city and the city of Vitoria suffered major, progressive and disturbing water restrictions which affected more than 1,200,000 consumers and major sectors of industrial production. These restrictions were the result of significantly lower than normal rainfall between August 1988 and November 1990, and beyond to a certain extent. This situation led to articles in the national newspapers: for instance the headline in *El País* on February 12th 1991, which translates as "*National Meteorological Office data shows the weather map in Spain upside down*". The practical absence of rain along the eastern Bay of Biscay coast and the headwaters of the river Ebro meant that there was not enough surface runoff even to maintain the water levels in the reservoirs in winter. On April 9th 1990 there was only twenty days' supply of water left, even allowing for savings in consumption of more than 30%. I myself saw the situation particularly clearly from my post in the *Consorcio de Aguas* (Water Board). Under the direction of Mr.. José Eizaguirre Basterrechea and in co-ordination with *Aguas Municipales de Vitoria (Amvisa)* (Vitoria Municipal Water Authority) I worked with the technical services of the *Consorcio* to determine what measures should be applied, and to organise, set in place, monitor and analyse those measures. Meanwhile, other working groups at the *Consorcio*, under the direction of the civil engineer Carlos García Marcos, also working jointly with *Amvisa*, designed and carried out seven thousand million pesetas' worth of emergency work in a record time of nine months. Between them they helped to overcome this critical situation. Now, several years later, these experiences have been used as the basis for an in-depth reflection on the events of that time, in the knowledge that some measures were more effective than others. We must learn from this. The purpose of this paper is therefore two-fold: first I wish to explain what measures were reasonable adopted and what results ensued from each of them. Secondly, I wish to present some conclusions and lay down the courses of action which we must take to start correcting our weak points and reviewing our assurance criteria with the ultimate aim of improving the service we provide to municipalities and industrial concerns.



*E. Cabrera and J. García-Serra (eds.), Drought Management Planning in Water Supply Systems,* 388–433.

## PART ONE. "GEOGRAPHICAL BACKGROUND: CAUSES OF THE DROUGHT AND INITIAL MEASURES"

### 1. Geography and Climate

#### 1.1. THE PHYSICAL SETTING

No two droughts are the same in terms of the areas in which they appear or of their duration, intensity and effects, all of which vary according to the stable climatic conditions of the area in question. This means that measures to combat them cannot be generalised. This paper concerns the action taken by *Consorcio de Aguas del Gran Bilbao* and *Aguas Municipales de Vitoria* in 1989, 1990 and 1991. It analyses the measures adopted and presents the courses of action taken to enable such a disturbing phenomenon to be overcome with the minimum discomfort to users and to industrial production.

Greater Bilbao occupies the lower reaches of the river Nervión. It comprises several towns with high levels of economic, demographic and industrial development. This area and the city of Vitoria have a joint population of 1,200,000, which is 55% of the total population of the Basque Autonomous Community.

*Figure 1. The physical setting. Source: Consorcio de Aguas del Gran Bilbao.*

Before the drought water consumption in Bilbao and Vitoria, including consumption in major industrial production, was between 170 and 180 $Hm^3$ per annum.

### 1.2. WATER SUPPLIES IN THE METROPOLITAN AREA OF BILBAO AND IN VITORIA.

The general water supply system for the metropolitan area of Bilbao and Vitoria is based on the following hydraulic infrastructure:

| System | River | Basin (km2) | Reservoir | Useful capacity in $Hm^3$ |
|---|---|---|---|---|
| Zadorra-Alegría | Sta. Engracia | 132 | Urrunaga | 65 |
| | Zadorra | 274 | Ullibarri | 127 |
| | Alegría | ~25 | - | - |
| Ordunte-Cerneja & Zalla | Ordunte | 48 | Ordunte | 22 |
| | Cerneja | ~10 | - | - |
| | Zalla | 2,7 | Zalla | 0,4 |
| Others | Various | ~20 | Various | 5 |

WHAT IS THE *CONSORCIO DE AGUAS*?
It is a local public body run under its own by-laws.Social purposes: supplying water & drainage
Company management systemsThe *CONSORCIO* comprises 36 municipalities

The Zadorra-Alegría system is a multiple-use system which supplies water to the aforementioned areas and is also used to generate electricity, combining the flows transferred to supply Bilbao and its surrounding area. It is also committed through concessions to maintaining various easements equivalent to around 30 $Hm^3$ per annum.

Before the drought the distribution of volumes was as follows:

| | BIZKAIA | | ALAVA | | TOTAL |
|---|---|---|---|---|---|
| | Greater Bilbao | Bilbao | Vitoria | Easements | |
| Zadorra-Alegría | 100 | 24 | 19 | 35 | 178 |
| Ordunte | 0 | 31 | 0 | 0 | 31 |
| Other northern systems | 10 | 2 | 0 | 0 | 12 |
| Other Ebro systems | 0 | 0 | 6 | 0 | 6 |
| TOTAL | 110 | 27 | 25 | 35 | 227 |

Figures in $Hm^3$ per annum

Figures 2a & 2b represent the general system of supply prior to the drought in 1988-1991.

a) ETAR = Waste water treatment plant
b) Cantabrian watershed

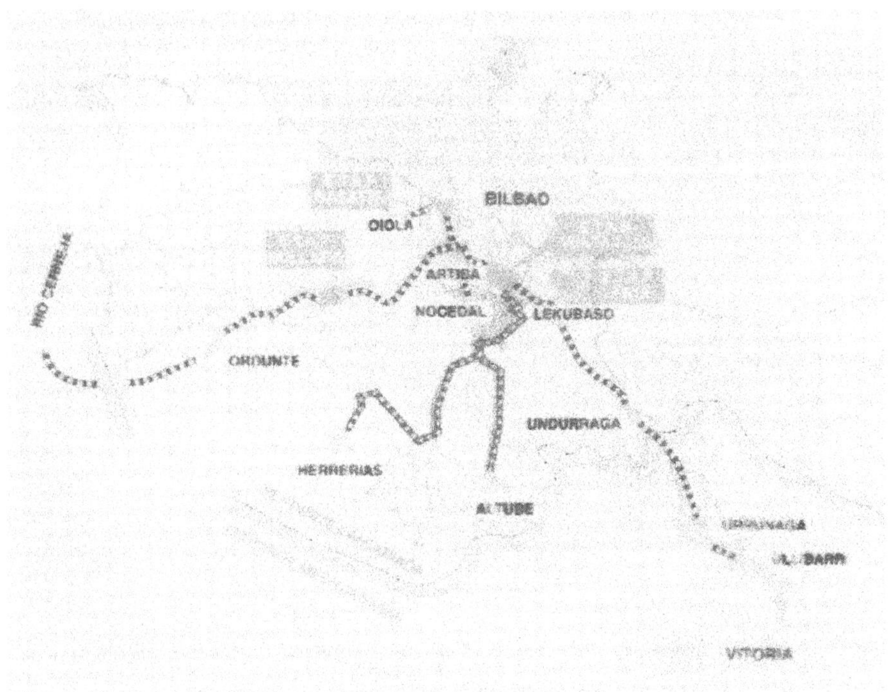

SUPPLY SYSTEM ETAR = Waste water treatment plant

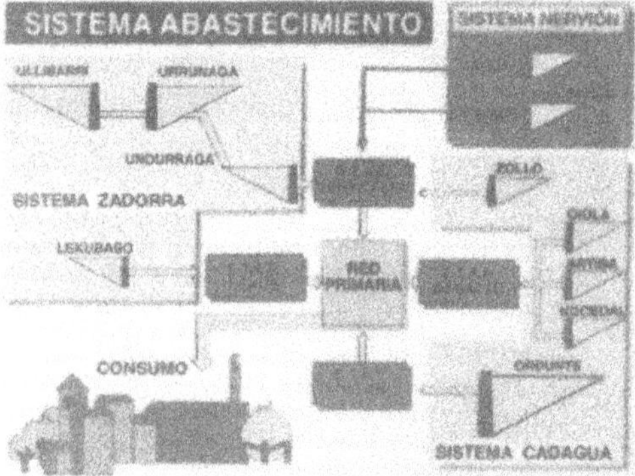


## 1.3. PRESENT CLIMATE IN THE BASQUE COUNTRY & EASTERN BAY OF BISCAY AREA

In the words of the illustrious climatologist Inocencio Font. Tullot in his work "*Climatología de España y Portugal*", published by the National Meteorological Institute in 1984, the climate is "The synthesis of the fluctuating set of atmospheric conditions in a particular area, over a period long enough to be geographically representative". In those terms climate is characterised by (a) the statistical figures for the different climate elements; and (b) the states and changes in weather conditions in accordance with meteorological maps.

Characterised in these terms, climate can be studied under two headings: "statistical climatology", which deals with (a), and "synoptic climatology", which deals with (b). (1) *Historia del Clima en España* by I. Font.

Statistical records and synoptic maps are available at most for the last 200 years, so our presentation of the present climate in the Basque Country and eastern bay of Biscay area is limited to that period. Figure 3 "Main Climatic Regions of the Iberian Peninsula" (1) shows that the Basque Country belongs to the so called "green area" of the peninsula.

*Fig. 3 Main Climatic Regions of the Iberian Peninsula.. Source: "Historia del Clima en España" by I. Font.*

This, with abundant, widespread rainfall, is what our historical memory also tells us. However, Figure 4 "Variation of the five-yearly rainfall rate" (1), giving five-yearly figures for rainfall expressed as percentages above and below the average figure for 1876-1985, shows that variations in the peninsula as a whole are less wide than for the green and brown areas individually. This proves that there is a certain compensation of rainfall in the two areas.

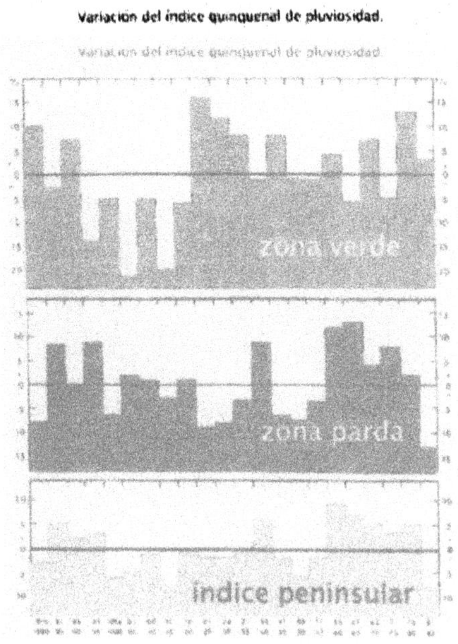


*Figure 4: Variation in five-yearly rainfall rates.*
*Source: "Historia del Clima en España" by I. Font.*

A look at these graphs shows that drops in rainfall of up to 20% have occurred in the green area: greater than those in the brown area. This means that droughts there have a greater impact.

There are four synoptic maps which depict how the climate of the Basque Country is formed. From the North West (NW) comes cool, rainy weather characteristic of autumn and winter. The North brings rainy, very cold weather and snow. From the south (S) comes dry, hot weather as a result of a stationary high pressure area to the west of the peninsula, sending warm winds into the green area which give rise to Foch effects along the Bay of Biscay coast. When this weather system settles in there are long periods with no rain. The anticyclone (A) is typical of summer, and results in pleasant temperatures along the coast. The north wind can sometimes bring light rain in summer.

Figure 5 shows the maps of these weather systems.

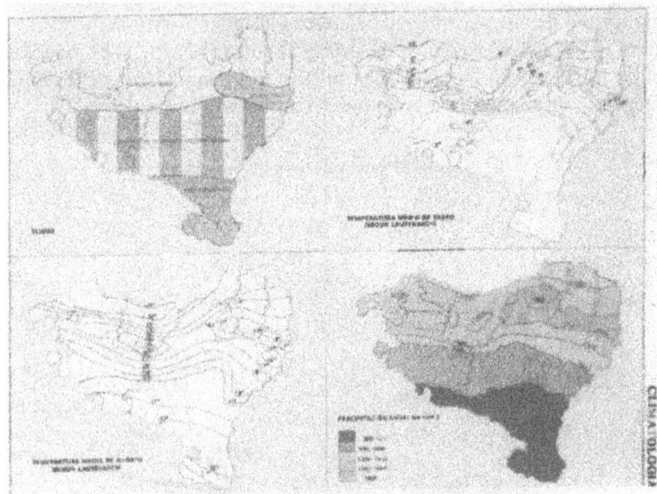


*Figure 5a Statistical climatology in the Basque Country. Source: "Atlas de Euskalherria Erein".*

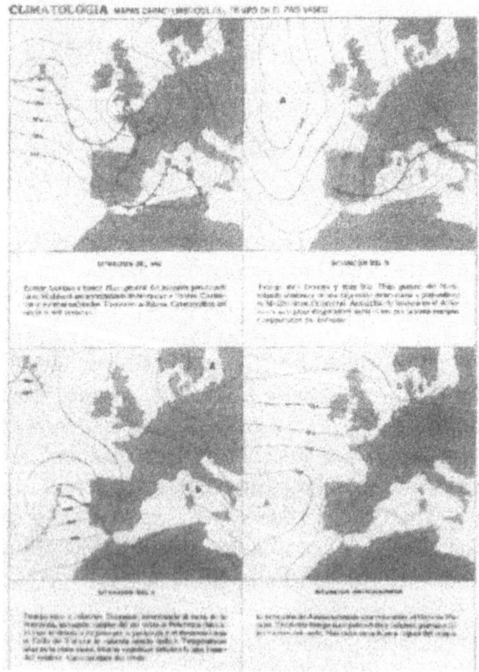


*Figure 5b Characteristic weather maps of the Basque Country. Source: "Atlas de Euskalherria Erein".*

Figure 6 corresponds to mid-winter 1990. This situation prevailed in the metropolitan area of Bilbao for 28 months, with two powerful anticyclones preventing any fronts from reaching the area.

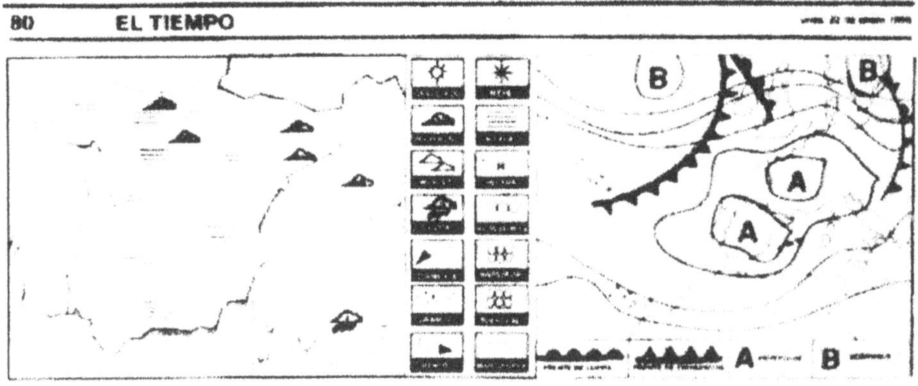


*Figure 6. Weather map for January 27th 1990. Source: " El Correo".*

## 1.4. AVERAGE ANNUAL RAINFALL IN THE THREE PROVINCIAL CAPITALS OF THE BASQUE AUTONOMOUS COMMUNITY.

To provide specific data on the type and extent of the events from Autumn 1988 to early 1991, we present the monthly rainfall figures for the three provincial capitals of the Basque Autonomous Community. Rainfall recorded at the Ullíbarri reservoir can be identified with Vitoria, that at Igeldo with San Sebastián and that at Sondika with Bilbao.

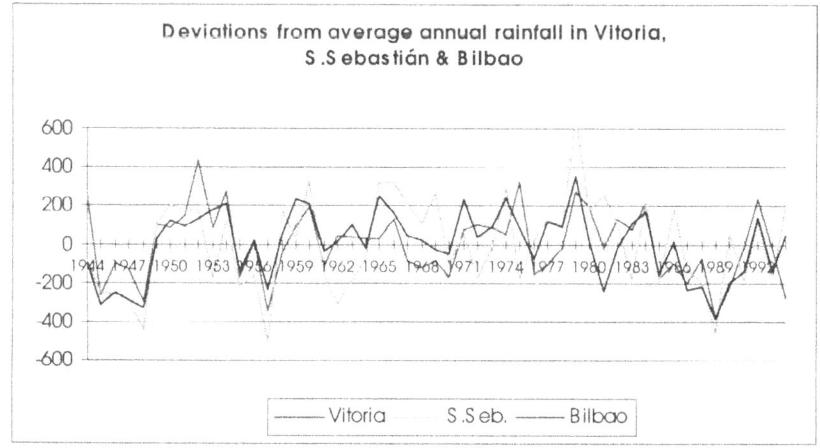


*Figure 7. Deviations from average annual rainfall in Vitoria, S. Sebastián & Bilbao. Source: "Consorcio de Aguas".*

| Rain in the Capital Cities of the Autonomous Community of the Basque Country. (in l/m$^2$). | | | |
|---|---|---|---|
| Year | Vitoria (Ullibarri) | S. Seb. (Igeldo) | Bilbao (Sondica) |
| 1944 | 1204 | 1449 | 1114 |
| 1945 | 676 | 1294 | 895 |
| 1946 | 844 | 1288 | 957 |
| 1947 | 815 | 1256 | 922 |
| 1948 | 647 | 1101 | 879 |
| 1949 | 1041 | 1655 | 1236 |
| 1950 | 1027 | 1743 | 1330 |
| 1951 | 1087 | 1718 | 1303 |
| 1952 | 1372 | 1721 | 1338 |
| 1953 | 1029 | 1367 | 1389 |
| 1954 | 1213 | 1665 | 1425 |
| 1955 | 769 | 1323 | 1069 |
| 1956 | 956 | 1383 | 1228 |
| 1957 | 599 | 1038 | 977 |
| 1958 | 899 | 1711 | 1260 |
| 1959 | 1033 | 1573 | 1444 |
| 1960 | 1135 | 1877 | 1421 |
| 1961 | 826 | 1379 | 1176 |
| 1962 | 984 | 1230 | 1221 |
| 1963 | 977 | 1358 | 1307 |
| 1964 | 974 | 1468 | 1193 |
| 1965 | 967 | 1865 | 1460 |
| 1966 | 1075 | 1864 | 1376 |
| 1967 | 861 | 1756 | 1257 |
| 1968 | 821 | 1649 | 1231 |
| 1969 | 856 | 1812 | 1183 |
| 1970 | 769 | 1442 | 1159 |
| 1971 | 1014 | 1594 | 1441 |
| 1972 | 1041 | 1367 | 1243 |
| 1973 | 1025 | 1532 | 1303 |
| 1974 | 994 | 1840 | 1450 |
| 1975 | 1264 | 1364 | 1297 |
| 1976 | 785 | 1567 | 1137 |
| 1977 | 838 | 1678 | 1327 |
| 1978 | 921 | 1630 | 1304 |
| 1979 | 1217 | 2206 | 1561 |
| 1980 | 1141 | 1719 | 1227 |
| 1981 | 924 | 1802 | 968 |
| 1982 | 1073 | 1652 | 1188 |
| 1983 | 1014 | 1365 | 1319 |
| 1984 | 1158 | 1765 | 1384 |
| 1985 | 763 | 1353 | 1057 |
| 1986 | 841 | 1733 | 1225 |
| 1987 | 743 | 1410 | 979 |
| 1988 | 869 | 1351 | 994 |
| 1989 | 565 | 1089 | 823 |
| 1990 | 717 | 1594 | 1008 |
| 1991 | 930 | 1358 | 1073 |
| 1992 | 1180 | 1669 | 1348 |
| 1993 | 918 | 1364 | 1071 |
| 1994 | 670 | 1749 | 1255 |
| Average | 942 | 1544 | 1210 |
| Median | 956 | 1573 | 1231 |
| Standard dev. | 179 | 238 | 1741 |

*Table 1: Annual rainfall rates in the three provincial capitals of the Basque Autonomous Community. Source: "Consorcio de Aguas".*

A glance at these curves shows how precipitation plummeted as from 1986, especially in Ullibarri.

## 2. Characterisation of the dry periods recorded.

### 2.1. DEFINITION OF DROUGHT FOR A WATER SUPPLY UTILITY.

For water supply utilities, a drought is considered to occur in absolute terms when the inflow over a certain period is significantly less than demand, resulting in an abnormal drop in water reserves. If this occurs habitually the problem is not so much one of drought as of a deficit in water infrastructure, or worse still of the establishing of the wrong exploitation strategies. In general terms traditionally wet areas are less likely to require large-scale adjustments, and therefore frequently have reservoir capacities of less than 50% of the average annual inflow. This makes them even more sensitive to extreme conditions: they are allowed almost to empty every year in the firm belief that "sometime soon" the "traditional" rains will come.

### 2.2. ANALYSIS OF THE DRY PERIODS RECORDED SINCE 1944.

To identify dry periods recorded prior to 1988-1991 we can consult the rainfall records for the reservoirs themselves and correlate them with more detailed records from other observation points, using them as a basis on which to infer events.

Table 2 shows the annual inflow figures for the Zadorra reservoirs, homogenised since 1994 with the inflow of the Alegría system. ( Figures in Hm$^3$):

TABLE N° 2.

ANNUAL INFLOW INTO THE ZADORRA SYSTEM IN HM$^3$

| | 1940 | 1950 | 1960 | 1970 | 1980 | 1990 |
|---|---|---|---|---|---|---|
| 0 | | 248 | 375 | 207 | 370 | 131 |
| 1 | | 237 | 263 | 271 | 290 | 296 |
| 2 | | 299 | 208 | 272 | 312 | 392 |
| 3 | | 218 | 202 | 232 | 241 | 279 |
| 4 | 270 | 442 | 181 | 334 | 311 | 213 |
| 5 | 181 | 118 | 295 | 313 | 201 | 210 |
| 6 | 189 | 285 | 301 | 238 | 248 | |
| 7 | 172 | 118 | 211 | 245 | 246 | |
| 8 | 119 | 302 | 222 | 359 | 297 | |
| 9 | 289 | 261 | 243 | 390 | 123 | |

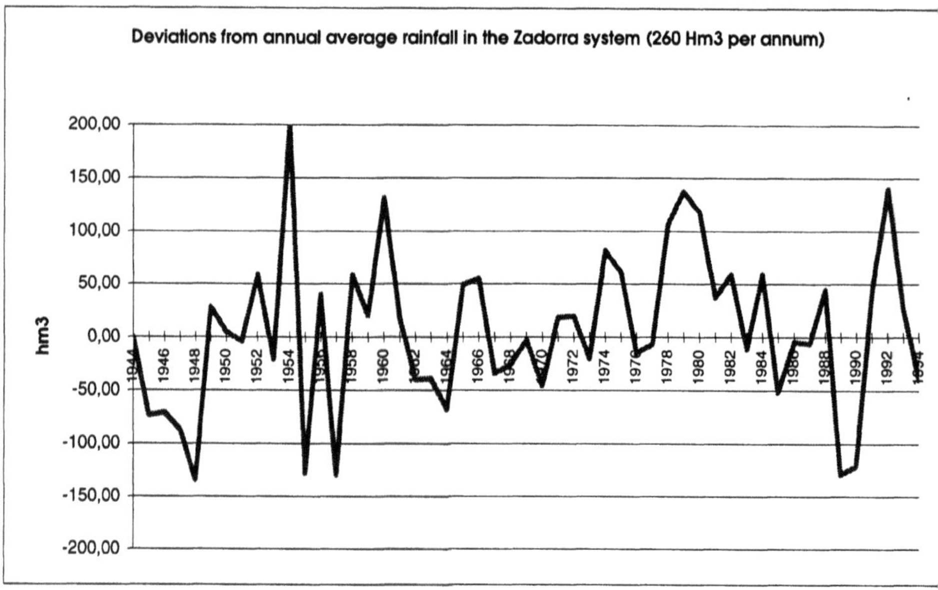


*Figure 8. Deviations from annual average with respect to the median. Source: "Consorcio de Aguas".*

Working with the relevant time series of rains and inflows with a month by month breakdown hydraulicity figures can be worked out via the moving summations of the inflows from 1, 2, 3, 4 ... n months. Taking into account the definition of a drought for water supply utilities, the following dry periods can thus be identified.

| DRY PERIOD | DURATION (MONTHS) | INFLOW $Hm^3$ | DEFICIT* $Hm^3$ | PARTIAL REFILLS |
|---|---|---|---|---|
| 4-1945 to 2-1949 | 47 | 543 | -163 | TWO |
| 4-1955 to 10-1957 | 31 | 390 | -74 | TWO |
| 8-1988 to 11-1990 | 28 | 244 | -175 | ONE |
| * Monthly consumption for all uses from Zadorra in 1988 ≅ 15 $Hm^3$. | | | | |

The first period is longer than the other two, but the deficit is less. There are two partial refills. The second period is only apparently dry, as the two refills practically fill the reservoirs. It is the third period that shows the greatest deficit, even though it is the shortest. The partial refill of April 1989 saved the situation, preventing the reservoirs from emptying if they start the dry period practically full.

## 2.3. DEFICITS OBSERVED: NEED FOR REGULATION.

Table 3 shows the inflows during the periods referred to above:

| TABLE Nº 3 | | | | | | |
|---|---|---|---|---|---|---|
| Nº MONTHS | series 47 | | series 31 | | series 28 | |
| 1 | Apr-45 | 4.150 | Apr-55 | 3.150 | Aug-88 | 5.674 |
| 2 | May-45 | 4.641 | May-55 | 1.000 | Sep-88 | 6.274 |
| 3 | Jun-45 | 0.000 | Jun-55 | 0.850 | Oct-88 | 3.231 |
| 4 | Jul-45 | 0.000 | Jul-55 | 1.089 | Nov-88 | 3.076 |
| 5 | Aug-45 | 0.000 | Aug-55 | 0.463 | Dec-88 | 12.048 |
| 6 | Sep-45 | 0.000 | Sep-55 | 0.369 | Jan-89 | 9.139 |
| 7 | Oct-45 | 0.564 | Oct-55 | 7.200 | Feb-89 | 13.177 |
| 8 | Nov-45 | 2.481 | Nov-55 | 3.300 | mar-89 | 14.783 |
| 9 | Dec-45 | 27.800 | Dec-55 | 7.500 | Apr-89 | 60.799 |
| 10 | Jan-46 | 15.072 | Jan-56 | 49.400 | May-89 | 10.637 |
| 11 | Feb-46 | 13.975 | Feb-56 | 33.400 | Jun-89 | 3.479 |
| 12 | mar-46 | 10.750 | mar-56 | 42.250 | Jul-89 | 2.199 |
| 13 | Apr-46 | 26.748 | Apr-56 | 49.150 | Aug-89 | 1.110 |
| 14 | May-46 | 46.000 | May-56 | 21.000 | Sep-89 | 1.456 |
| 15 | Jun-46 | 6.304 | Jun-56 | 13.600 | Oct-89 | 0.642 |
| 16 | Jul-46 | 0.000 | Jul-56 | 1.505 | Nov-89 | 5.669 |
| 17 | Aug-46 | 0.000 | Aug-56 | 0.956 | Dec-89 | 0.688 |
| 18 | Sep-46 | 0.000 | Sep-56 | 0.458 | Jan-90 | 6.101 |
| 19 | Oct-46 | 0.256 | Oct-56 | 3.200 | Feb-90 | 4.209 |
| 20 | Nov-46 | 6.109 | Nov-56 | 56.300 | mar-90 | 2.654 |
| 21 | Dec-46 | 64.000 | Dec-56 | 14.500 | Apr-90 | 55.302 |
| 22 | Jan-47 | 38.900 | Jan-57 | 24.400 | May-90 | 6.624 |
| 23 | Feb-47 | 32.400 | Feb-57 | 17.400 | Jun-90 | 2.335 |
| 24 | mar-47 | 21.750 | mar-57 | 2.250 | Jul-90 | 0.957 |
| 25 | Apr-47 | 4.150 | Apr-57 | 7.150 | Aug-90 | 0.562 |
| 26 | May-47 | 5.200 | May-57 | 13.000 | Sep-90 | 1.592 |
| 27 | Jun-47 | 0.850 | Jun-57 | 13.600 | Oct-90 | 1.701 |
| 28 | Jul-47 | 0.000 | Jul-57 | 1.375 | Nov-90 | 8.704 |
| 29 | Aug-47 | 0.000 | Aug-57 | 0.625 | | 244.722 |
| 30 | Sep-47 | 3.933 | Sep-57 | 0.355 | | |
| 31 | Oct-47 | 2.256 | Oct-57 | 0.200 | | |
| 32 | Nov-47 | 10.109 | | 390.995 | | |
| 33 | Dec-47 | 52.700 | | | | |
| 34 | Jan-48 | 57.200 | | | | |
| 35 | Feb-48 | 9.801 | | | | |
| 36 | mar-48 | 0.625 | | | | |
| 37 | Apr-48 | 6.395 | | | | |
| 38 | May-48 | 18.000 | | | | |
| 39 | Jun-48 | 4.000 | | | | |
| 40 | Jul-48 | 0.000 | | | | |
| 41 | Aug-48 | 0.180 | | | | |
| 42 | Sep-48 | 10.540 | | | | |
| 43 | Oct-48 | 3.323 | | | | |
| 44 | Nov-48 | 5.676 | | | | |
| 45 | Dec-48 | 3.422 | | | | |
| 46 | Jan-49 | 19.800 | | | | |
| 47 | Feb-49 | 2.575 | | | | |
| | | 542.635 | | | | |

*Table 3. Identification of dry periods at the Zadorra reservoirs. Monthly inflows Source: "Consorcio de Aguas".*

Figure 9 shows the monthly deficits as regards consumption from the Zadorra reservoirs in each of the three dry periods identified, lasting 47, 31 & 28 months respectively.

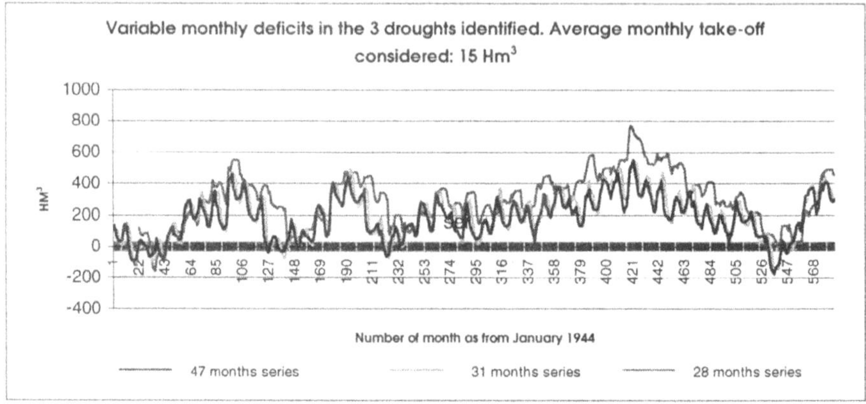


*Figure 9. Variable monthly deficits for each dry period studied. Source: "Consorcio de Aguas".*

By integrating these monthly deficits we obtain the water reserves required at the beginning of the period to ensure that the reservoirs do not empty.

It can be observed that these cumulative deficits are variable, and depend clearly on the intensity and duration of the dry series.

The evolvent of the highest figures for each month over all known periods determines the guarantee curve. The effectiveness of this is discussed below.

## 2.4. EXPLOITATION OF THE ZADORRA SYSTEM UP TO 1990.

Since the Zadorra system is a multiple use system, covering both water supplies and electricity generation, as indicated in section 1.2 above, a frontier of interests must be drawn to determine when hydroelectric customers can use the reservoir without endangering supplies.

Up to 1990 hydroelectric users and the *Consorcio* set this frontier of interests by determining the "guarantee curve" on the basis of the hydraulicity figures known at that time. The safety coefficient was the dead volume of the reservoir which could not be used to generate electricity. This gave a reserve of just two months even with heavy restrictions. Guarantees of supply were therefore based on deterministic procedures with a 100% guarantee rate, as shown by a historical series of inflow records covering 51 years, with a month by month breakdown. The changes in water reserves in 1988 are shown in figure 10.

Changes in the Zadorra system reservoirs in 1988

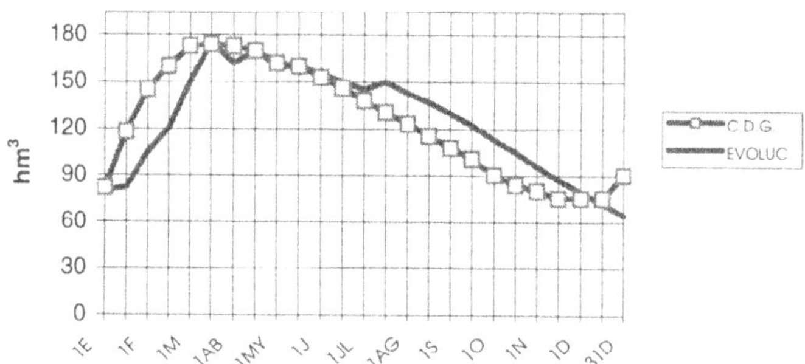


*Figure 10. Changes in water reserves in the Zadorra system in 1988. Source: "Consorcio de Aguas".*

The guarantee curve used at the time is drawn on the basis of hydraulicity from 1944 to 1987, a period which includes two dry series. The reservoir could reach its maximum level in April, so that water would need to be released and it was not possible to store any autumn or summer rains.

The summer of 1988 was exceptionally wet, with a runoff of more than 40 $Hm^3$, which was not stored, in line with agreed practices.

However the wet summer was followed by a dry autumn and a dry winter.

So it was that on December 1st the free exploitation curve, or guarantee curve, then set at 70 $Hm^3$ , was intersected. There was no rain in December, when the weather was much warmer than normal for the time of year.

## 3. Events leading up to the crisis: January 1989 - April 1989.

### 3.1. SITUATION & ANALYSIS OF WATER RESERVES AT THE END OF 1988.

On December 28th 1988 the Exploitation Section of the *Consorcio* sent the following analysis of the situation to its superiors: "The lack of rain in the Autumn has resulted in the water reserves in the Zadorra system being at their lowest level at this time of year for the last 12 years. The reservoirs contain a total of 65 $Hm^3$, and there is no runoff from the latest rains because the basin is in summer-like condition. Rainfall in January will be crucial for our reserves".

At the beginning of 1989 things failed to change. Public awareness, in the context of a collective historical memory of a very wet climate (1960-1980),was low. It was not until

February that the first headlines appeared in the press referring to the problem facing us. On January 15th 1989 the reservoirs of the Zadorra system held 60 $Hm^3$, a figure 62 $Hm^3$ below the guarantee curve. Reserves at Ordunte were a little higher in percentage terms at 12 $Hm^3$, a little over 50% of total capacity.

## 3.2. DRAWING UP THE FIRST PROGRAMME OF CORRECTIVE MEASURES.

The lack of surface runoff was evident. Around that time the first programme of corrective measures was submitted. This was to be deployed in three stages as follows (February 6th 1990):

Stage 1:

- Limiting of rights of use of Zadorra.
- Gathering of information & involvement of other institutions in regulating different levels of alarm due to lack of water.
- Study of the difficulties involved in restricting water supplies to the population.
- Launching of an advertising campaign to call for voluntary water savings.
- Banning of watering by night and by day and of non-essential water consumption.
- Stepping up of the location and repair of leaks in municipal distribution networks. Inventory of possible savings.

Stage 2: (if there was no rain in February)

- Forming of a drought monitoring committee.
- Stepping up of the advertising campaign.
- Organising of working meetings with heads of municipal services to put restrictions into action.
- Penalising of consumption higher than average established rates.

Stage 3: (to follow stage 2 if matters continued to worsen)

- Limiting of water also to large industrial users.
- Gradual extension of restrictions.
- Study of how to use "back-up" reserves located below elevation 533
- Stepping up of water monitoring at source and in the network.

By that time the problem was public knowledge. On Sunday February 5th the newspaper *El Correo* published an article under the headline "*Driest winter of the decade: outlook gloomy*"

At the same time the Head of the bay of Biscay Weather Office, José Ignacio Usabiaga, explained in the newspaper *Deia* that "*the currents from the west which formerly flowed through the Bay of Biscay have now moved up towards Great Britain, which removes their moderating effects and jams the situation, giving rise to long periods of settled weather*". He also added that he doubted whether there had been a similar climatic situation in the last 60 years.

### 3.3. INVENTORY OF CAPABILITY TO MAKE SAVINGS AND INITIAL MEASURES

The first stage of this plan envisaged the drawing up of an inventory of possible water savings.

In cities with large irrigated areas, watering restrictions are a good source of savings. The main source of savings available to us was to restrict the easement rights to the Zadorra, and that we did.

After analysing all the possibilities we concluded that there were only very limited possibilities of savings, and that we should concentrate on attempting to change they way people used water. Restrictions could be used in two areas where there was no recent experience: one was to set an example to encourage saving and show solidarity with other institutions, and the other was to "recover" some of the water lost through leaks by cutting supplies at night, provided that such hypothetical savings were not offset by possible pipeline breaks as a result of the manipulation of the network, so that the water saved was not used after all. Such cuts were also assumed to be useful in bearing witness to the situation.

However it was believed that large scale industry could restrict its consumption voluntarily via undertakings to save such percentages as might be advisable as the situation developed.

Even with all these ideas the belief at the time was that savings could not exceed 20% of the figures for consumption in September 1988, which were used as the basis for calculation.

Initial simulations were also drawn up to determine how far we could go taking into account inter-annual inflows and savings in consumption.

The first meeting between institutions was held with the *Confederación Hidrográfica del Ebro* on February 22nd 1989, with just 48 $Hm^3$ of water in the Zadorra system. The meeting was held in Vitoria, and it was agreed to reduce the easements involving the rivers in the Zadorra "*to the minimum compatible with the needs and commitments of other users downstream*". At the same time it was undertaken to launch a campaign to save water, with municipal authorities being asked to co-operate by restricting watering.

The first measures under the plan established 16 days earlier were thus set in motion.

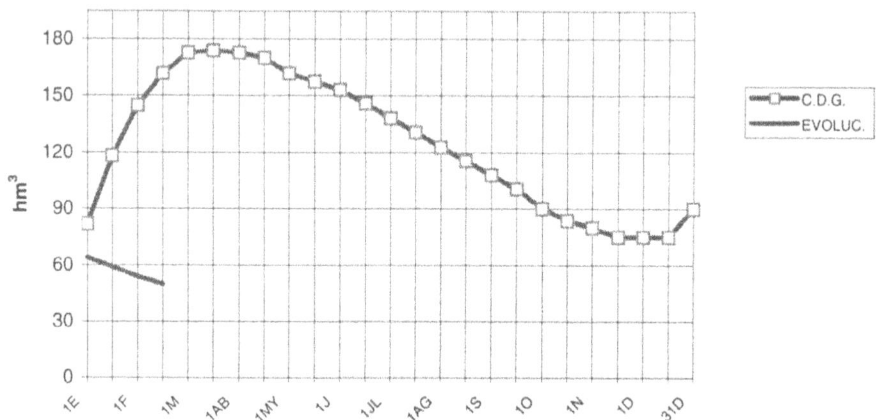


*Figure 11. Changes in water reserves up to February 15th 1989.*

3.4. THE RAIN IN MARCH & APRIL 1989 AND FURTHER ANALYSIS OF THE SITUATION.

The measures taken heightened awareness and led altogether to a saving (not counting the reduction in easements, which was not official) of barely 5%. Since August 1988 we had been through two seasons which were to all intents and purposes dry summers. What would the Spring bring?

In late February and early March there was rain, which reversed the trend. A total of 30 $Hm^3$ was recorded, bringing the reserves in Zadorra back up to 60 $Hm^3$. However we could not relax, as the situation was still difficult, with a further 110 $Hm^3$ needed to reach the level where supplies were guaranteed. April was an exceptionally good month, with 60 $Hm^3$ being added to the reservoirs so that by the end of the month Zadorra held almost 100 $Hm^3$ and Ordunte was full, when it had held only 8 $Hm^3$ on February 25th.

We knew that we now had to face the summer with reserves which were more typical of September, so action was stepped up, in the belief that restrictions would have to be imposed in October.

It was also confirmed that a new pattern of water resources was appearing, and that the guarantee curve would have to be revised in the short term, not only as regards statistics but also as regards the very concepts underlying it.

Situation in the reservoirs of the Zadorra system up to May 1st
1989

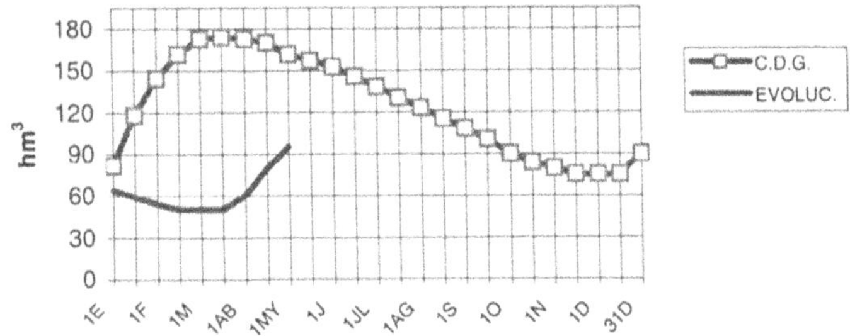


*Figure 12. Changes in water reserves up to April 30th 1989.*

## PART TWO.
## "THE DROUGHT AND HOW IT WAS HANDLED IN 1989, 1990 & 1991"

### 4. Measures Taken in May - December 1989.

#### 4.1. ADVERTISING CAMPAIGNS ARE STEPPED UP.

Several slogans were drawn up to bring home the message that water consumption should be cut back. Leaflets were delivered door to door, posters were placed on telephone booths and bus-stops and commercials were placed on TV, radio, in the local press, etc.

The campaign was completed with the distribution of stickers with the same messages in schools and major centres of production. All these actions were aimed at encouraging better use of water, without being unduly pessimistic. Friendly, not too alarmist forms were used. The savings resulting from this campaign did not exceed 2%, though the campaign was continued throughout April, May and June.
A more aggressive campaign was launched and maintained in the Autumn of 1989, based on a phrase which was to become famous: "*CUIDEMOS EL AGUA HASTA LA ÚLTIMA GOTA*" ["let's take care of every last drop"]. Enormous numbers of telephone calls were received at the *Consorcio's* switchboards reporting misuse of water by consumers and offering solutions, though most of those offered were impractical or of very little use.

Some of the posters designed for this campaign are shown below.

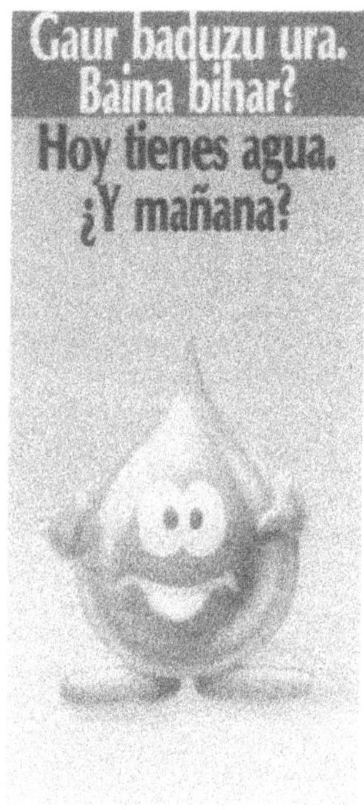


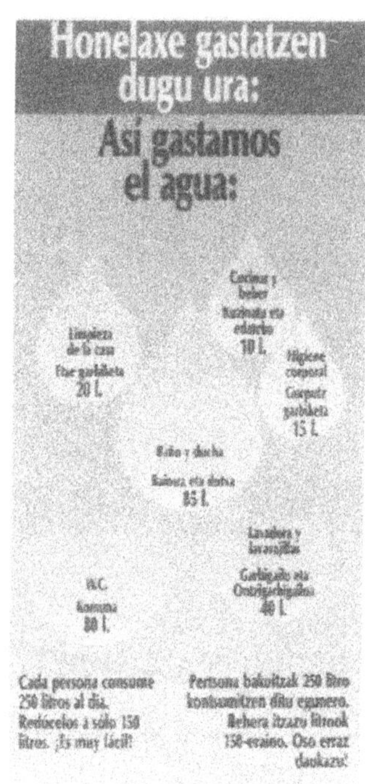


a) You have water today, but what about tomorrow? b) How we use water
c) Cooking & drinking 10 l d) House cleaning 20 l
e) Bodily hygiene 15 l f) Baths & showers 85 l
g) Washing machine & dishwasher 40 l h) Toilet 80 l
i) We use 250 litres per person per day: it could easily be cut to 150

## 4.2. RESTRICTIONS ON USE BY CONSUMERS

As the advertising campaign was drawing to a close in June 1989 the *Consorcio* agreed to ask all town halls to adopt measures which would not directly affect users but would still allow water consumption to be reduced. This means limiting the watering of allotments, parks & gardens and the washing down of roads.

The washing of cars, trucks and other vehicles with mains water was banned except by forms specialising in this activity. These measures, together with the advertising campaign, resulted in savings of around 4%.

A common sight during the period when restrictions were in force was that of town council vehicles loading up with water from streams and wells, and operators painting "non potable water" signs on their trucks.

At the same time the *Consorcio* began to organise a service of water tankers and to provide technical assistance to answer enquiries from private individuals concerning the possibility of installing water tanks (it being assumed by then that restrictions would be introduced). This was to become a burning issue subsequently.

## 4.3. REDUCTION OF EASEMENT RIGHTS TO THE ZADORRA SYSTEM

These reductions had been going on discreetly ever since they were agreed in the first meeting with the CHE on February 22nd 1988. The continued drought in several river systems forced the Ministry of Public Works and Town Planning to table Royal Decree 798 of June 30th 1989, under which exceptional measures were authorised to make use of water resources as covered by article 56 of the Water Act.

This Royal Decree affected the Duero, Tagus, Guadiana, Guadalquivir, Segura & Ebro river systems, and showed just how widespread the drought was. The governing boards of the affected water authorities were empowered to introduce reductions in water provisions with a view to ensuring a rational description of their scant resources, and to carry out emergency work, etc.

This provision enabled the Ebro water authorities to reduce the easement flows from the Zadorra reservoirs by 500 l/sec.

## 4.4. RESTRICTIONS BEGIN TO AFFECT USERS.

On October 4th 1989 the General Assembly of the *Consorcio* agreed to apply restrictions in the water supply to the municipalities and industrial concerns connected to the primary networks, and the Chairman was empowered to increase them progressively and take any other action which might be in order to implement measures.

At the same time, Aguas Municipales de Vitoria, S.A. adopted a similar agreement. These measures did not affect large scale industrial concerns, which imposed their own restrictions, but were monitored daily by equipment set up for that purpose. The initial restriction of six hours a day, from 00.00 to 06.00, remained in place until November 5th. On that date the restrictions were extended to 8 hours, starting at 22.00, though the earlier timetable was maintained for non-working days.

To manipulate, adjust, close, open, monitor and measure consumption at the primary network meters, etc. around 300 sites, some almost inaccessible, had to be visited every day. Special operatives groups were hired and trained. A total of 38 operators working on timetables adapted to the requirements of each period, handled the actual manipulation work. They filled in daily incident reports, and covered an area of 18*10 km. Their main jobs were the following.

Tank outlets: regulate the inlet valve to the reduced level agreed on. Difficulties arose here due to lack of adjustment.

Direct outlets: shut off at the agreed time and re-open long enough in advance for high points of the primary network, which might have emptied, to fill. This made for a much more effective resumption of service to municipal networks.

Pump outlets: limit times so that pumps operated in accordance with the savings agreed on.

In all cases these groups had to co-ordinate their work with municipal workers in charge of cutting off supplies in their respective local networks.

The Venta Alta treatment plant, which is the supply control centre, treated the average flow rates imposed by the cuts agreed on. Wide variations were observed in output from tanks. There was a drastic increase in cuts in late 1989.

Savings were as follows, calculated on the basis of consumption in September 1989, duly homogenised.

| TYPE | CUTS | % SAVED |
|---|---|---|
| Domestic | 00.00 - 06.00 October 1989 | 10 – 15 |
| Domestic | 22.00 - 06.00 Nov. 1989 | 15 – 20 |
| Domestic | 21.00 - 06.00 Dec. 1989 | 20 – 22 |
| Industrial | Voluntary in October | 20 |
| Industrial | Voluntary Nov. - Dec. | 25 – 30 |

*Figure 13* shows the water reserves in the Zadorra system in 1989. On December 31st the system held only 15 $Hm^3$.

Situation in the reservoirs of the Zadorra system in 1989

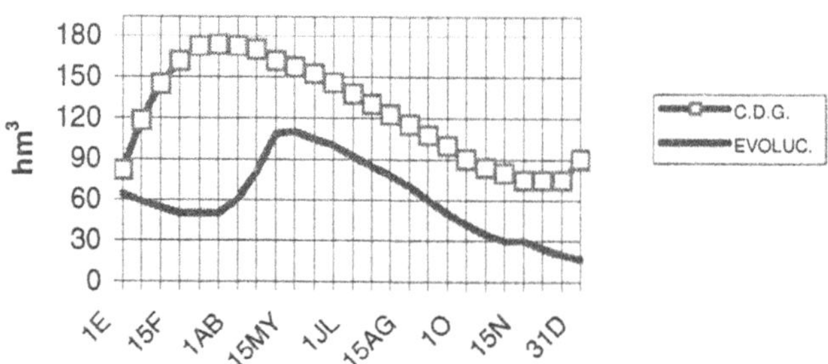


*Figure 13. Changes in the Zadorra reservoirs in 1989*
*Source: "Consorcio de Aguas"*

On the commencement of these measures in October 1989 the *Consorcio* felt it necessary to set up a department to handle complaints and deal professionally with customers, co-ordinating the actions of the exploitation sections. This department worked well and with very little recognition of its merits, assuming the responsibility for "giving explanations", though they sometimes failed to convince people. It has now become the *Consorcio's* Communications and Image Office.

## 4.5. CAMPAIGN TO DETECT AND REPAIR OF WATER LEAKS: STAGE 1

The *Consorcio de Aguas* is directly responsible for the primary network but not for local distribution networks, which are run and maintained by the respective municipal authorities. However in this situation a supra-municipal policy needed to be established to intensify actions to detect and repair water leaks in distribution pipes as one more factor in combating the drought.

The *Consorcio* agreed to use the funds available under its "Aid Plan" for this purpose, and a two-year campaign was set in motion with the aim of recovering the equivalent of 10% of the total volume of supply. This plan went into action in November 1989.

## 4.6. EMERGENCY WORK AND HOW IT WAS APPROACHED.

From October 1989 onwards the technical services of the *Consorcio* looked into alternative supplies from river courses other than those included in the present systems. At least 15 water catchment projects were carried out, at an estimated cost of 4,000 million pesetas.

Water collected in the northern river systems was fed directly into the Undúrraga - Venta Alta and Ordunte - Elejabarri mains pipelines, owned by Bilbao City Hall. This would have positive effects in that with demand being met by non regulated water the take-off from the reservoirs would be lower. The water collected in the Ebro system was fed into the Zadorra reservoirs to increase their infeed capacity.

These projects helped to normalise the situation, but for them to be effective it would have to start to rain so that the larger surface area used for water catchment could be taken advantage of. The emergency projects carried out are listed below.

| **PROJECTS** | **RIVER SYSTEMS** |
|---|---|
| ARRATIA RUNOFF | NORTH |
| PUMPING FROM CADAGUA | NORTH |
| COTORRIO BARBADUN | NORTH (WATER FOR INDUST. USE) |
| BOLINTXU | NORTH |
| PUMPING FROM ZAYAS | EBRO |
| ARAYA HEADWATERS | EBRO |
| PUMPING FROM NANCLARES | EBRO |
| PUMPING FROM BAYAS | EBRO |
| PUMPING FROM EGA | EBRO |
| TAKE-OFF FROM UYAR | EBRO |
| GROUND WATER | EBRO (PROSPECTING) |
| PUMPING FROM NERVIÓN | NORTH |
| RE-USE OF WASTE WATER | GALINDO |

## 5. Measures Taken in January - April 1990.

### 5.1. RESTRICTIONS ARE STEPPED UP.

At the beginning of 1990 the *Consorcio* was forced to introduce very severe restrictions. On January 2nd water supplies began to be cut off from 18.00 to 06.00, i.e. 12 hours a day. Such long restrictions came close to the technical limits compatible with the time needed to cut off and restore service, especially in municipalities with large diameter mains. Many areas suffered drops in water pressure and hydrants were left open, with the resulting threat to water quality. On February 2nd the *Consorcio* approved a document titled "Standards of application for water users in *Consorcio* member and associate municipalities". The simulations run to check out various hypotheses and distributions of inflows shoed that it was advisable to aim for restrictions of more than 45% to ensure that the reservoirs did not empty (there was a reserve of 10 $hm^3$ below elevation 533).

These heavy, prolonged restrictions had the following effects:

| TYPE | CUTS | % SAVED |
|---|---|---|
| Domestic | 18.00 - 06.00 Jan. 1990 | 25 – 30 |
| Domestic | Idem incl. non-working days Feb. 1990 | 30 – 32 |
| Domestic | Idem. March - April 1990 | 20 - 25 (rain in April) |
| Industrial | Voluntary Jan. 1990 | 40 |
| Industrial | Voluntary Feb - April | 44 - 50 |

Figure 14 shows how levels changed in the Zadorra system reservoirs up to April 1st 1990, when there was only 5 $Hm^3$ left above elevation 533.

Situation in the reservoirs of the Zadorra system up to April 1st 1990

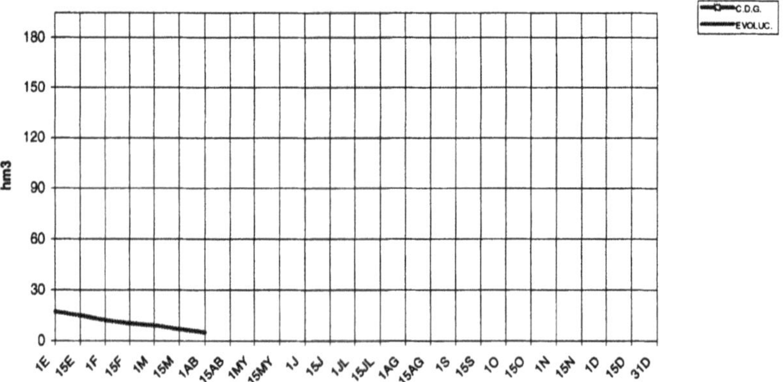


*Figure 14. Changes in the Zadorra system up to April 1st 1990*
*Source: "Consorcio de Aguas"*

This figures does not include the "guarantee curve", which was undergoing an in-depth revision. It is worth pointing out that throughout this complex process there were less breakdowns than expected, though the first fortnight was discouraging to the point where it was thought that it might prove technically impossible to impose restrictions. But after the initial breaks at the weakest points the number settled down to an average of 10 per day after the mains valves were opened, in an overall system estimated to contain around 2000 km of piping. Manipulation was carried out carefully and professionally.

There were difficulties in measuring the dose of sterilising agents used, and there was a considerable dispersal in the chlorine content in the network. The savings achieved with 12 hour cuts were considered as being the most that could be achieved, and these cuts were believed to be the most that the patience of users would tolerate. Stepping up to 16 hour cuts would have been very difficult, and the increase in savings would have been negligible in many sectors of the community.

The consumption time modulation curves at the outlet from the plant are revealing: the extent of stocking up in the period before the cut-off time could be used to estimate the extent of saving on the following day.

## 5.2. CAMPAIGN TO DETECT AND REPAIR LEAKS: STAGE 2

In 1990 882 km of piping was inspected and a total of 1,278 leaks were located and repaired. Three pipeline auscultation companies and three firms of experts in urgent repairs of malfunctions in cities were used.

By the time the plan came to an end in 1991 a volume equivalent to over 200 litres per second had been saved, i.e. almost 5% of total water consumption in the metropolitan area.

## 5.3. EMERGENCY WORK AND HOW IT WAS CARRIED OUT

Legal backing for the emergency work was provided by the Ministry of Public Works and Town Planning in the shape of Royal Decree 296 of March 2nd 1990, under which exceptional measures were adopted as envisaged in Article 56 of the Water Act to meet the demand for water supplies in the Basque Country.

This decree envisaged the following actions:

* Authorisation from the water authorities in the Ebro and Northern systems to collect surface and ground water.

* Permission to go on doing this up to December 31st 1991.

* Supplying utilities benefit from these measures, but it is agreed that compensation will be paid when appropriate.

* Ownership of the work done is to remain with the water authorities.

* Setting up of exceptional contracting arrangements.

Perfect co-ordination was required between the *Consorcio de Aguas, Aguas Municipales de Vitoria S.A.*, Bilbao City Hall and the Cadagua area fire service to put these measures into operation throughout 1990.

Meeting the targets set as quickly as possible with all the inconveniences of the prevailing circumstances required tremendous efforts in planning, co-ordination, management and performance of projects and work on the part of all the people and organisations involved. The firm support of the different water authorities, the co-operation and understanding of the municipalities and private individuals affected by work and the dedication of the people performing that work were all highly praiseworthy. More than 7,000 million pesetas' worth of work was done, funded by the central government, the Basque Government, the provincial councils of Bizkaia and Alava, the *Consorcio de Aguas & Aguas Municipales de Vitoria S.A.*, as follows (figures are given in millions of pesetas):

| | Central Govt. | Basque Govt. | Bizkaia Prov. C. | Álava Prov. C. | *Consorc. Aguas* | *Amvisa, S.A.* | TOTAL |
|---|---|---|---|---|---|---|---|
| Block I | 650,000 | 650,000 | 520,000 | 130,000 | | | 1.950,000 |
| Block II | 350,000 | 322,660 | 258,080 | 64,520 | 800,000 | 200,000 | 1.995,260 |
| Block III | 774,311 | 801,651 | 641,368 | 160,342 | 619,448 | 154,862 | 3.151,984 |
| TOTAL | 1.774,311 | 1.774,311 | 1.419,448 | 354,862 | 1.419,448 | 354,862 | 7.097,243 |

The emergency work on the northern river system was highly stable thanks to the behaviour of the river Cadagua.

The emergency work in the Ebro watershed was more sensitive to rainfall, with effective inflows being obtained with very little rain.

Catchments from the northern river systems contributed around 25 $hm^3$, of which 12 $hm^3$ was pumped from the Cadagua.

Catchments from the Ebro system contributed around 19 $hm^3$, of which 8 $hm^3$ was obtained by pumping. *Figure 15* shows where emergency work took place.

## 5.4. OTHER MEASURES

Among the most outstanding emergency measures taken were the pumping of water from the river Nervión, the re-use of industrial water and ground water. The use of the Nervión was proposed by the inter-institutional committee as an extreme measure only, because this river is highly polluted, establishing that it would only be used if water quality could be maintained at grade A2 and/or A3.

a) MEASURES TAKEN DURING THE 89-90 DROUGHT: GENERAL VIEW
b) Shed line
c) Venta Alta treatment plant
d) ... dam
e) ... reservoir
f) LEGEND
- Arratia runoff
- Pumping from Cadagua
- Cotorrio-Barbadun
- Take-off from river Zayas
- Bolintxu
- Fuente Iturriotz headwaters (Araya)
- Fuente Turbaz (Opakua)
- Take-off from river Bayas
- Take-off from river La Torca (Nanclares)
- Take-off from river Ega (Maeztu)
- Take-off from stream at Vicuña
- Take-off from river Uyar (Zalduendo)
- Pumping P-16
- Re-use of industrial water (Galindo)
- Take-off from river Nervión (Arrigorriaga)

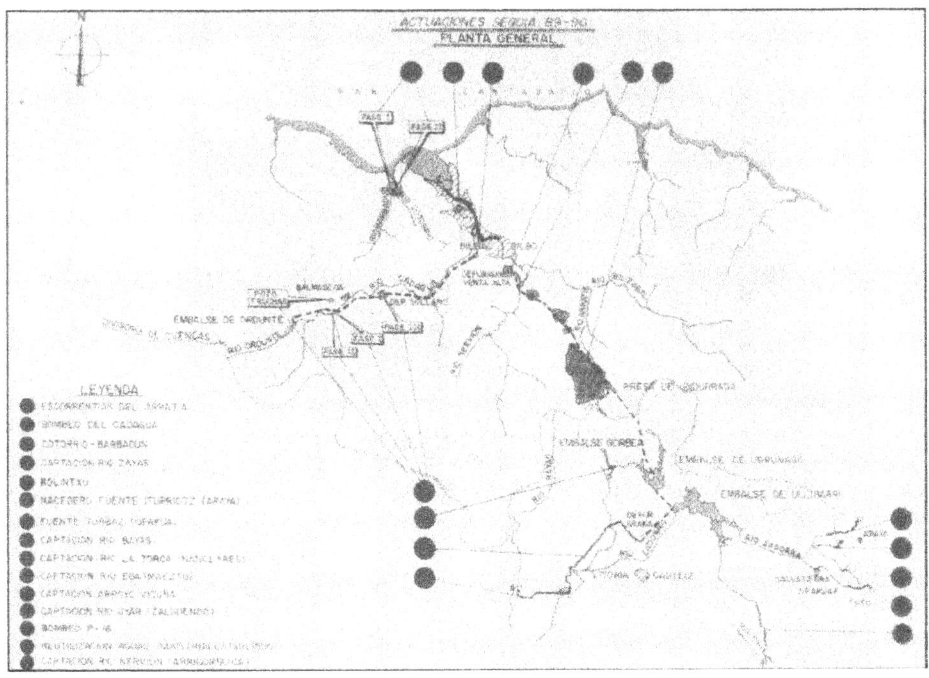


*Figure 15. Emergency work in the Northern and Ebro systems.*
*Source: "Consorcio de Aguas"*

The Venta Alta treatment plant is designed to deal with extreme water qualities, as it has a treatment line based on ozone and active carbon. A committee set up at the Basque Government Department of the Environment laid down the conditions for the use of water from this river, setting minimum flow rates and a biological control system at the entry to the plant.

The re-use of industrial water was proposed as a drought palliation measure, but as part of a scheme designed for other reasons. This consisted in building a tertiary connection for the tributaries of the Galindo waste treatment plant to supply up to 10,000 m3 per day to Altos Hornos de Vizcaya. This plan never went into operation.

Ground water was seen as another emergency option. This scheme was led by the *Ente Vasco de la Energía* ("Basque Power Organisation"), the Geology Dept. of the MOPU (Ministry of Public Works) and the Provincial Council of Álava. Except for the major spring at La Torca in Nanclares, ground water did not prove to be a significant solution. None of the facilities planned actually went into operation.

Un sacerdote leridano dirige la oración en demanda de agua en pleno centro de la capital vizcaína

Varios centenares de personas rezaron ayer en Bilbao para que llueva

**Rosarios para el agua**

La céntrica plaza bilbaína de Federico Moyúa fue escenario en la tarde de ayer de esperanzadas rogativas a la Virgen para pedir que llueva abundantemente y en breve en todo el País Vasco. El acto fue organizado por un sacerdote de la provincia de Lérida, que aseguraba haber recibido un mensaje divino a través del cual se le comunicaba que si el día 3 de marzo esa plaza se llenaba de gente en «íntima oración» se aliviaría la pertinaz sequía. Una imagen de la Virgen de Begoña, otra de Fátima y un Corazón de Jesús presidieron las rogativas, desde un altar improvisado y frente a unas trescientas personas provistas de blancos rosarios de cera y bastante fe.

*Figure 16. Prayers offered up in Bilbao, March 2nd 1990. Source: "El Correo".*

Other, more exotic ideas were also put forward, such as carrying water by sea in tankers and inducing rain artificially. It comes as no surprise that these last, extremist ideas were those seized upon most eagerly by the newspapers, in detriment to serious, truthful information.

Figure 16 illustrates a curious anecdote. It is a cutting from the local press dated March 4th 1990 reporting that prayers for rain had been offered up in the centre of Bilbao.

## 5.5. SUPPLY SITUATION FROM THE APRIL RAINS TO DECEMBER 1990

April 1990 was a providential month: starting on the 5th, the reservoirs began to recover, reaching 60 $hm^3$, and reserves then held steady until June 1st, as shown in Figure 17.

Situation in the reservoirs of the Zadorra system up to June 1st 1990

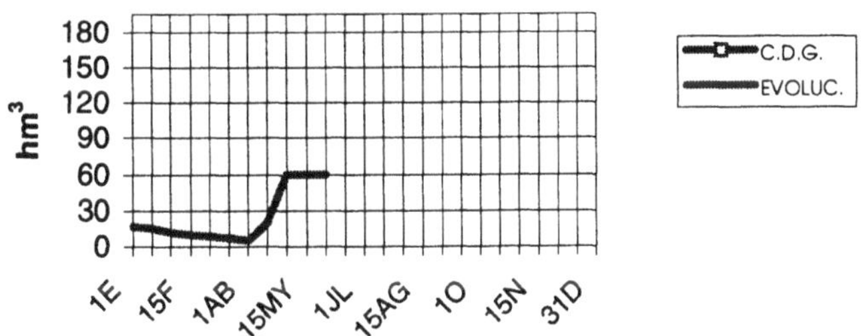


*Figure 17. Changes in water reserves in the Zadorra system up to June 1st 1990 Source: "Consorcio de Aguas".*

The level of restrictions was maintained, and emergency work continued to bring water in from both river systems. On June 1st the cut-off time was put back to 22.00, and on July 1st to 00.00, with re-connection at 06.00. This timetable was then maintained until cuts were ended.

*Figure 18* shows the changes in reserves throughout 1990.

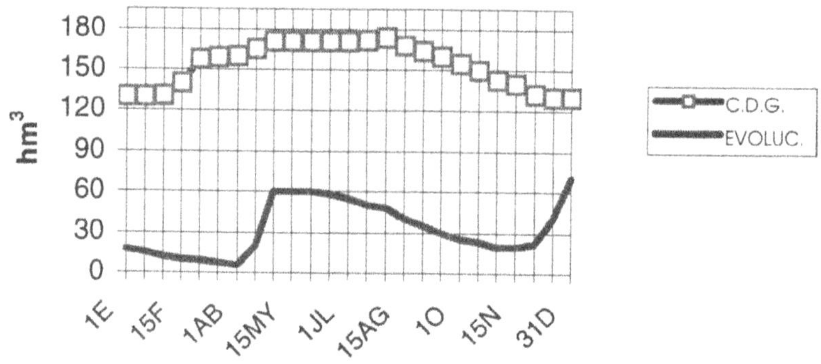


*Figure 18. Changes in water reserves in the Zadorra system in 1990. Source: "Consorcio de Aguas".*

## 5.6. PROCESS MONITORING & QUALITY CONTROL METHODS

### 1. WATER SIMULATION MODEL

To monitor changes in water reserves hydraulic models were set up which enabled short-term strategies to be decided. The variables considered were inflow (random), demand (defined), actual stocks in reservoirs and the desired horizon. The worst case scenario envisaged restrictions of 50% with an inflow of only 40 $hm^3$. Actual inflow in April 1990 exceeded expectations, so different hydraulicity levels were simulated for 1990 and 1991, taking into account in each of them two different distribution systems for rainfall throughout the year: concentrated at the beginning or end, and evenly distributed throughout the year. The simulations revealed that with the 60 $hm^3$ obtained in April and with the extraordinary inflow from emergency work, restrictions could be done away with early in 1991. In fact they were lifted in February 1991.

### 2. MONITORING CONSUMPTION IN THE PRIMARY NETWORK

This consisted of daily readings of primary network meters at municipalities and industrial concerns. The operative groups handling regulation & cutting of supplies took these readings after they had completed their operations. A team working exclusively with major industry monitored consumption day to day, as this sector was not subject to restrictions and made its savings through self-discipline. Daily data were processed and compared with reference flow rates.

3. QUALITY CONTROL

The number of analyses taken of water in the distribution networks was increased by 35%. In compliance with the provisions of Royal Decree 353/1987 on the monitoring and control of water for public consumption, the *Consorcio de Aguas* carried out twice as many analyses as were strictly necessary, and increased the number of overall analyses by 15%. Resources which were out of service were recovered, which required additional efforts. Weather variations increased the turbidity of water and made for lower microbiological quality because the chlorine systems in place were insufficient. Even so, the public health authorities only had to declare as "not fit for drinking" the water from a single rehabilitated spring. In parallel to the work described above quality control checks were increased on the deepest levels of the water in the reservoirs, and break-point treatment was envisaged with chlorine doses of between 6.5 and 8.5 ppm. Water quality studies were carried out on the lower reaches of rivers in Sta. Ageda (Cadagua), Sangróniz (Asúa), Arbina (Butrón), etc.

## 5.7. THE INTER-INSTITUTIONAL COMMITTEE

The Basque Government was deeply concerned at the situation created by the drought, and passed Decree 32 of February 13th 1990 to set up a committee to co-ordinate action to palliate its effects.

This co-ordinating body was answerable to the Lehendakari, or President of the Basque Government. It was chaired by the Secretary to the President's Office, and its members included representative of the different government departments. There were also representatives of the provincial councils, associations of municipalities, the *Consorcio de Aguas, Aguas Municipales de Vitoria* and the Añarbe District Council.

The job of this committee was "to correct any distortions which the drought might bring about in the normal development of activities in the different sectors of the economy and society". The seriousness of the situation made it almost a parallel government.

# 6. The Situation After 1990.

## 6.1 WATER CONSUMPTION AFTER THE DROUGHT

A new culture of water use grew up in Greater Bilbao. Even now, six years after the drought, consumption remains stable in spite of slight increases in population and provisions.

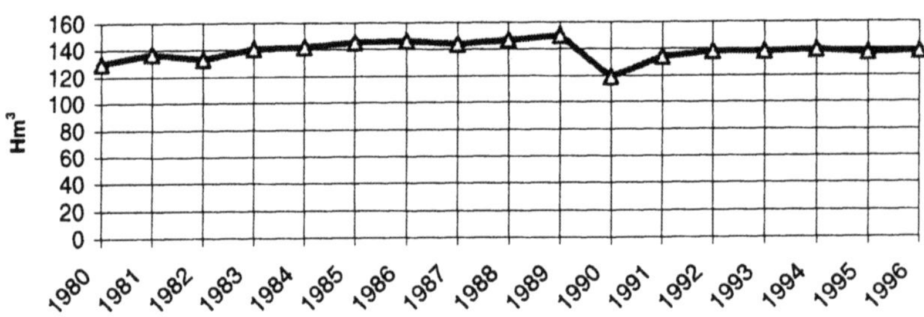


*Figure 19. Water consumption in Bilbao and its metropolitan area*
*Source: "Consorcio de Aguas"*

## 6.2. THE NEW GUARANTEE CURVE

At the end of this drought, a new, harsher type of drought situation was defined: the "critical drought". This was not only a dry period but also an atypical one, as it was preceded by an extremely wet summer in 1988. This led to an equally atypical guarantee curve in which 174 $hm^3$ were available in August when in July 171 $hm^3$ had been required. This happened because the inflow in July was sufficient to meet demand. It is easy to see that this guarantee curve is out of line with reality: reservoirs should fill in accordance with the climate of their catchment areas, reaching their highest point in Spring.

## 6.3. INCORPORATION OF EMERGENCY WORK INTO GENERAL SUPPLY SYSTEM

When the drought was over we had new supply systems owned by the water authorities. The beneficiaries of this were the supply utilities themselves. The period of exceptional circumstances envisaged by Royal Decree 296 of March 2nd 1990 had ended on December 31st 1991.

In the Northern river system the pumping facilities from the Cadagua were now a consolidated reality, and could be used to increase the supply systems for Bilbao and its metropolitan area. Pumping from the Nervión provided another highly interesting additional resource. The Ebro system had also seen major work, including especially the highly stable pumping facility at Nanclares connected to the Araca waste water treatment plant. Vitoria could thus be said to have acquired a whole new alternative supply. Other catchments such as Araya and Ega were also still available. All these sites were owned by the respective water authorities, which could use them as they saw fit at any time.

On March 23rd 1992 the water authorities of the Ebro and Northern systems, the *Consorcio de Aguas & Aguas Municipales de Vitoria* signed an agreement under which suppliers could continue to use the facilities provided by emergency work, provided that third party rights were respected and water reserves were below certain operating levels.

The idea was for catchments to come on line gradually, starting with those in the Northern system. Those in the Ebro system would follow in a predetermined order if the reservoir levels continued to fall. Maintenance and upkeep expenses would be paid by suppliers in proportion to the use of the facilities.

The agreement prohibited the taking of water from the Ebro system between July and October. At present some of these facilities are used in that period by irrigation co-operatives on the plains of Alava, which have obtained the relevant permission from the authorities.

Figure 20 shows the guarantee curve characteristic of free turbination (1996 situation) and the use curves of the emergency facilities.

## 6. 4. COST AND BENEFITS OF THE DROUGHT

The drought had positive consequences as well as negative ones: it forced us to learn from our errors. What was the ultimate cost and what were the ultimate benefits? The cost cannot be measured only in terms of money: there was a cost in terms of the morale of those people who felt insecure in a situation which looked as if it might become chaotic. There was a cost in terms of loss of image by the *Consorcio*, which had previously appeared as a colossus wielding inexhaustible resources. Three was a cost in terms of the inconvenience of being without water at home, in sports centres, etc. There was a cost in the loss of quality of service in the catering, hairdressing and other sectors. There as a cost in the increased risk of fire or contamination of water because the mains networks were almost empty. There was the cost of having to install tanks in hospitals and other premises. There was a cost in terms of the high risk of supply failure with absolutely unforeseeable consequences. All these and other sensitive areas created a solid mass of opinion which certainly had a cost in terms of the political image and loss of face of the many authorities responsible for water management. There were also costs in the material used up every day by the operative groups as they did their job, and there were other costs which although necessary did not result in any payback.

However some costs did pay dividends, such as the 200 million pesetas invested in pipeline repairs and the 7,000 million spent on catchments which remained in place to benefit water authorities later. We have no records of the production sector suffering losses or losing contracts as a result of the lack or limited supply of water.

a) SITUATION & GUARANTEE CURVE OF ZADORRA SYSTEM RESERVOIRS
b) Year: 1997
c) Total volume
d) Useful volume
e) Total surface area of catchment area
f) Reference elevation for useful volume
g) Maximum normal reservoir
h) Guarantee curve
i) Spillway threshold
j) Northern curve
k) Gravity
l) Elevation
m) Pumping
n) Useful supplies (gravity + pumping)
o) Millions of cubic metres
p) Reservoir elevation
q) January-February-March-April-May-June-July-August-September-October-November-December

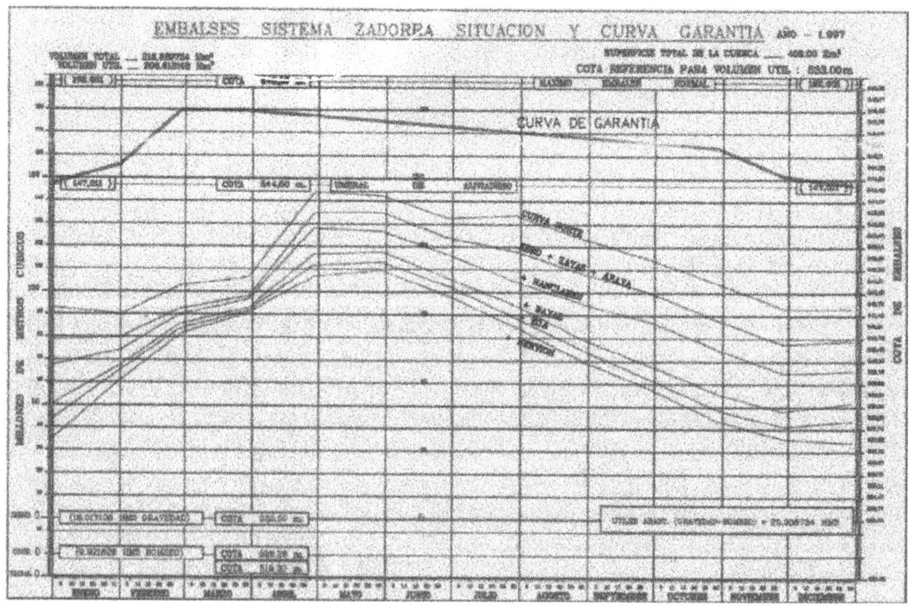


*Figure 20. Free turbination curve (1996) & use curves for emergency facilities*
*Source: "Consorcio de Aguas"*

There were also direct benefits: people learned to handle droughts, to make better use of water. Companies set up recovery systems and water consumption stabilised. We learned not to trust our calculation systems, and to manage water more in terms of assuring supply and perhaps less in terms of mathematical rationales. Nature put us in our place in terms of the new resources available to us. We also learned to reassess earlier agreements on free turbination at reservoirs and we began to design new exploitation systems which were simpler but surer from the viewpoint of suppliers. What we did not learn, however, is how to predict a drought far enough in advance.

## PART THREE
## "REFLECTIONS AND CONCLUSIONS DRAWN AFTER THE DROUGHT"7.

**Review of Criteria of Exploitation after the Critical Drought Recorded**

### 7.1. REFLECTIONS AND SELF-CRITICISM

Now that we have looked at events from 1988 to 1991 in Bilbao and its metropolitan area and in Vitoria, we must look back and with hindsight reflect on various points.

First of all we must state firmly that the advent of a dry period cannot be predicted with our present level of knowledge and techniques. A dry period could start at any time and cause disturbances in supplies as a result of even very slight changes in weather patterns. The most effective action is to remain alert so that recurrence periods are longer. If the drought does arrive anyway, effective measures must be taken.

Suppliers must apply procedures and strategies which prioritise the filling of reservoirs over other uses, as laid down in the Water Act. These rights must prevail, and must be defended firmly.

The implementing of "tough" restrictions in our case not only slowed the drop in water levels at reservoirs but also worked as an example of solidarity. Overall savings in the 14 months of restrictions were spectacular in absolute terms at 27 $hm^3$, but not so in relative terms. If this volume is spread over the whole period it works out to 8% of total consumption (domestic and industrial), equivalent to an aggressive campaign to heighten awareness.

Emergency work provided an added value of great interest for supplies in both river systems, The resulting facilities should not be regarded as temporary, but as consolidated. Even they could not work miracles, however: the miracle was provided by nature with timely, gratefully received rain. Emergency work did however help to normalise the situation in the shortest possible time.

The measures adopted and their timing were "text-book" solutions applied correctly, even though some of them (e.g. the re-use of waste water for industry) were never actually put into practice.

Another point of debate was the lack of alternative resources held in reserve. Such resources are now being worked on as part of the Northern Hydrological Plan. They are a fundamental part of that plan, and are seen as resources to be used in case of malfunctions in the main facilities or during planned overhauls. Such alternative resources were sorely missed during the drought.

## 7.2. HYDROLOGICAL ASPECTS OF THE REVIEW OF CRITERIA OF EXPLOITATION

In his 1993 work "Applied Hydrology" the late Vent Te Chow, an illustrious hydrologist, said " *Hydrological design is considered to be satisfactory if water can be supplied at the required rates during an equivalent critical* period". This reference extends the possibility of using critical records of the same severity but with different rainfall distributions.
The critical drought involved here was 28 months long, and the monthly inflow figures for the period add up to 244 $hm^3$. The distribution of rainfall was atypical, as indicated, with a normal spring and an extremely wet summer.

We could simulate a less serious drought with a larger total inflow, but with a rainfall distribution more in accordance with the climate of the area. It would be logical to expect less demanding results, but this is not the case: evolvent curves could be found which would involve larger volumes at the start of the summer, on the side of logic.

To compensate for the lack of historical records, we recommend synthetically generating long duration series to confirm probabilistic bases for designing drought prevention measures. *(Hirsch, 1979, Salas, 1980).*

Figure 21 below is an example not of a synthetic series by of a real one, which reinforced these recommendations. It shows a series of rains in Bilbao investigated between 1865 and 1995, with the reference point being the Sondica observatory. More or less dry periods can be seen not only in 1944, 1957 & 1989, but also in 1881, 1891, 1989, 1910, 1920, 1925 & 1934.

Considering the 108 years between the earliest and latest droughts (1881 & 1989), the average recurrence period works out at 12 years, practically equivalent to 1 period every 10 years. The minimal deviations in rainfall timing and quantity required to cause these effects on runoff into river systems should make us especially cautious, and we should assume these recurrence times for the sake of safety.

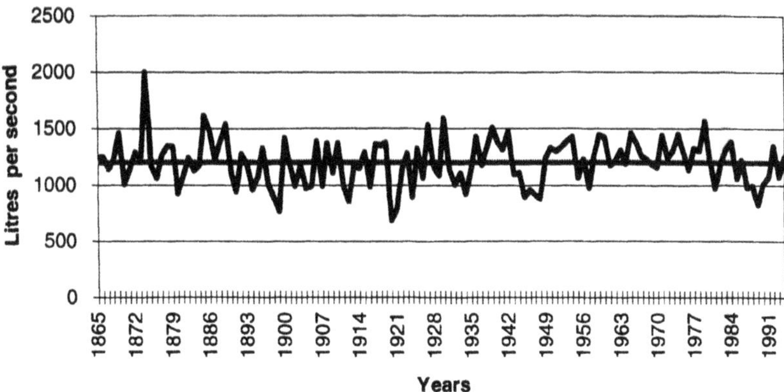


*Figure 21. Long rainfall series investigated in Bilbao (Sondica) between 1865 & 1995. Source: Author's doctoral thesis (in preparation)*

Figure 22 gives a spectrum of rainfall records in Bilbao of 950 litres per square meter or less. Although this is an isolated figure it is considered sufficient to cause appreciable changes in runoffs in Northern rivers.

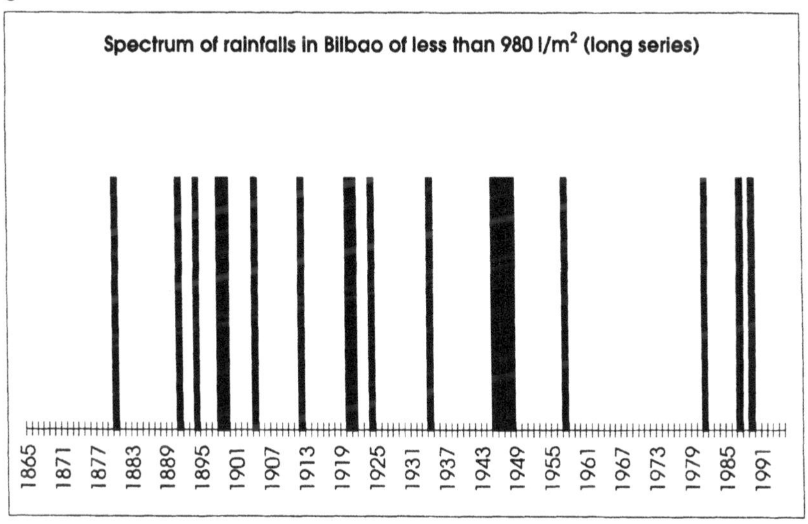


*Figure 22. Spectrum of rainfall records investigated in Bilbao (Sondica) between 1865 & 1995 of 980 l/m$^2$ or less. Source: Author's doctoral thesis (in preparation)*

Similarly, Figure 23 shows a spectrum of records on inflows into the Zadorra system of 180 $hm^3$ per annum or less. This figure is equivalent to the annual take-off for all uses. The average recurrence time also indicates the annual shortfall

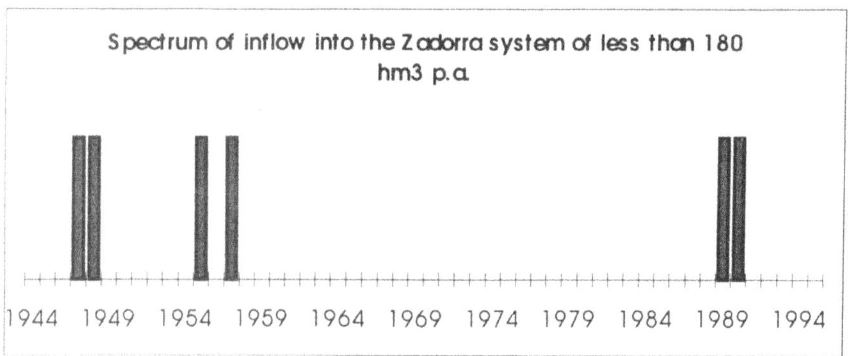


*Figure 23. Spectrum of records on inflows into the Zadorra system between 1944 and 1995 of 180 $hm^3$ p.a. or less.*
*Source: Author's doctoral thesis (in preparation)*

The same results can be reached by adjusting annual hydrological variables to probability distributions to estimate whether their behaviour is random and independent (central limit theorem), though successions of dry years appear in many cases, so there may be dome dependence.

The table below shows the statistical parameters of the variables studied. The low level of consistency of the results obtained with respect to the rain in Bilbao is significant, in both the long and short series. The same deductions can be made when calculating averages, etc. for subseries of the long series. This shows that this inconsistency is "congenital" to this hydrological variable and that the period from 1962 to 1982 was, as we have stated, an exceptionally wet one.

| | Long series of rains in Bilbao in l/m2 p.a. | Short series of rains in Bilbao in l/m2 p.a. | Inflow to Zadorra in $hm^3$ p.a. |
|---|---|---|---|
| Average | 1195 | 1210 | 257 |
| Median | 1193 | 1231 | 249 |
| Standard deviation | 189 | 174 | 73 |
| Coeff. of assymetry | -0.1620 | -0.3704 | 0.1306 |

Figure 24 shows the fit of the rains in Bilbao from 1944 to 1995 with the normal and log-normal distribution.

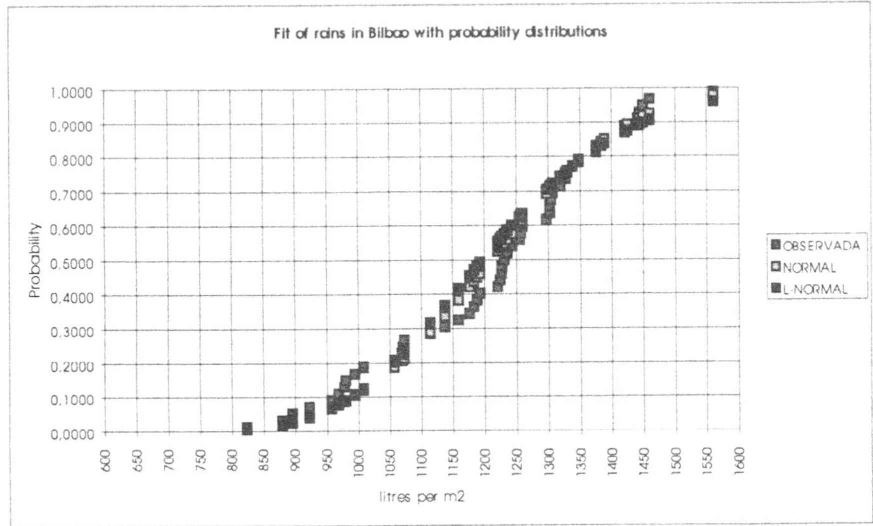


*Figure 24. Fit with probability distributions of the rain in Bilbao between 1944 & 1995Source: Author's doctoral thesis (in preparation)*

Figure 25 shows the fit of the annual inflow into the Zadorra between 1944 and 1995 with normal and log-normal distribution.

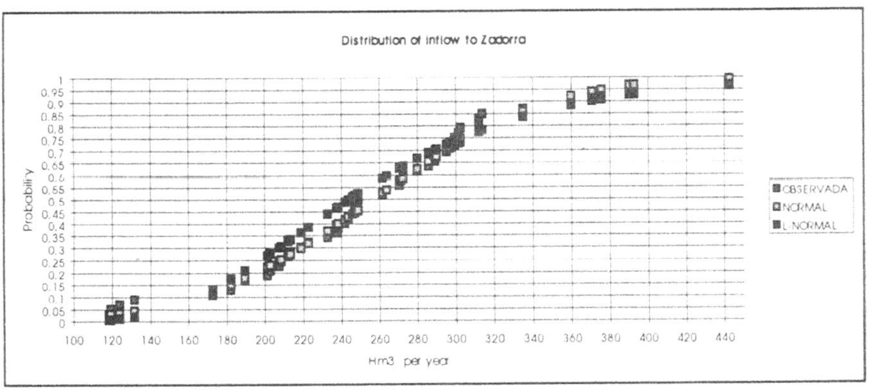


*Figure 25. Fit with probability distributions of inflow to Zadorra between 1944 & 1995 Source: Author's doctoral thesis (in preparation).*

It can be seen that there is a 20% probability of inflow being lower than take-off, coinciding, obviously, with the recurrence time. This is the probability of reservoirs failing to recover within one year.

Annual inflow series are formed from monthly inflows using a 12th moving summation so that a matrix is formed of 12 times the number of years studied. This matrix contains all the possible values for inflows accumulated from January of the earliest year and the eleven subsequent months, from February of the earliest year and the eleven subsequent months, and so on. Of all the series formed, if we take the ones beginning in January and ending on December 31st we will obtain the series of inflows for the calendar years as adjusted above (see Figure 25).

The same process was followed for 28th order accumulated inflows (length of the critical drought) and 47th order (length of the drought prior to the critical one). These series fit well into normal distribution, as shown in Figures 25 and 26.

Figure 26 shows the fit of the 28th order accumulated inflows (recorded drought) into the Zadorra system. The probability of reaching the critical inflow figure of 244 $hm^3$ is 2%. This distribution does not take into account the monthly rainfall distribution (inflow), so that more serious situations may arise with the same inflow levels.

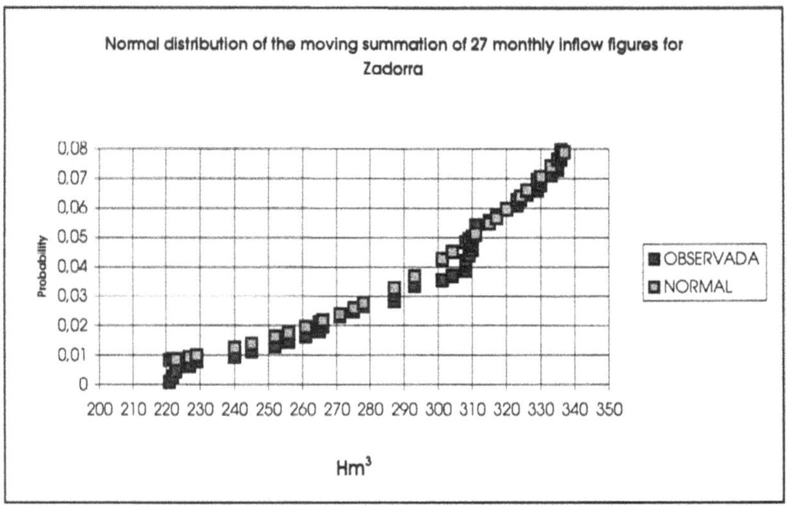


*Figure 26. Normal fit of moving accumulated inflows for 28 months into the Zadorra. Source: Author's doctoral thesis (in preparation).*

Figure 27 shows the moving accumulated inflow figures for 47 months into the Zadorra system.

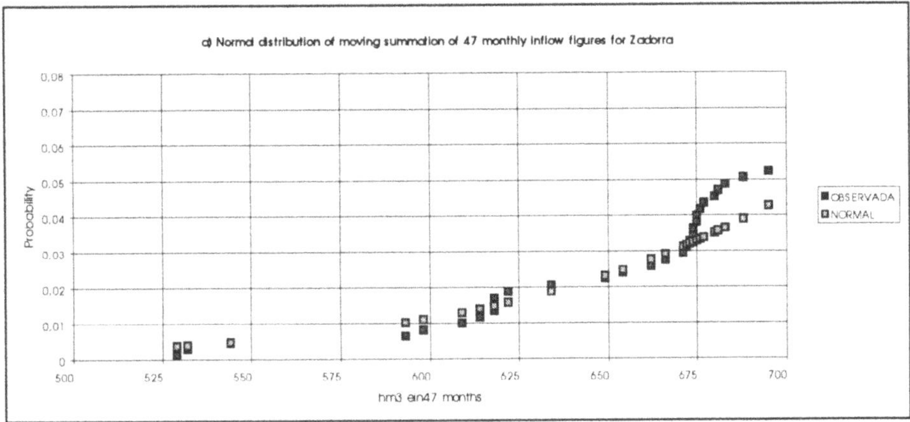


*Figure 27. Normal fit of moving accumulated inflows for 47 months into the Zadorra system. Source: Author's doctoral thesis (in preparation).*

## 7.3. DETERMINING SUPPLY GUARANTEE AND QUALITY LEVELS

Levels for guaranteeing supplies were established prior to the drought by analysing the n° of months in which the reservoirs were "just empty" and establishing monthly volumes by means of that equality. Clearly only one period was found in the simulation in which the reservoirs emptied (leaving a reserve of 2 months with restrictions of over 50%) and then began to recover straightaway. The deterministic guarantee was 100%, or 98% if the sample size (52 years) is taken into account.

This form of assessing safety levels is clearly not valid, and indeed it is not used in large supplies, which draw up exploitation strategies conducive to the filling of reservoirs an the end of the traditional rainy season. The singular nature of the multiple use catchment on which supply is based in our case should not be an obstacle in this.

We can set up an operating band *above* the curve which guarantees that the reservoirs will not empty (former guarantee curves), and curves which show the likelihood of the reservoirs filling by May 1st. These curves, which have been plotted recently, are known as "filling guarantee curves". They are being discussed at present with the hydroelectric power company which uses this water. In this way, exploiting the reservoirs in this area, we can considerably reduce the number of failures of service, which concept has also been reviewed. We must therefore obviously talk in terms of "service quality levels" rather than guaranteed levels.

Service must be considered to have failed not when the reservoirs are almost empty but when water consumption has to be limited. The *Consorcio* has established the following guarantee level:

*"Limitations in service will not be admitted in more than 5% of years, and will in no case exceed 10% reductions in all types of consumption from reservoirs".*

The figure of 10% is set because it is felt that this level can be reached without major upsets for users, and without cutting off water supplies in the distribution network in any case.

To determine how many times 10% restrictions need to be imposed, we must consider time criteria, i.e. when measures have to be applied. This is being reviewed at present, but for the moment the criterion is the moment in which reservoirs hold less water than is necessary to reach the next but one traditional rainy season with 10% restrictions, except in months in which shorter periods offer a greater safeguard.

Figure 28 shows the minimum safeguard curve or corrective measure curve mentioned above.

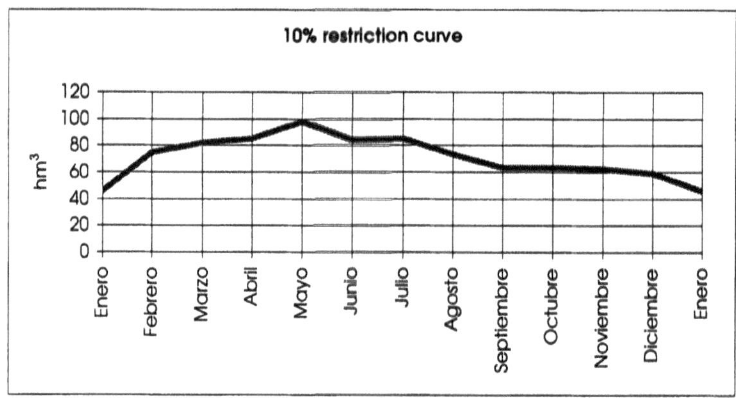


*Figure 28. Corrective measure curve for limiting take-off from the Zadorra reservoirs by 10% (now being reviewed). Source: "Consorcio de Aguas"*

With these criteria the "service quality level" is considerably improved.

The filling guarantee curves for periods of 1, 2 and 3 years are indicated below, meaning that after between 2 and 3 years the reservoirs empty and the operating band must therefore lie between the filling curves for 2 or 3 years, the free turbination curve and the probability of our targeting May 1st for filling. As can be seen, reservoir capacity is exceeded in many cases.

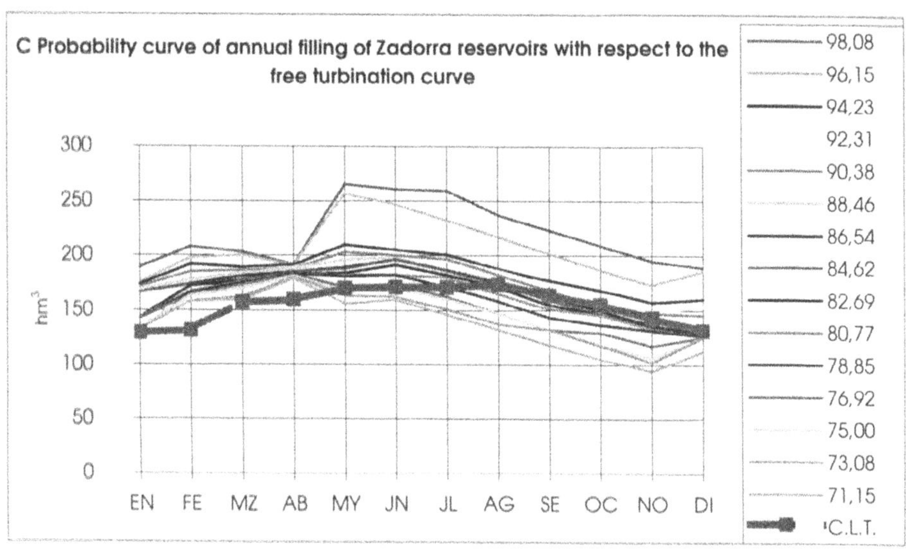


*Figure 29. Filling guarantee curve (180 hm³ on May 1st) in 1 year (Zadorra)*
*Source: "Consorcio de Aguas"*

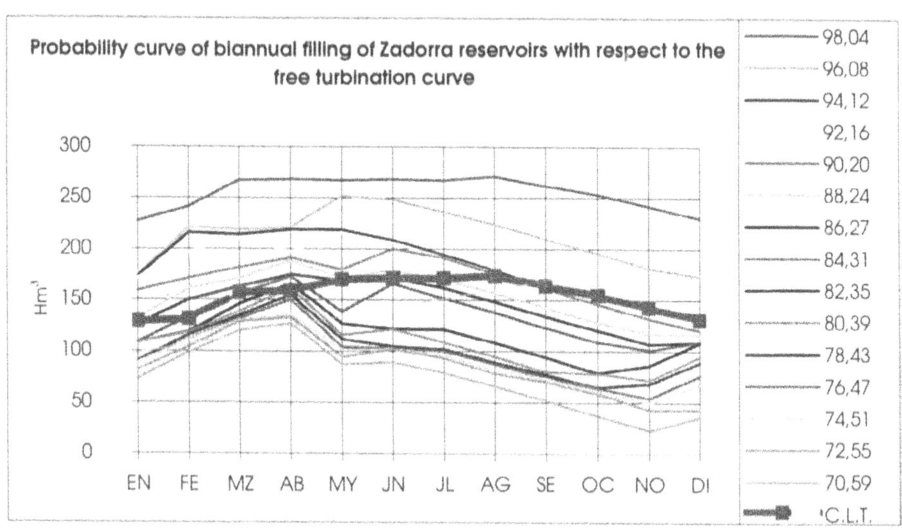


*Figure 30. Filling guarantee curve (180 hm³ on May 1st) in 2 years (Zadorra)*
*Source: "Consorcio de Aguas"*

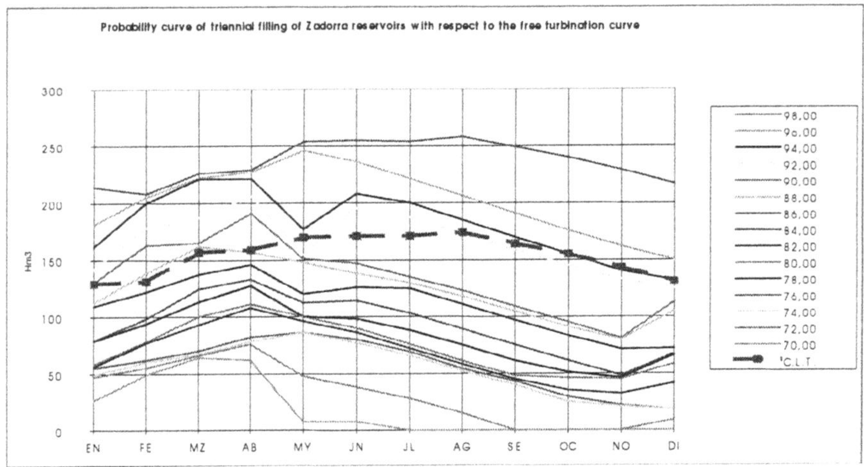


*Figure 31. Filling guarantee curve (180 hm³ on May 1st) in 3 years (Zadorra)*
*Source: "Consorcio de Aguas"*

## /.4. OVERALL OPTIMISATION OF A SUPPLY SYSTEM AND CONTINUOUS IMPROVEMENT

Once these targets are set, optimisation measures must not stop. In a seminar held in San Sebastián on supply management, organised by the Fluid Mechanics Dept. of the Polytechnic University of Valencia and the Nuclear Engineering and Fluid Mechanics Dept. of the University of the Basque Country, I was invited to present a paper. I will reproduce below with slight modifications my comments in that paper on optimising water supplies.

What is the best way to achieve continuous improvement? The answer is obviously that which leads us towards higher service quality levels and therefore to greater guarantees. How can we increase the guaranteed supply level and therefore rest easier? How much will it cost and where is the point of equilibrium? To answer these questions arguments and justifications clearly have to be put forward.

An increase in the guaranteed supply level leads to an increase in the satisfaction of users with their suppliers. This can be expressed as follows:

Pr(G)=(N° statements analysed - N° statements with deficit) / (N° statements analysed)

The n° of statements with deficit is obtained by a balance between system inputs and outputs. Guaranteed supply is therefore a random, probabilistic concept, and the relevant index measures the degree of assurance and protection of users against the risk of restrictions, supply failure or any other catastrophic situation.

It is random because one of the variables involved is random. The other variable (take-off or consumption) can be human controlled. By decreasing it, the guarantee is increased. "Consumption and water requirements are not comparable terms, because we consume more than we need". When water consumption levels are consistent with reasonable needs, with no discomfort or wastage, management comes close to optimisation

One way of encouraging lower consumption is to promote service organisation measures which not only handle fast, efficient repairs of chance breakdowns but also include auscultation plans to detect new leaks in urban water mains. With this we would recover added value, as the hypothetical new facilities needed to supply the amount saved would not be needed. Furthermore, if the extra cost involved in these measures is lower than the cost of producing an equivalent amount of water, management is optimised at all levels. This is not always possible: organisational systems must be set up to manage networks of a suitable size so that economies of scale can be achieved.

Single urban mains management units are an interesting proposition for populations of 75,000 - 100,000, depending on the type of network and its initial condition. It is only above this size that economic aspects affecting the price of water due to the establishing of additional services of this type become significant.

a) Consumption b) High reaches c) Treatment d) Storage
e) Urban mains f) Leaks g) Filtration h) Cleaning treatment
i) Middle reaches j) Lower reaches k) Sea

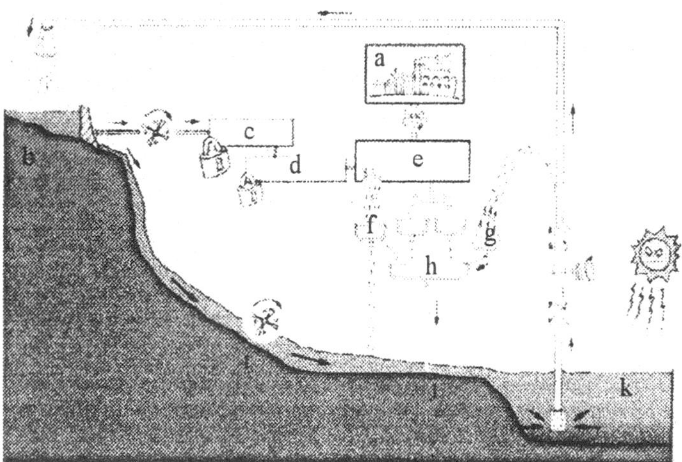


*Figure 32. Supply with headwater regulation. Source: Author's sketch*

Figure 32 shows supplies based on regulation via a reservoir of the high reaches of an irregular river, in conditions very similar to those with which the *Consorcio de Aguas* and many other suppliers work. By reducing leaks in the mains and encouraging rational use of water we can improve supply guarantees in an area which is probably highly sensitive to the random variable. In this way we increase the recurrence time to very low probability levels.

Reliability is a probabilistic concept which covers the ability of a system to remain up and running over time. If the probability of failure drops, reliability increased. There are failures which are not associated with the hydrological variable, and whose likelihood of occurrence may be linked to other equally random aspects of different kinds, such as accidental breakages of mains pipes or even earthquakes.

Redundant, alternative or substitute systems are the tools which must be used to increase reliability of major supplies in these cases. In analyses of this type there must be hydroeconomic studies to compare investment to increase reliability with the decrease in the likelihood of risk.

These are the methods undertaken by the *Consorcio de Aguas*.

## 8. Summary & Conclusions

Part one of this study discussed the origin and antecedents of the drought suffered by much of the eastern Bay of Biscay coast, which affected water supplies to Bilbao and its metropolitan area especially seriously. There were natural causes and problems involving the way in which our exploitation systems were considered, linked to energy uses and a lack of alternative or back-up systems.

Part two is descriptive, and gives a chronological account of events and the measures put in place to mitigate the effects of the drop in water stocks. All possible remedies were employed: campaigns, restrictions and emergency work. The prevailing impression left by these events is that the people who worked to implement measures, to carry out emergency work and to operate the resulting emergency facilities achieved a friendly, spontaneous co-ordination which did not extend to politicians, who were more concerned with debating the proposals which were put forward by the governing bodies of the *Consorcio.*

Part three gives a series of references to help assess droughts, evaluating the likelihood that they will recur and presenting long-duration hydrological data sets. Then an explanation is given of how we believe quality of service should be designed for a supply system catering for more than 1,250,000 people. Finally, procedures for exploiting water resources are presented.

As a last reflection, let us all recognise that there should be a single authority over this resource, which we all say has no frontiers but around which we all try to set up barriers. Could a single authority for such an important management task be set up? We have an

*Figure 33. The Valencia Water Court. Source: "El agua en España"*

## REFERENCES


Experiencias de una sequía. XII Jornadas Técnicas de la AEAS. Abril 1991.
José M. Eizaguirre Basterrechea & Angel Silveiro Gª-Álzórriz.

Annual Reports of the Consorcio de Aguas de 1988, 1989, 1990 & 1991.
(Consorcio de Aguas)

Hidrología Aplicada. Mc. Graw Hill.1993.
Vent Te Chow, David R. Maidment & Larry W. Mays
(Spanish translation)

Historia del Clima en España
Inocencio Font

## Water Science and Technology Library

1. A.S. Eikum and R.W. Seabloom (eds.): *Alternative Wastewater Treatment.* Low-Cost Small Systems, Research and Development. Proceedings of the Conference held in Oslo, Norway (7–10 September 1981). 1982 ISBN 90-277-1430-4
2. W. Brutsaert and G.H. Jirka (eds.): *Gas Transfer at Water Surfaces.* 1984 ISBN 90-277-1697-8
3. D.A. Kraijenhoff and J.R. Moll (eds.): *River Flow Modelling and Forecasting.* 1986 ISBN 90-277-2082-7
4. World Meteorological Organization (ed.): *Microprocessors in Operational Hydrology.* Proceedings of a Conference held in Geneva (4–5 September 1984). 1986 ISBN 90-277-2156-4
5. J. Němec: *Hydrological Forecasting.* Design and Operation of Hydrological Forecasting Systems. 1986 ISBN 90-277-2259-5
6. V.K. Gupta, I. Rodríguez-Iturbe and E.F. Wood (eds.): *Scale Problems in Hydrology.* Runoff Generation and Basin Response. 1986 ISBN 90-277-2258-7
7. D.C. Major and H.E. Schwarz: *Large-Scale Regional Water Resources Planning.* The North Atlantic Regional Study. 1990 ISBN 0-7923-0711-9
8. W.H. Hager: *Energy Dissipators and Hydraulic Jump.* 1992 ISBN 0-7923-1508-1
9. V.P. Singh and M. Fiorentino (eds.): *Entropy and Energy Dissipation in Water Resources.* 1992 ISBN 0-7923-1696-7
10. K.W. Hipel (ed.): *Stochastic and Statistical Methods in Hydrology and Environmental Engineering.* A Four Volume Work Resulting from the International Conference in Honour of Professor T. E. Unny (21–23 June 1993). 1994
    10/1: Extreme values: floods and droughts ISBN 0-7923-2756-X
    10/2: Stochastic and statistical modelling with groundwater and surface water applications ISBN 0-7923-2757-8
    10/3: Time series analysis in hydrology and environmental engineering ISBN 0-7923-2758-6
    10/4: Effective environmental management for sustainable development ISBN 0-7923-2759-4
    Set 10/1–10/4: ISBN 0-7923-2760-8
11. S.N. Rodionov: *Global and Regional Climate Interaction: The Caspian Sea Experience.* 1994 ISBN 0-7923-2784-5
12. A. Peters, G. Wittum, B. Herrling, U. Meissner, C.A. Brebbia, W.G. Gray and G.F. Pinder (eds.): *Computational Methods in Water Resources X.* 1994 Set 12/1–12/2: ISBN 0-7923-2937-6
13. C.B. Vreugdenhil: *Numerical Methods for Shallow-Water Flow.* 1994 ISBN 0-7923-3164-8
14. E. Cabrera and A.F. Vela (eds.): *Improving Efficiency and Reliability in Water Distribution Systems.* 1995 ISBN 0-7923-3536-8
15. V.P. Singh (ed.): *Environmental Hydrology.* 1995 ISBN 0-7923-3549-X
16. V.P. Singh and B. Kumar (eds.): *Proceedings of the International Conference on Hydrology and Water Resources* (New Delhi, 1993). 1996
    16/1: Surface-water hydrology ISBN 0-7923-3650-X
    16/2: Subsurface-water hydrology ISBN 0-7923-3651-8

## Water Science and Technology Library

16/3: Water-quality hydrology ISBN 0-7923-3652-6
16/4: Water resources planning and management ISBN 0-7923-3653-4
Set 16/1–16/4 ISBN 0-7923-3654-2

17. V.P. Singh: *Dam Breach Modeling Technology.* 1996 ISBN 0-7923-3925-8
18. Z. Kaczmarek, K.M. Strzepek, L. Somlyódy and V. Priazhinskaya (eds.): *Water Resources Management in the Face of Climatic/Hydrologic Uncertainties.* 1996 ISBN 0-7923-3927-4
19. V.P. Singh and W.H. Hager (eds.): *Environmental Hydraulics.* 1996 ISBN 0-7923-3983-5
20. G.B. Engelen and F.H. Kloosterman: *Hydrological Systems Analysis.* Methods and Applications. 1996 ISBN 0-7923-3986-X
21. A.S. Issar and S.D. Resnick (eds.): *Runoff, Infiltration and Subsurface Flow of Water in Arid and Semi-Arid Regions.* 1996 ISBN 0-7923-4034-5
22. M.B. Abbott and J.C. Refsgaard (eds.): *Distributed Hydrological Modelling.* 1996 ISBN 0-7923-4042-6
23. J. Gottlieb and P. DuChateau (eds.): *Parameter Identification and Inverse Problems in Hydrology, Geology and Ecology.* 1996 ISBN 0-7923-4089-2
24. V.P. Singh (ed.): *Hydrology of Disasters.* 1996 ISBN 0-7923-4092-2
25. A. Gianguzza, E. Pelizzetti and S. Sammartano (eds.): *Marine Chemistry.* An Environmental Analytical Chemistry Approach. 1997 ISBN 0-7923-4622-X
26. V.P. Singh and M. Fiorentino (eds.): *Geographical Information Systems in Hydrology.* 1996 ISBN 0-7923-4226-7
27. N.B. Harmancioglu, V.P. Singh and M.N. Alpaslan (eds.): *Environmental Data Management.* 1998 ISBN 0-7923-4857-5
28. G. Gambolati (ed.): *CENAS. Coastline Evolution of the Upper Adriatic Sea Due to Sea Level Rise and Natural and Anthropogenic Land Subsidence.* 1998 ISBN 0-7923-5119-3
29. D. Stephenson: *Water Supply Management.* 1998 ISBN 0-7923-5136-3
30. V.P. Singh: *Entropy-Based Parameter Estimation in Hydrology.* 1998 ISBN 0-7923-5224-6
31. A.S. Issar and N. Brown (eds.): *Water, Environment and Society in Times of Climatic Change.* 1998 ISBN 0-7923-5282-3
32. E. Cabrera and J. García-Serra (eds.): *Drought Management Planning in Water Supply Systems.* 1999 ISBN 0-7923-5294-7

Kluwer Academic Publishers – Dordrecht / Boston / London

GPSR Compliance
The European Union's (EU) General Product Safety Regulation (GPSR) is a set of rules that requires consumer products to be safe and our obligations to ensure this.

If you have any concerns about our products, you can contact us on

ProductSafety@springernature.com

In case Publisher is established outside the EU, the EU authorized representative is:

Springer Nature Customer Service Center GmbH
Europaplatz 3
69115 Heidelberg, Germany


www.ingramcontent.com/pod-product-compliance
Ingram Content Group UK Ltd.
Pitfield, Milton Keynes, MK11 3LW, UK
UKHW021859190726
13853UKWH00003B/1341